Oxidative Stress and Diseases:
Advanced Topics

Oxidative Stress and Diseases: Advanced Topics

Edited by **Nick Gilmour**

New York

Published by Callisto Reference,
106 Park Avenue, Suite 200,
New York, NY 10016, USA
www.callistoreference.com

Oxidative Stress and Diseases: Advanced Topics
Edited by Nick Gilmour

International Standard Book Number: 978-1-63239-502-3 (Hardback)

Printed in the United States of America.

Contents

Preface

Information related to oxidative stress and diseases has been examined in this book. The formulation of hypothesis of oxidative stress in the 1980s aroused the interest of biomedical and biological sciences which exists in the present day scenario as well. The contributions in this book educate the readers with the knowledge on the role of reactive oxygen species in distinct pathologies in animals as well as humans. The book is designed in such a way that it covers all the particular groups of pathologies like cancer, hormonal, neuronal, and systemic ones. It also elucidates the potential of antioxidants to safeguard organisms against harmful effects of reactive species. This book is a useful source of information for readers including researchers and students.

The information contained in this book is the result of intensive hard work done by researchers in this field. All due efforts have been made to make this book serve as a complete guiding source for students and researchers. The topics in this book have been comprehensively explained to help readers understand the growing trends in the field.

I would like to thank the entire group of writers who made sincere efforts in this book and my family who supported me in my efforts of working on this book. I take this opportunity to thank all those who have been a guiding force throughout my life.

Editor

Section 1

Systemic, Neuronal and Hormonal Pathologies

Effects of Oxidative Stress on the Electrophysiological Function of Neuronal Membranes

Zorica Jovanović
Department of Pathological Physiology,
Faculty of Medicine, Kragujevac,
Serbia

1. Introduction

Numerous experimental and clinical observations suggest that reactive oxygen species (ROS) play a significant role in several pathological conditions of the central nervous system where they directly injure tissue and where their formation may also be a consequence of tissue injury. Reactive oxygen metabolites are particularly active in the brain and neuronal tissue, and they are involved in numerous cellular functions, including cell death and survival. In comparison with other organs of the body, the brain may, for a number of biochemical, physiological and anatomical reasons, be especially vulnerable to oxidative stress and ROS mediated injury. A high metabolic rate (the brain consumes approximately 20% of total-body oxygen) and an abundant supply of transition metals, make the brain an ideal target for free radical attack (Facchinetti et al., 1998; Gutowicz, 2011). In addition, the brain has high susceptibility to oxidative stress due to high polyunsaturated fatty acid content and relatively lower regenerative capacity in comparison with other tissues. On the other hand, the brain is poor in catalytic activity and has moderate amounts of glutathion peroxidase and superoxyde dismutase. Of all the cell types in the body, neuronal cells may be among the most vulnerable to oxidative stress. These cells are continuously exposed to ROS. Accumulating evidence demonstrating that the defense of nerve cells against ROS-mediated oxidative damage is essential for maintaining functionality of nerve cells. Because hydrogen peroxide (H_2O_2) is the peroxide generated in the highest quantity in the brain, the defense against the oxidative stress appears to be particularly important. When production exceeds antioxidant protection, oxidative stress leads to molecular damage. An important component of the cellular detoxification of ROS is the antioxidant glutathione (GSH) (Dringen & Gutterer, 2002; Dringen & Hirrlinger, 2003). Because neurons have limited antioxidant capacity, they rely heavily on their metabolic coupling with astrocytes to combat oxidative stress. Evidence is growing that glutathione plays an important role in the detoxification of H_2O_2 and organic hydroperoxides in the brain and that glutathione is the main antioxidant molecule in neurons (Aoyama et al., 2008; Haskew-Layton et al., 2010; Limon-Pacheco & Gonsebat, 2010). Ongoing studies have shown that neuron-glial compartmentalization of antioxidants is critical for neuronal signaling by H_2O_2, as well as neuronal protection. The neurons are more vulnerable to oxidative stress than astrocytes, due to an insufficient detoxification of ROS via their glutathione system (Dringen et al., 1999; Martin & Teismann, 2009). But, the concentration of glutathione is in relatively lesser quantities in the brain in comparison to the other organs of the body (Skaper et al., 1999). In contrast to other ROS, H_2O_2 is neither a free radical nor an ion, which limits its reactivity (Cohen, 1994). However, in the presence of transition metals such as iron or copper, H_2O_2 can give rise to the indiscriminately reactive and toxic hydroxyl radical ($HO^•$) by Fenton chemistry. H_2O_2 is able to diffuse across biological membranes, and therefore can diffuse freely from a site of generation (Bienert et al., 2007; Makino et al., 2004) so that it is well-suited as a diffusible messenger. Increasing evidence indicates that H_2O_2 is a particularly intriguing candidate as an intracellular and intercellular signaling molecule because it is neutral and membrane-permeable (Nistico et al., 2008; Forman et al., 2010). Recent research into mechanisms of ROS-induced modifications in ion transport pathways involves: oxidation of sulfhydryl (SH) groups on the ion transport proteins, lipid peroxidation, and alterations of calcium (Ca^{2+}) homeostasis, a major second messenger system (Kourie, 1998). Increases in Ca^{2+} initiate inappropriate activation of several enzyme systems e.g., nitric oxide synthase and phospholipase A_2. Overactivation of these enzymes results in the breakdown of proteins and phospholipids and initiates several cascades that damage cells (Lee et al., 1999). It has been described that elevation in cytoplasmic Ca^{2+} levels activates the mitogen-activated protein kinase (MAPK) cascade (Liu & Templeton, 2008; Son et al., 2011) and the phosphatidylinositol 3'-kinase (PI3K)-Akt pathway (Cheng et al., 2003). ROS produce cell damage through multiple mechanisms, including excitotoxicity, metabolic dysfunction and disturbance of intracellular homeostasis of Ca^{2+} (Halliwell & Gutteridge, 1984; Del Maestro et al., 1980; Bracci, 1992). Activation of glutamate ionotropic receptors promptly triggers membrane depolarization and Ca^{2+} influx, resulting in the activation of several different protein kinases (Ca^{2+}-calmodulin-dependent kinase, protein kinase C and MAPK) and transcription factors, such as cyclic AMP response element binding protein (CREB). Neurons efficiently repair glutamate-induced oxidative DNA damage by a process involving CREB-mediated up-regulation of apurinic endonuclease 1 (APE1) (Yang et al., 2011).

Studies have demonstrated that ROS can induce or mediate the activation of the MAPK pathways (McCubrey et al., 2006). The mechanisms by which ROS can activate MAPK pathways are unclear. Because ROS can alter protein structure and function by modifying critical amino acid residues of proteins (Thannickal & Fanburg, 2000), the oxidative modification of signaling proteins by ROS may be one of the plausible mechanisms for the activation of MAPK pathways. However, the precise molecular target(s) of ROS is unknown. The prevention of oxidative stress by antioxidants blocks MAPK activation after cell stimulation with cellular stimuli indicating the involvement of ROS in activation of MAPK pathways. The recent observations provide a strong argument for activation of MAPK pathways by direct exposure of cells to exogenous H_2O_2 (Ruffels et al., 2004; Son et al., 2011).

The cell membrane would seem of special interest because of its large surface area and because of the susceptibility of membrane unsaturated fatty acids and proteins containing oxidizable amino acids (such as cysteine and methionine) to oxidant attack. Oxidative stress affects cellular membrane lipids and proteins. Cell membranes are either a source of neurotoxic lipid oxidation products or the target of pathogenic processes that cause permeability changes or ion channel formation (Axelsen, 2011). Reactive oxygen metabolites modify ion transport mechanisms either directly via ion transport pathway proteins and regulatory proteins or indirectly via peroxidation of membrane lipids. The nature and sequence of events that lead to the disruptions of these ion transport pathways are not fully understood. Early studies revealed that the effects of ROS on membrane properties could be deduced from electrophysiological parameters of the membrane. These include changes in membrane potential and current, ionic gradients, action potential duration and amplitude, spontaneous activity and excitability (Tarr et al., 1995; Tarr & Valenzeno, 1989; Beleslin et al., 1998). Oxygen-derived free radicals are thought to induce alterations in nervous electrical activity, however, the underlying membrane ionic currents affected by ROS and the mechanisms by which ROS induce their effects on ion channels in the nerve cells are not well defined. Considering neuronal function, ROS can attack ion channels and transporters directly, or indirectly by causing lipid peroxidation (Kourie, 1998; Carmeliet, 1999) and affecting associated signaling molecules (Hool, 2006). The mechanism of initiation of ROS peroxidation is not understood completely. The hydroxyl radical (HO•), a highly reactive oxidant, has been proposed as the initiating species. The ability of the HO• to initiate lipid peroxidation has been questioned by some investigators. In addition to initiating lipid peroxidation, the HO• has been implicated in direct cellular damage. Peroxidation of membrane phospholipids has been demonstrated to affect various transmembrane processes, such as receptor activation and formation of intracellular second messengers and Ca^{2+} homeostasis. Ca^{2+} ions also play a central role in the control of neuronal excitability. ROS oxidatively modify numerous membrane-bound proteins including ion channels. ROS can also react with proteins directly and in this case seem to have a prevalence for SH groups or disulfide bridges on the ion transport proteins (Van der Vliet & Bast, 1992). Oxidative sensitivity of ion channels is often conferred by amino acids containing sulfur atoms (Su et al., 2007). The mechanism of ROS-induced modifications in ion transport pathways involves the inhibition of membrane-bound regulatory enzymes and modification of the oxidative phosphorylation and ATP levels.

Studies have demonstrated that oxidative stress, perturbations in the cellular thiol level and redox balance, affects many cellular functions, including signaling pathways. In the CNS, cells respond to oxidative stress by initiating endogenous protective cascades often regulated at the transcriptional level. The transcription factor, nuclear factor erythroid 2-related factor 2 (Nrf2) plays an integral role in astrocyte-mediated protection of neurons from oxidative stress. Previous studies have reported that MAPKs may play a role in the induction and regulation of the Nrf2 system in the brain (Clark & Simon, 2009). When cells are exposed to oxidative stress, the Nrf2 binds to the antioxidant responsive element (ARE). The Nrf2–ARE pathway elicits transcriptional activation of antioxidant genes and detoxifying genes that protect cells and organisms from oxidative stress. Activation of this pathway protects cells from oxidative stress-induced cell death (Hur and Gray, 2011). The NRF2/KEAP1 signaling pathway is the main pathway responsible for cell defense against oxidative stress and maintaining the cellular redox state (Stepkowski & Kruszewski, 2011). The Nrf2-mediated GSH biosynthesis and release from astrocytes protects neurons from oxidative stress (Shih et al., 2003). Increased levels of GSH may be a major component of the protection observed by Nrf2 activation. In the CNS, Nrf2 plays an integral role in astrocyte-mediated protection of neurons from oxidative stress. Neuronal viability is enhanced significantly by an increased supply of GSH precursors from Nrf2-overexpressing glia. Thus Nrf2-dependent enhancement of glial GSH release appears to be necessary and sufficient for neuronal protection. The observations of Correa et al. (2011) suggest that activated microglia can stimulate or reduce the GSH-related anti-oxidant defense in cultured astrocytes. Recently, Zou et al. (2011) reported that overexpression of Nrf2 increased GSH content and efficiently protected t-BHP-induced mitochondrial membrane potential loss and apoptosis in cultured human retinal pigment epithelial cells.

Ion channels and transporters are susceptible to oxidative stress. For example, voltage-dependent Na^+, K^+ and Ca^{2+} channels, Ca^{2+}-activated K^+ channels, and K_{ATP} channels have all been identified as targets for ROS (Hool, 2006). Several previous studies indicate that H_2O_2 alters energy metabolism, ATP- sensitive K^+currents, L-type Ca^{2+}currents (Goldhaber & Liu, 1994; Racay et al, 1997), as well as delayed rectifier K^+ currents (Goldhaber et al., 1989). However in literature the data concerning the effect of ROS on potassium current are controversal. For example, Cerbai et al. (1991) and Ward & Giles (1997) did not observe any effect, in contrast to Tarr & Valenzeno (1989) who obtained a decrease in the outward, delayed rectifier potassium current. The results of Hasan et al. (2007) suggest that oxidative stress, which inhibits the delayed-rectifier current, can alter neural activity. Ward and Giles (1997) studied the effects of H_2O_2 on action potentials and underlying ionic currents in isolated rat ventricular myocytes. They showed that H_2O_2 caused no significant changes in either the Ca^{2+}-independent transient outward K^+ current (Ito) or the inwardly rectifying K^+ current (I_{K1}). The most prominent effect of H_2O_2 on the ionic currents which underlie the action potential is a slowing of inactivation of the I_{Na}. Potassium channels constitute a highly diverse class of ion channel and thus participate in multiple modulatory functions. Although altered potassium dynamics play a major role in this type of neuronal activity (Dudek et al., 1998) the role of K^+ channels is still incompletely understood. The voltage-activated K^+ channels are responsible for the establishment of the resting membrane potential, repolarization during action potentials and regulation of action potential frequency (Vacher et al., 2008). Three principal K^+ currents were identified in LRNC (Stewart et al., 1989). These differed in their time courses of activation and inactivation and in their responses to Ca^{2+} channel blockers. K^+ currents of the A-type (I_A) with rapid activation and inactivation kinetics, were not affected by Ca^{2+} channel blockers. The A-type K^+ currents were a minor component of the outward current in LRNC. A Ca^{2+} activated K^+ current (I_{Ca}), that inactivated more slowly and was reduced by Ca^{2+} channel blockers, constituted the major outward current in LRNC. The third K^+ current resembled the delayed rectifier currents (I_{K1} and I_{K2}) of squid axons, activated and inactivated slowly. Modifications of K^+ channel activity by ROS in the brain would lead to drastic changes in the electrical excitability of neuronal cells and could easily explain a tendency to brain hyperexcitability, or even neuronal death.

2. Materials and methods

All experiments were carried out at room temperature (22-25°C) on the Retzius nerve cells of isolated abdominal segmental ganglia of the ventral nerve cord of the horse leech *Haemopis saguisuga*. The dissection procedure, the recording method and point voltage clamp technique were employed as described previously (Beleslin et al., 1988). Dissected segments of 4 ganglia were immediately transferred to a 2.5 ml plastic chamber with a leech Ringer and fixed by means of fine steel clips. The plastic chamber was then placed in a grounded Faraday's cage mounted on a fixed table in a manner that prevents vibrations. Identification and penetration of the cells was performed in the cage under a stereomicroscope. Retzius neurons were identified based on the position within the ganglion, the size and the bioelectrical properties of the cells. The Retzius cells, are the largest neurons (40-60 μm in diameter) situated on the ventral side of the ganglia. It is well known that the resting potential of Retzius nerve cells of medical and horse leeches are lower than in other neurons (Hagiwara & Morita, 1962; Beleslin, 1985). Theoretically, this can be due to the low resting potassium permeability or the high membrane permeability to sodium. The Retzius cell is spontaneously active and responds to depolarization with an increased firing rate proportional to the depolarization (Lent, 1977). In leech Retzius nerve cells three classes of K+ channels (fast, slow calcium-activated and late voltage-regulated) have been identified (Beleslin et al., 1982; Beleslin, 1985). To change the solution, the chamber was flushed continuously with a volume of fluid at least five times that of the chamber volume. The perfusion rate was kept low so that implanted microelectrodes remained inside the cells during the perfusion. The bath volume was 2 ml and the solution changes were completed within 30 sec.

2.1 Electrical methods

In this study, we investigated the time-dependent changes in action potential configuration and changes in steady-state membrane currents in leech Retzius nerve cells. The spontaneous spike activity was recorded with a conventional 3 M KCl microelectrode. Membrane voltage and current were recorded using voltage-clamp techniques. This was shown in voltage-clamped neurons by long depolarizing steps (to 500 ms) from the holding potential which was more negative than –40 mV in a sodium free leech Ringer, in order to induce fast and slow K+ outward currents. The recording electrodes were prepared from 1.5 mm borosilicate capillaries (Clark Electromedical Instruments, UK) and filled with a 3 M KCl-containing solution. The pipette resistance ranged from 5 to 10 MΩ (when filled with 3M KCl solution). Microelectrodes with a tip potential less than 5 mV in an artificial solution, were selected for use. Usually the microelectrode was connected through a Ringer bridge and Ag-AgCl electrode via a negative capacitance high input resistant amplifier "Bioelectric Instrument DS2C" to a computer. Command pulses were derived from a "Tektronix 161" pulse generator. The signals were digitized by the use of an analog-to-digital converter (Digidata 1200; Axon Instruments) and saved in a computer for off-line analysis.

2.2 Solutions

Pharmacological agents were prepared and dissolved immediately before application in the physiological salt solution at the concentrations stated. H_2O_2, cumene hydroperoxide (CHP) and reduced glutathione (GSH), were added to the leech or Tris-Ringer. All drugs were administered sufficient to reach a steady-state response (up to 30 min). The bath volume was 2 ml and the solution changes were completed within 30 sec. The ganglia were bathed in a leech Ringer containing (mM): 115 NaCl, 4 KCl, 2 $CaCl_2$, 1.2 Na_2HPO_4, 0.3 NaH_2PO_4 (pH 7.2). In the Na+-free Ringer, 115 mM NaCl was completely replaced with an equal amount of Tris (hydroxy-methyl)aminomethane-Cl (Tris Ringer) and Na_2HPO_4 and NaH_2PO_4 were omitted.

Solutions containing H_2O_2 were prepared freshly before their use from 30% H_2O_2 solution (Zorka Pharma, Sabac) and added to the Ringer solution (or Tris-Ringer) at final concentrations of 1, 5 and 10 mM. CHP and GSH were added to the normal or Tris-Ringer solution. The CHP was obtained from Sigma (St. Louis, U.S.A.), dissolved in 0.01% dimethyl sulfoxide (Sigma, St. Louis, U.S.A.) and added to the Ringer solution (or Tris-Ringer) in a concentrations of 0.25, 1 and 1.5 mM. A GSH (Sigma, St. Louis, U.S.A.) was added to the Ringer solution to produce a final concentration of 0.2 mM.

2.3 Data analysis

Data are expressed as mean ± SEM. Comparisons between the mean values were made with a Student's t-analysis. P values <0.05 were considered significant.

3. Results

3.1 Effects of cumene hydroperoxide on the spontaneous spike potential of leech Retzius nerve cells

In our work we used CHP to stimulate lipid peroxidation as the mechanism of free radical-induced cell membrane damage. We investigated the time-dependent changes in action potential configuration and changes in steady-state membrane currents in LRNC. Superfusion of leech abdominal ganglia with CHP (0.25, 1 and 1.5 mM) caused an extreme change to the shape and action potential duration (APD) in LRNC. Exposure of LRNC to CHP prolonged the duration of the action potentials of the LRNC in a concentration-dependent manner. Figure 1 illustrates the representative record obtained 15, 20, 25 and 30 min after the exposure of an isolated ganglia to 1 mM of CHP. A cardiac-like action potential with a rapid depolarization, followed by a sustained depolarization or plateau and fast repolarization was recorded. During the 20 min of exposure with leech Ringer containing 1 mM CHP, early after depolarization was recorded. Higher concentration of CHP led to appearance of repetitive firing only a few minutes after application of CHP, which was followed by loss of excitability of leech Retzius nerve cells.

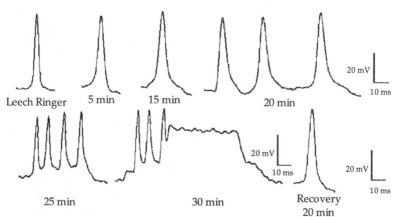

Fig. 1. Early after depolarization and repetitive firing recorded in LRNC after exposure of isolated ganglion to 1 mM CHP

Table 1 summarizes the values of the APD in a leech Ringer and after adding CHP (0.25, 1 and 1.5 mM). Table 1 shows that CHP caused a concentration-dependent increase in APD.

	Leech Ringer	5 min	15 min	20 min	30 min	Recovery 20 min
0.25 mM CHP n= 6	10.45±0.98	12.33±1.74	13.50±2.13	13.50±1.39	14.4±0.80 P>0.05	15.66±1.05
1 mM CHP n= 6	9.66±0.52	13.17±4.02	25.17±2.91	31.05±5.96	35.33±13.60 P≤0.01	26.50±8.38
1.5 mM CHP n= 11	9.66±2.18	16.09±3.15	41.64±8.27*	68.72±13.4*	127.80±15.95* P≤0.01	23.43±4.61

Table 1. Values of the APD (in ms) of LRNC before, 5, 15, 20 and 30 minutes after the adding of CHP (0.25, 1 and 1.5 mM) and during the recovery. Data are expressed as mean ± SEM; n=number of cells. * *repetitive firing*

3.2 Effects of hydrogen peroxide on the spontaneous spike potential of leech Retzius nerve cells

Previous investigations have shown that H_2O_2 is involved in cascades of pathological events affecting neural cells. The background of this study were the findings that 1 mM CHP treatment caused an extreme change in the duration of the action potential and suppression of Ca^{2+} activated outward K^+ currents of LRNC. The aim of the present experiments was to examine the effect of the higher concentrations of H_2O_2 on LRNC. H_2O_2 in concentrations up to 10 mM in the reaction mixture had no effect on spontaneous spike potential. Extracellular application of H_2O_2 (1, 5 and 10 mM) did not significantly change (P > 0.05) the duration of the action potential of the LRNC. H_2O_2 is ineffective in generating either cardiac-like action potential or early after depolarization in LRNC.

3.3 Effects of glutathione on the cumene hydroperoxide-induced change of the spontaneous spike potential of leech Retzius nerve cells

The background of this study were the findings that the hydroxyl radical scavenger, mannitol (5 mM) significantly reduced neurotoxic effect of CHP on the spontaneous spike electrogenesis of LRNC (Jovanović, 2010).

Taking in to consideration that it has been proved that CHP affects the lasting action potentials of Retzius nerve cells, the possibility of recovering the changes by the effect of antioxidant, GSH was also examined. Firstly, Retzius cells were exposed to the effect of CHP (1 mM) then GSII was added in a concentration of 0.2 mM to the Leech Ringer solution. The application of GSH, a free radical scavenger, to a bathing solution reverses the CHP effects. The GSH, largely inhibited the effects of CHP on the APD. In the presence of the GSH the APD has been extended by 9.22±1.14 ms (in controlled conditions, before the application of

CHP and GSH) to 12.45±1.56 ms (30 min after adding CHP and GSH to the Leech Ringer solution), which did not have any significant statistical result (Table 2). In the presence of GSH repetitive firings has not been registered in any examined cells.

	Leech Ringer	5 min	15 min	20 min	30 min	Recovery 20 min
CHP+GSH n=10	9.22±1.14	8.87±2.34	9.67±1.44	10.76±2.32	12.45±1.56 P>0.05	10.34±1.21

Table 2. Effects of GSH on CHP-induced prolongation of the APD in LRNC

The APD of LRNC before and after the adding of CHP (1 mM) and GSH (0.2 mM). Data are expressed as mean ± SEM; n=number of cells

The application of 0.2 mM GSH solution significantly decreased the bursting frequency, duration and amplitude of depolarization plateaus, and the number of spikes per plateau. These observations point out the significance of GSH in the protection of SH groups of membrane proteins as well as lipids in oxidative stress caused by CHP.

3.4 Modulation of Ca^{2+} activated K^+ current in leech Retzius nerve cells by cumene hydroperoxide

The elongation of action potentials by CHP suggested that CHP decreased the magnitude of ion currents needed for the repolarization phase of action potentials. The action potentials of **leech** Retzius nerve cells elongated after the exposure to 1mM of CHP, suggested that CHP modified the outward K^+ currents that form action potentials together with the Na^+ current. In order to explore the ionic mechanism by which CHP prolongs spike potential, we examined its effects on membrane K^+currents. Several studies have reported the possibility that ROS alter ionic channel function. The K^+ channels, key regulators of neuronal excitability, are targets of ROS. It is well known that outward K^+ current are essential for maintaining normal APD. The K^+ currents contributing to the resting membrane potential and repolarization of the action potential were studied in voltage-clamped Retzius neurones. Modifications of voltage-sensitive K^+ channel activity by ROS would lead to changes in APD and the electrical excitability of neuronal cells. To test this hypothesis, the effect of CHP on Ca^{2+} activated K^+ current was studied.

This was shown in voltage-clamped neurons by long depolarizing steps (to 500 ms) from the holding potential which was more negative than –40 mV in the sodium free leech Ringer in order to induce fast and slow K^+ outward currents. Figure 2 illustrates the typical outward currents elicited in a CHP responsive neuron, depolarized in steps from a holding potential of -56 mV to +4 mV. Both components of the delayed outward K^+ current, Ikr and Iks were inhibited by external CHP.

Application of CHP (in a concentration of 1 mM) caused suppression of fast and slow Ca^{2+} activated outward K^+ currents. The fast and slow steady part of the K^+ outward current was reduced by 40% and 31%, respectively. Figure 2 shows K^+ current amplitudes measured before and during exposure to CHP.

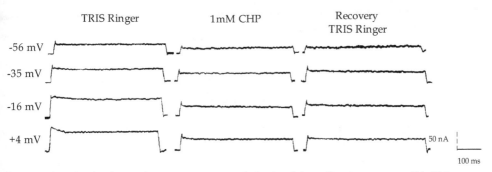

Fig. 2. Patterns of voltage clamp current record obtained from Retzius nerve cell in Tris-Ringer, after adding 1 mM CHP and again in an Na-free fluid (recovery) during displacement of holding potential from -56 mV to +4 mV.

Figure 3 shows the current-voltage relationship, separately, for the peak and established a steady level of depolarizing K^+ outward current. At the test potential of +4 mV the fast and late steady part of the K^+ outward current dropped from 60 to 36 nA (40%) and from 33 to 23 nA (31%). These results demonstrate the marked electrophysiological effects of CHP in leech Retzius nerve cells. Upon washout of the CHP, the fast and slow steady part of the K^+ outward current recovered by approximately 30% and 40% within 15 min.

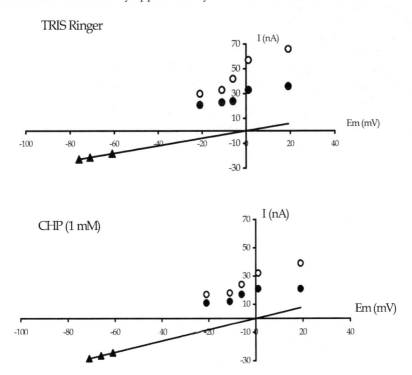

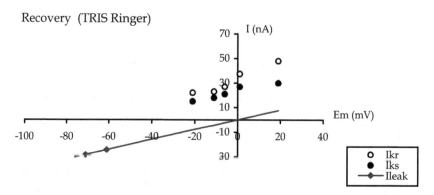

Fig. 3. The current-voltage relationship for the same cell, measured at the peak of the K⁺ outward current (open circles) and at the end of stimulation (solid circles) in the Tris-Ringer, 25 min after adding 1 mM CHP and during the recovery.

3.5 Effects of hydrogen peroxide on the Ca²⁺ activated K⁺ current of leech Retzius nerve cells

Modification of on membrane potassium currents by H_2O_2, a membrane-permeable form of ROS, in LRNC was examined using the voltage clamp technique. Using a two-electrode voltage clamp, we examined the H_2O_2 effect on the K⁺ outward current. In contrast to the effect of CHP, application of the H_2O_2 failed to inhibit fast and slow outward K⁺ currents in leech Ringer. At the test potential of -14 mV from holding potential of -77 mV, the fast and late steady part of K⁺ outward current dropped by 7.48% and 6.07%, respectively. In the current-voltage relationship (Fig. 4) there were no significant changes on the early or late part of the K⁺ outward current in the presence of H_2O_2. Voltage clamp experiments using double microelectrode methods revealed that H_2O_2 reduced a fast and slow K⁺ outward current by 7.48% and 6.07% respectively at the test potential of -14 mV, after 25 min.

Ikr- rapid Ca²⁺ activated K⁺ current; Iks- slow Ca²⁺ activated K⁺ current; Ileak - passive leak current

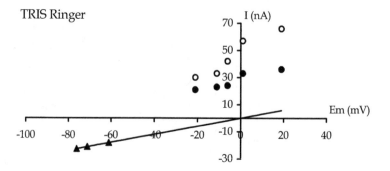

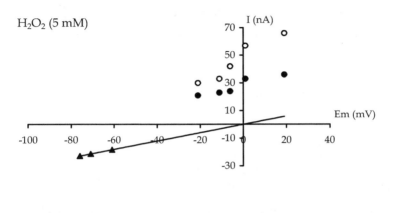

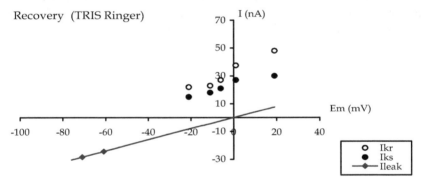

Fig. 4. Current-voltage relationship for the same cell measured at the peak of the K+ outward current (open circles) and at the end of stimulation (solid circles) in Tris-Ringer, 25 min after adding 5 mM H_2O_2 and during the recovery.

Ikr- rapid Ca^{2+} activated K+ current; Iks- slow Ca^{2+} activated K+ current; Ileak - passive leak current

3.6 Effects of glutathione on cumene hydroperoxide-induced suppression of the Ca^{2+} activated K+ current of leech Retzius nerve cells

Reduced glutathione applied in a concentration of 0.2 mM partially blocked the effect of CHP on Ca^{2+} activated outward K+ currents. Figure 5 illustrates the effect of GSH on Ca^{2+} activated K+ currents. The application of the GSH reduced fast and slow K+ outward currents in the leech Ringer. At the test potential of -17 mV from the holding potential of -57 mV, the fast and late steady part of the K+ outward current dropped by 21% and 12%, respectively.

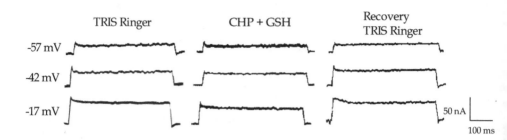

Fig. 5. Patterns of the voltage clamp current record obtained from Retzius nerve cell in the Tris-Ringer, after adding 1 mM CHP and 0.2 mM GSH, and in Na-free fluid (recovery) during displacement of the holding potential from -57 mV to -17 mV.

In the corresponding current-voltage relationship (Fig. 6) there were no significant changes on the early or late part of the K^+ outward current in the presence of 1 mM CHP and 0.2 mM GSH. At the test potential of -17 mV the fast and late steady part of the K^+ outward current dropped from 65 to 51 nA (21%) and from 46 to 38 nA (12%).

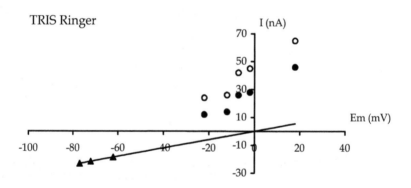

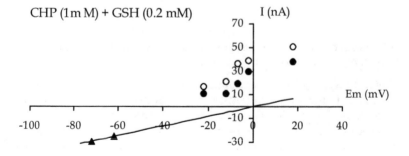

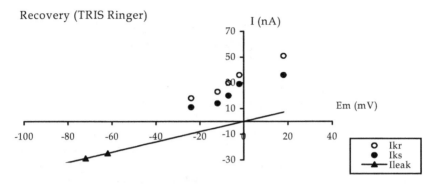

Fig. 6. Current-voltage relationship for the same cell measured at the peak of the potassium outward current (open circles) and at the end of stimulation (solid circles) in the Tris-Ringer, 25 min after adding 1 mM CHP and 0.2 mM GSH, and during the recovery.

Ikr- rapid Ca^{2+} activated K^+ current; Iks- slow Ca^{2+} activated K^+ current; Ileak - passive leak current

4. Discussion

The interest in H_2O_2 as a biologically active oxygen-derived intermediate is evident, because it is associated to a series of alterations and effects in different types of cells. The present data show that H_2O_2 did not significantly change, within 30 min, the shape of the amplitude of spontaneous spike potentials of LRNC. In the voltage clamp experiments, H_2O_2 was ineffective in the supression of fast and slow Ca^{2+} activated K^+ currents. The background of this study were the findings that a 1 mM H_2O_2 treatment with or without $FeCl_2$ did not significantly change the resting membrane potential of LRNC (Jovanović & Beleslin, 1996; Jovanović & Beleslin, 1997; Jovanović & Beleslin, 2004).

The present results suggest that leech ganglia have an efficient system against oxidative stress. There are several explanations why leech Retzius nerve cells should be resistant to H_2O_2. The simplest could be that leech Retzius nerve cells have a low concentration of polyunsaturated fatty acids, which are very sensitive to radical injury. This possibility is unlikely since neuronal membrane are rich in lipids (Whittemore et al., 1995; Wilson, 1997). Another explanation could be that they have an efficient scavenging enzyme system which reacts rapidly with H_2O_2. Peroxidation of lipids that inactivates membrane-associated enzymatic protein, increases membrane permeability. However, since we have insignificant changes in action potential with H_2O_2, it is reasonable to suppose that lipid membrane peroxidation did not occur. On the other hand, since changes in the action potential were not significant, it could further be expected that H_2O_2 was decomposed by a number of enzymatic and nonenzymatic antioxidant systems. A possible explanation for weak responses of LRNC to H_2O_2 is that the extracellular H_2O_2 application, results in an intracellular concentration some 7–10-fold below that of extracellular (Stone, 2004). It is well known that concentration as well as time of exposure plays an important role in the response generated by ROS. The range of [H_2O_2] used by different authors varies from 0.1 to 50 mM (Kourie, 1998).

When cells are exposed to external H_2O_2, the intracellular consumption of H_2O_2 catalyzed by anti-oxidants and enzymes is able to generate a gradient of H_2O_2 across membranes, which makes the intracellular H_2O_2 concentration lower than the extracellular one. Previous studies have reported that H_2O_2 did not affect channel activity when added to the extracellular side (Soto, 2002). In particular, oxidation of free SH groups of cysteines, present in a greater proportion at the intracellular side, could explain the observed difference. These results provide evidence for an intracellular site(s) of H_2O_2 action. It has been recently demonstrated that H_2O_2 activates TRPC6 channels via modification of thiol groups of intracellular proteins (Graham et al., 2010). H_2O_2 is a weaker oxidizing agent than other ROS. H_2O_2 is not by itself reactive enough to oxidize organic molecules in an aqueous environment. Nevertheless, the biological importance of H_2O_2 stems from its participation in the production of more reactive chemical species such as $HO^{\bullet}$ and its role as a source of free radicals has been emphasized rather than its chemical reactivity. However, because of its extremely short half-life, $HO^{\bullet}$ is effective only near the locus of its production.

The results reported in this paper show that an alkyl-hydroperoxide, CHP is a more efficient oxidant than H_2O_2. In contrast to H_2O_2, CHP, induced dose and time dependent membrane depolarization with a marked prolongation of spontaneous repetitive activity. These actions appear to underlie the prolongation of the action potential, and contribute to repetitive firing. Several mechanisms have been proposed for the plateau of the prolonged action potential, such as sustained inward Na^+ current, block of Ca^{2+} activated K^+ channels, modification of Ca^{2+} channels or its transformation in Na^+ channels. Our findings support the second proposal. A possible explanation is that CHP form free radicals that are more stable than the $HO^{\bullet}$. H_2O_2 and organic hydroperoxides, generate distinct ROS. $HO^{\bullet}$ generated by one-electron reduction of the H_2O_2, damages adjacent molecules at diffusion controlled rates. By contrast, the organic hydroperoxide triggers the generation of the free radical intermediates peroxyl and alkoxyl radicals, which can cross cellular membranes and evoke the production of the $HO^{\bullet}$ (Hwang et al., 2009). It was well known that $HO^{\bullet}$ generated from H_2O_2 could cause peroxidation of lipids that inactivates membrane-associated protein, increasing membrane permeability. This metabolic and physico-chemical alteration of a cell membrane would produce intracellular Ca^{2+} overload. CHP is more hydrophobic than H_2O_2. The most important finding of the present study is that CHP modulates Ca^{2+} activated K^+ channels in leech Retzius nerve cells. In the voltage clamp experiments, fast and slow Ca^{2+} activated outward K^+ currents were suppressed with CHP. The present results support the view that CHP stimulates lipid peroxidation, as the mechanism of ROS-induced cell membrane injury.

Although several previous investigations have described electrophysiological effects of H_2O_2 and CHP, the literature describing these effects is sometimes contradictory. For example, Cerbai et al. (1991) and Ward and Giles (1997) did not observe any effect, in contrast to Tarr and Valenzeno (1989) who obtained a decrease in the rectifying current. Vega-Saenz de Miera and Rudy (1992) reported that H_2O_2 inhibited three cloned voltage-gated K^+ channels expressed in Xenopus oocytes. A recent paper reported that ROS donors (H_2O_2 and t-BHP) reduced the voltage operated Ca^{2+} current but increased the amplitude of the delayed rectifier K^+ current in adult rat intracardiac ganglion neurons (Dyavanapalli et al., 2010; Whyte et al., 2009). Nakaya et al. (1992) examined the mechanism of membrane depolarization induced by CHP in guinea-pig papillary muscles, using ion-selective microelectrode and patch clamp techniques. They demonstrated that the depolarization of the resting membrane is, at least in part, due to the inhibition of inward rectifier K+ channel activity, and may play an important role in the genesis of reperfusion-induced arrhythmias.

There are conflicting descriptions of the current changes induced by ROS, and an incomplete understanding of which is responsible for the observed changes in action potential duration. For example, the inward rectifying K+ current has been reported to be either unaffected (Cerbai et al., 1991) or decreased (Tarr & Valenzeno, 1991). The electrophysiological effects of ROS generally consist of a reduction in action potential amplitude and an increase in action potential duration followed by a reduction (Tarr and Valenzeno, 1989; Barrington, 1994; Satoh and Matsui, 1997), although either only a reduction (Goldhaber et al., 1989; Hayashi et al., 1989; Coetzee et al., 1990) or only an increase in action potential duration (Barrington, 1994) have also been reported. Nakaya et al. (1991) reported that ROS-induced shortening of the action potential duration of guinea-pig isolated ventricular myocytes. The underlying mechanisms of the action potential shortening are undoubtedly complex, and changes in membrane currents other than the outward current through the ATP-sensitive K+ channels may also contribute to the action potential shortening. Matsuura and Shattock (1991) demonstrated that oxidant stress induces a decrease in the resting potassium conductance and an increase in Ca^{2+} activated membrane conductance. Both factors may underlie the depolarization of resting membrane potential, prolongation of the action potential and automaticity. Tokube et al. (1996) reported biphasic changes in the action potential duration, with initial lengthening of the action potential and subsequent shortening. In voltage-clamp experiments, ROS suppressed the L-type calcium current, the delayed rectifier K+ current and the inward rectifier K+ current. A recent paper reported that relatively low concentrations of CHP (100 µM) led to a significant decrease in the cellular content of ATP and reduced glutathione (Vimard et al., 2010).

K+ channels are a family of ion channels that govern the intrinsic electrical properties of neurons in the brain (Lujan, 2010). K+ currents control action potential duration, Ca^{2+} dependent synaptic plasticity, the release of neurotransmitters and epileptiform burst activity. Enhanced excitability of K+ channels (via downmodulation, or changes in biophysical properties such as inactivation kinetics and voltage dependence), could all result in enhanced Ca^{2+} responses to firing activity (Pongs, 1999). Ca^{2+} activated K+ channels are a large family of K+ channels that are found throughout the central nervous system and in many other cell types, but its *in vivo* physiological functions have not been fully defined. Ca^{2+} dependent K+ channels are activated by both depolarization and increases in intracellular free $[Ca^{2+}]$. Ca^{2+} dependent K+ currents contribute to the repolarization of neurons to resting membrane potential (Hille, 2001). Thus, Ca^{2+} dependent K+ channels determine the shape of the action potential and help in regulating cell excitability (Goodman, 2008). Outward currents play a principal but not a sole role in repolarization in many types of excitable cells. The excitability of neurons is governed by the input they receive from their neighbours and the intrinsic excitability of the neuron. Electrophysiological studies have revealed that the voltage gated ion channels are important in determining the intrinsic excitability of neurons in the CNS. The voltage gated ion channels are critical in producing hyperexcitability such as that associated with seizure discharges (Errington, 2005) and causing epilepsy in humans (Du et al., 2005; Ez-Sampedro et al., 2006; Jorge et al., 2011). Ca^{2+} activated K^+ channels are essential for the production of bursting activity in mammalian cortical neurons (Jin et al., 2000) and they can also influence rhythmic firing patterns and bursting output (Gu et al., 2007). Voltage gated K^+ currents play crucial roles in modifying neuronal cellular and network excitability and activity (Muller & Connor, 1991).

Results of our study demonstrate that SH reducing agent, GSH, incompletely inhibited the effect of CHP on calcium-activated potassium channels in LRNC. The SH groups are known to be important for the function of many membrane transport systems. These include also various potassium channels (Lee et al., 1994; Han et al., 1996). Redox modification of cysteine SH groups may also be an important mechanism of controlling ion channel function. There are several explanations for the incomplete recovery of calcium-activated potassium channels. The simplest could be that CHP treatment must be modifying other amino acid residues, e.g., as methionine or tryptophan, besides cysteine. Cysteine and methionine residues are particularly sensitive to oxidation by almost all forms of ROS. In addition, the localization of the critical SH groups (responsible for the inhibitory action of the oxidative agents), could explain the observed partly protective effects of glutathione against cumene hydroperoxide-induced toxicity. The part of changes in channel properties depend on cysteine residues located on the cytoplasmic domains of the calcium-activated potassium channels in LRNC. The activity of potassium channels is dependent on the redox status of one or more SH groups on the channel protein, or an associated regulatory protein. Of course, it is possible that the oxidant agent affects other components associated to the membrane or to the channel (the target of the oxidizing agent could be a β-subunit or some membrane-bound enzyme able to promote channel phosphorylation).

5. Conclusions and implications

The present data show that the CHP is a more efficient neurotoxin and oxidant than H_2O_2, as well as that the suppression of Ca^{2+} activated outward K^+ currents is responsible for the prolongation of spike potential in leech Retzius nerve cells. Here we discuss the implications that free radicals can have a significant role in the appearance of spontaneous repetitive activity in Retzius nerve cells by interrupting the process of repolarization.

What is the pathophysiological relevance of a block of voltage-gated K^+ channels? In the past few years it has become more appreciated that a block of voltage-gated K^+ channels by ROS contributes to increased neuronal excitability and repetitive firing. These data indicate that a block of voltage gated K^+ channels contributes to an increase in neuronal excitability such as that associated with seizure discharges and causing epilepsy in humans. The interaction of ROS with K^+ channels may cause modifications of membrane currents and potentials thereby leading to neuronal dysfunction. The suppression of Ca^{2+} dependent K^+ currents proposed in this paper might have a broader significance, pertaining not only to leeches, but mammalian neurons as well. Leech ganglia are good models for studying the cellular basis for epileptiform activity. The largest neurons in the leech nervous system are Retzius cells which exhibit stable resting membrane potential and which are non-bursting neurons with a low spontaneous firing rate. An understanding of ion mechanisms epilepsy will provide insight into the molecular events of epileptogenesis, improve molecular diagnostic utility, and identify novel therapeutic targets for improved treatment of human epilepsy.

Electrophysiological analyses showed that oxidative modification of K+ channels might represent a fundamental pathogenic mechanism in the mammalian brain during normal aging, as well as in neurodegenerative diseases such as Alzheimer's and Parkinson's. Therefore, it is probable that the action of ROS on K+ channels might play a role in changes in electrical identity of neurons produced by ischemia and of course in neuronal death.

Considering that K+ channels and ROS are universal players in the biological game, we put forward the hypothesis that the oxidation of voltage-gated K+ channels may represent a general pathogenic mechanism in biological organisms. Obviously, more work is needed to establish the possible involvement of ion channels and of their modulation by ROS as important mechanisms in several pathological conditions in the brain. In addition, such knowledge may help to explain pathophysiological alterations in neurological diseases and to develop new strategies for therapeutic intervention that aim at preventing oxidative stress in the brain.

6. References

Aoyama, K.; Watabe, M. & Nakaki, T. (2008). Regulation of neuronal glutathione synthesis. *Journal of Pharmacological Sciences*, Vol.108, No.3, pp. 227-238

Axelsen, P.H.; Komatsu, H. & Murray, I.V. (2011). Oxidative stress and cell membranes in the pathogenesis of Alzheimer's disease. Physiology (Bethesda), Vol. 26, No.3, pp. 54-69

Barrington, P.L. (1994). Interactions of H2O2, EGTA and patch pipette recording methods in feline ventricular myocytes. *Journal of Molecular and Cellular Cardiology, Vol.26*, pp. 557–568

Bienert, G.P.; Moller, A.L.; Kristiansen, K.A.; Schulz, A.; Moller, I.M.; Schjoerring, J.K. & Jahn, T.P. (2007). Specific aquaporins facilitate the diffusion of hydrogen peroxide across membranes. *Journal of Biological Chemistry*, Vol.282, No.2, pp. 1183-1192

Beleslin, B.B. (1982). Membrane physiology of excitable cells in annelids. In: *Membrane physiology of invertebrates*. R. B. Podesta, (Ed.), 199-260, Marcel Dekker, ISBN 0824715039, New York, U.S.A

Beleslin, B.B. (1985). Sensitivity of the resting sodium conductivity of leech Retzius nerve cells to tetrodotoxin. *Comparative Biochemistry and Physiology*, Vol.81, No.2, pp. 323-328

Beleslin, B.B.; Ristanovic, D. & Osmanovic, S. (1988). Somatic outward currents in voltage clamp leech Retzius nerve cell. *Comparative Biochemistry and Physiology* Part A: Physiology, Vol.89, No.2, pp. 187-196

Beleslin, B.B.; Dekleva, N.M.; Jovanovic, D.Z.; Beleslin-Cokic, B.B.; Cokic, V. & Cemerikic, D. (1998). Resistivity of different tissues to oxidative stress. *Iugoslavica Physiologica et Pharmacologica Acta*, Vol.34, No.2, pp. 39-45

Bracci, R. (1992). Calcium involvement in free radical effects. *Calcified Tissue International*, Vol.51, No.6, pp. 401-405

Carmeliet, E. (1999). Cardiac ionic currents and acute ischemia: from channels to arrhythmias. Physiological Reviews, Vol.79, No.3, pp. 917–1017

Cerbai, E.; Ambrosio, G.; Porciatti, F.; Chiariello, M.; Giotti, A. & Mugelli, A. (1991). Cellular electrophysiological basis for oxygen radical-induced arrhythmias. A patch-clamp study in Guinea pig ventricular myocytes. *Circulatio*, Vol.84, pp. 1773-1782

Cheng, A.; Wang, S.; Yang, D.; Xiao, R. & Mattson, M.P. (2003). Calmodulin mediates brain-derived neurotrophic factor cell survival signaling upstream of Akt kinase in embryonic neocortical neurons. *Journal of Biological Chemistry*, Vol.278, No.9, pp. 7591-7599

Clark, J. & Simon, D.K. (2009). Transcribe to survive: transcriptional control of antioxidant defense programs for neuroprotection in Parkinson's disease. *Antioxidants and Redox Signaling*, Vol.11, No.3, pp. 509-528

Cohen, G. (1994). Enzymatic/nonenzymatic sources of oxyradicals and regulation of antioxidant defenses. *Annals of the New York Academy of Sciences*, Vol.73, pp. 8-14

Coetzee, W.A.; Owen, P.; Dennis, S.C.; Saman, S. & Opie, L.H. (1990). Reperfusion damage: Free radicals mediate delayed membrane changes rather than early ventricular arrhythmias. *Cardiovascular Research*, Vol. 24, pp. 156-164

Correa, F.; Mallard, C.; Nilsson, M. & Sandberg M. (2011). Activated microglia decrease histone acetylation and Nrf2-inducible anti-oxidant defence in astrocytes: restoring effects of inhibitors of HDACs, p38 MAPK and GSK3β. *Neurobiology of Disease*, Vol.44, No.1, pp. 142-151

Goldhaber, J.I.; Ji, S.; Lamp, S.T. & Weiss, J.N. (1989). Effects of exogenous free radicals on electromechanical function and metabolism in isolated rabbit and guinea pig ventricle. Implications for ischemia and reperfusion injury. *Journal of Clinical Investigation*, Vol.83, No.6, pp. 1800-1809

Goldhaber, J.I. & Liu, E. (1994). Excitation-contraction coupling in single guinea-pig ventricular myocytes exposed to hydrogen peroxide. *Journal of Physiology*, Vol.477, No.1, pp. 135-147

Goodman, B.E. (2008). Channels active in the excitability of nerves and skeletal muscles across the neuromuscular junction: basic function and pathophysiology. *Advances in Physiology Education*, Vol.32, No.2, pp. 127-135

Graham, S.; Ding, M.; Ding, Y.; Sours-Brothers, S.; Luchowski, R.; Gryczynski, Z.; Yorio, T.; Ma, H. & Ma, R. (2010). Canonical transient receptor potential 6 (TRPC6), a redox-regulated cation channel. *Journal of Biological Chemistry*, Vol.285, No.30, pp. 23466-23476

Gu, N.; Vervaeke, K. & Storm, J.F. (2007). BK potassium channels facilitate high-frequency firing and cause early spike frequency adaptation in rat CA1 hippocampal pyramidal cells. *Journal of Physiology*, Vol.580, No.3, pp. 859-882

Gutowicz, M. (2011). The influence of reactive oxygen species on the central nervous system. Postepy Hig Med Dosw (Online), ISSN:1732-2693 (Electronic), Vol.65, pp. 104-113

Del Maestro, R.; Thaw, H.H.; Björk, J.; Planker, M. & Arfors, K.E. (1980). Free radicals as mediators of tissue injury. *Acta Physiologica Scandinavica*, Vol.492, pp. 43-57

Dringen, R. & Gutterer, J.M. (2002). Glutathione reductase from bovine brain. *Methods in Enzymology*, Vol. 348, pp.281-288

Dringen, R. & Hirrlinger, J. (2003). Glutathione pathways in the brain. *Journal of Biological Chemistry*, Vol. 384, No.4, pp. 505-516

Du, W.; Bautista, JF.; Yang, H.; ez-Sampedro, A.; You, SA.; Wang, L.; Kotagal, P.; Lüders, H.O.; Shi, J.; Cui, J.; Richerson, G.B. & Wang, Q.K. (2005). Calcium-sensitive potassium channelopathy in human epilepsy and paroxysmal movement disorder. *Nature Genetics*, Vol.37, No.7, pp. 733-738

Dudek, E.; Yasumura, T. & Rash, J.E. (1998). 'Non-synaptic' mechanisms in seizures and epileptogenesis. *Cell Biology International*, Vol.22, pp. 793-805

Dyavanapalli, J.; Rimmer, K. & Harper, A.A. (2010). Reactive oxygen species alters the electrophysiological properties and raises [Ca2+]i in intracardiac ganglion neurons.

The American Journal of Physiology - Regulatory, Integrative and Comparative Physiology, Vol. 299, No.1, pp. 42-54

Errington, A.C.; Stohr, T. & Lees, G. (2005). Voltage gated ion channels: targets for anticonvulsant drugs. *Current Topics in Medicinal Chemistry,* Vol.5, No.1, pp. 15-30

Ez-Sampedro, A.; Silverman, W.R.; Bautista, J.F. & Richerson, G.B. (2006). Mechanism of increased open probability by a mutation of the BK channel. *Journal of Neurophysiology,* Vol.96, pp. 1507–1516

Facchinetti, F.; Dawson, V.L. & Dawson, T.M. (1998). Free radicals as mediators of neuronal injuri. *Cellular and Molecular Neurobiology,* Vol.18, No.6, pp. 667-682

Forman, H.; Maiorino, M. & Ursini, F. (2010). Signaling Functions of Reactive Oxygen Species. *Biochemistry,* Vol. 49, No.5, pp. 835-842

Hagiwara, S. & Morita, H. (1962). Electrotonic transmission between two nerve cells in leech ganglion. *Journal of Neurophysiology,* Vol.25, pp. 721-731

Halliwell, B. & Gutteridge, J.M.C. (1984). Oxygen toxicity, oxygen radicals, transition metals and disease. *Biochemical Journal,* Vol. 219, pp. 1-14

Han, J.; Kim, E.; Ho, W.K. & Earm, Y.E. (1996). Sulfhydryl redox modulates ATP-sensitive K+ channels in rabbit ventricular myocytes. *Biochemical and Biophysical Research Communications,* Vol. 219, No.3, pp. 900-903

Hasan, S.M.; Joe, M. & Alshuaib, W.B. (2007). Oxidative stress alters physiological and morphological neuronal properties. *Neurochemical Research,* Vol.32, No.7, pp. 1169-1178

Haskew-Layton, R.E.; Payappilly, J.B.; Smirnova, N.A.; Ma, T.C.; Chan, K.K.; Murphy, T.H.; Guo, H.; Langley, B.; Sultana, R.; Butterfield, D.A.; Santagata, S.; Alldred, M.J.; Gazaryan, I.G.; Bell, G.W.; Ginsberg, S.D. & Ratan, R.R. (2010). Controlled enzymatic production of astrocytic hydrogen peroxide protects neurons from oxidative stress via an Nrf2-independent pathway. *Proceedings of the National Academy of Sciences of the United States of America,* Vol. 107, No.40, pp. 17385-17390

Hayashi, H.; Miyata, H.; Watanabe, H.; Kobayashi, A. & Yamazaki, N. (1989). Effects of hydrogen peroxide on action potentials and intracellular Ca2+ concentration of guinea pig heart. *Cardiovascular Research,* Vol.23, pp. 767–773

Hille, B. (2001). *Ion Channels of Excitable Membranes,* Third edition, Sinauer Associates, Inc. (Ed.), ISBN 0-87893-321-2, Massachussetts, U.S.A

Hool, L.C. (2006). Reactive oxygen species in cardiac signalling: from mitochondria to plasma membrane ion channels. *Clinical and Experimental Pharmacology and Physiology,* Vol.33, pp. 146–151

Hur, W. & Gray, N.S. (2011). Small molecule modulators of antioxidant response pathway. *Current Opinion in Chemical Biology,* Vol.15, No.1, pp. 162-173

Hwang, Y.P.; Yun, H.J.; Chun, H.K.; Chung, Y.C.; Kim, H.K.; Jeong, M.H.; Jeong, M.H.; Yoon, T.R. & Jeong, H.G. (2009). Protective mechanisms of 3-caffeoyl, 4-dihydrocaffeoyl quinic acid from Salicornia herbacea against tert-butyl hydroperoxide-induced oxidative damage. *Chemico-Biological Interactions,* Vol.181, pp. 366–376

Jin, W.; Sugaya, A.; Tsuda, T.; Ohguchi, H. & Sugaya, E. (2000). Relationship between large conductance calcium-activated potassium channel and bursting activity. *Brain Research,* Vol.860, pp. 21–28

Jovanović, Z. & Beleslin, B.B. (1996). Resistivity of leech Retzius nerve cells to long-lasting oxidant. *Journal of Neurochemistry,* Vol.66, pp. S32, ISSN:0022-3042

Jovanović, Z. & Beleslin, B.B. (1997). Resistivity of leech Retzius nerve cells to long-lasting oxidant. In: *Neurochemistry, Cellular, Molecular and Clinical Aspects*, A. Teelken & J. Korf, (Ed.), 983-986, ISBN 0-306-45705-9, New York, U.S.A.

Jovanović, Z. & Beleslin, B.B. (2004). Effects of long lasting oxidants on the electrophysiological properties of leech Retzius nerve cells. *Iugoslavica Physiologica et Pharmacologica Acta*, Vol.40, No.1-3, pp. 55-64

Jovanović, Z. (2010). Neuroprotective effect of the mannitol on oxidative stress induced by cumene hydroperoxide. *Medicinski časopis*, Vol.44, No.3, pp. 9-14, ISSN:0350.1221.UDK.61.

Jorge, B.S.; Campbell, C.M.; Miller, A.R.; Rutte, E.D.; Gurnett, C.A.; Vanoye, C.G.; George, A.L. & Kearney, J.A. (2011). Voltage-gated potassium channel KCNV2 ($K_v8.2$) contributes to epilepsy susceptibility. *Proceedings of the National Academy of Sciences of the United States of America*, Vol.108, No.13, pp. 5443-5448

Kourie, J.I. (1998). Interaction of reactive oxygen species with ion transport mechanisms. (1998). *American Journal of Physiology - Cell Physiology*, Vol. 275, pp. C1–C24

Lee, M.Y.; Bang, H.W.; Lim, I.J.; Uhm, D.Y. & Rhee, S.D. (1994). Modulation of large conductance calcium-activated K+ channel by membrane-delimited protein kinase and phosphatase activities. *Pflügers Archiv*, Vol.429, No.1, pp. 150-152

Lee, J.W.; Miyawaki, H.; Bobst, E.V.; Hester, J.D.; Ashraf, M. & Bobst, A.M. (1999). Improved functional recovery of ischemic rat hearts due to singlet oxygen scavengers histidine and carnosine. *Journal of Molecular and Cellular Cardiology*, Vol. 31, pp. 113–121

Lent, C.M. (1977). The Retzius cells within the central nervous system of leeches. *Progress in Neurobiology*, Vol. 8, pp. 8l- 117

Limon-Pacheco, J.H. & Gonsebatt, M.E. (2010). The glutathione system and its regulation by neurohormone melatonin in the central nervous system. *Current Medicinal Chemistry - Central Nervous System Agents*, Vol.10, No.4, pp. 287-297

Liu, Y. & Templeton, D.M. (2008). Initiation of caspase-independent death in mouse mesangial cells by Cd2+: involvement of p38 kinase and CaMK-II. *Journal of Cellular Physiology*, Vol.217, pp. 307–318

Lujan, R. (2010). Organisation of potassium channels on the neuronal surface. *Journal of Chemical Neuroanatomy*, Vol.40, No.1, pp. 1-20

Makino, N.; Sasaki, K.; Hashida, K. & Sakakura, Y. (2004). A metabolic model describing the H_2O_2 elimination by mammalian cells including H_2O_2 permeation through cytoplasmic and peroxisomal membranes: comparison with experimental data. *Biochimica et Biophysica Acta*, Vol.1673, No.3, pp. 149-159

Matsuura, H. & Shattock, M.J. (1991). Effects of oxidant stress on steady-state background currents in isilated ventricular myocytes. *American Journal of Physiology*, Vol. 261, pp. H1358-H1365

McCubrey, J.A.; Lahair, M.M. & Franklin, R.A. (2006). Reactive oxygen species-induced activation of the MAP kinase signaling pathways. *Antioxidants and Redox Signaling*, Vol.8, No.9-10, pp. 1775-1789

Muller ,W. & Connor, J.A. (1991). Dendritic spines as individual neuronal compartments for synaptic Ca2+ responses. *Nature*, Vol. 354, pp. 73–76

Nakaya, H.; Takeda, Y.; Tohse, N. & Kanno, M. (1991). Effects of ATP-sensitive K + channel blockers on the action potential shortening in hypoxic and ischaemic myocardium. *British Journal of Pharmacology*, Vol.103, pp. 1019-1026

Nakaya, H.; Takeda, Y.; Tohse, N. & Kanno, M. (1992). Mechanism of the membrane depolarization induced by oxidative stress in guinea-pig ventricular cells. *Journal of Molecular and Cellular Cardiology*, Vol.24, No.5, pp. 523-534

Nistico, R.; Piccirilli, S.; Cucchiaroni, M.L.; Armogida, M.; Guatteo, E.; Giampa, C.; Fusco, FR.; Bernardi, G.; Nistico, G. & Mercuri, N.B. (2008). Neuroprotective effect of hydrogen peroxide on an *in vitro* model of brain ischaemia. *British Journal of Pharmacology*, Vol.153, No.5, pp. 1022-1029

Pongs, O. (1999). Voltage-gated potassium channels: from hyperexcitability to excitement. *FEBS Letters*, Vol.452, pp. 31–35

Racay, P.; Kaplan, P.; Mezesova, V. & Lehotsky, J. (1997). Lipid peroxidation both inhibits $Ca^{(2+)}$-ATPase and increases Ca^{2+} permeability of endoplasmic reticulum membrane. *Biochemistry and Molecular Biology International*, Vol.41, No.4, pp. 647-655

Ruffels, J.; Griffin, M. & Dickenson, J.M. (2004). Activation of ERK1/2, JNK and PKB by hydrogen peroxide in human SH-SY5Y neuroblastoma cells: role of ERK1/2 in H2O2-induced cell death. *European Journal of Pharmacology*, Vol.483, No.2-3, pp. 163-173

Satoh, H. & Matsui, K. (1997). Electrical and mechanical modulations by oxygen-derived free-radical generating systems in Guinea-pig heart muscles. *Journal of Pharmacy and Pharmacology*, Vol.49, pp. 505–510

Shih, A.Y.; Johnson, D.A.; Wong, G.; Kraft, A.D.; Jiang, L.; Erb, H.; Johnson, J.A. & Murphy, T.H. (2003). Coordinate regulation of glutathione biosynthesis and release by Nrf2-expressing glia potently protects neurons from oxidative stress. *Journal of Neuroscience*, Vol.23, pp. 3394–3406

Skaper, S.D.; Floreani, M.; Ceccon, M.; Facci, L. & Giusti, P. (1999). Excitotoxicity, Oxidative Stress, and the Neuroprotective Potential of Melatonin. *Annals of the New York Academy of Sciences*, Vol.890, pp. 107–118

Son, Y.; Cheong, Y.K.; Kim, N.H.; Chung, H.T.; Kang, D.G. & Pae, H.O. (2011). Mitogen-Activated Protein Kinases and Reactive Oxygen Species: How Can ROS Activate MAPK Pathways? *Journal of Receptors and Signal Transduction*, 2011;2011:792639. Epub 2011 Feb 6

Son ,Y.O.; Wang, X.; Hitron, J.A.; Zhang, Z.; Cheng, S.; Budhraja, A.; Ding, S.; Lee, J.C. & Shi, X. (2011). Cadmium induces autophagy through ROS-dependent activation of the LKB1-AMPK signaling in skin epidermal cells. *Toxicology and Applied Pharmacology*, Vol.255, No.3, pp. 287-296

Soto, M.A.; Gonzalez, C.; Lissi, E.; Vergara, C. & Latorre, R. (2002). Ca2+-activated K+ channel inhibition by reactive oxygen species. *American Journal of Physiology - Cell Physiology*, Vol.282, pp. C461-C471

Stepkowski, T.M. & Kruszewski, M.K. (2011). Molecular cross-talk between the NRF2/KEAP1 signaling pathway, autophagy, and apoptosis. *Free Radical Biology & Medicine*, Vol.50, No.9, pp.1186-1195

Stewart, R.R.; Nicholls, J.G. & Adams, W.B. (1989). Na+, K+ and Ca2+ currents in identified leech neurones in culture. *Journal of Experimental Biology*, Vol.141, pp. 1-20

Stone, J.R. (2004). An assessment of proposed mechanisms for sensing hydrogen peroxide in mammalian systems. *Archives of Biochemistry and Biophysics*, Vol.422, No.2, pp. 119-124

Su, Z.; Limberis, J.; Martin, R.L.; Xu, R.; Kolbe, K.; Heinemann, S.H.; Hoshi, T.; Cox, B.F. & Gintant, G.A. (2007). Functional consequences of methionine oxidation of hERG potassium channels. *Biochemical Pharmacology*, Vol.74, No.5, pp. 702-711

Takeuchi, K. & Yoshii, K. (2008). Superoxide modifies AMPA receptors and voltage-gated K⁺ channels of mouse hippocampal neurons. *Brain Research*, Vol.21, No.1236, pp. 49-56

Tarr, M. & Valenzeno, D.P. (1989). Modification of cardiac action potential by photosensitizer-generated reactive oxygen. *Journal of Molecular and Cellular Cardiology*, Vol.21, pp. 539–543

Tarr, M. & Valenzeno, D.P. (1991). Modification of cardiac ionic currents by photosensiter-generated reactive oxygen. *Journal of Molecular and Cellular Cardiology*, Vol.23, pp. 639-649

Tarr, M.; Arriaga, E. & Valenzeno, D. (1995). Progression ofcardiac potassium current modification after brief exposure to reactive oxygen. *Journal of Molecular and Cellular Cardiology*, Vol.27, pp. 1099–1109

Thannickal, V.J. & Fanburg, B.L. (2000). Reactive oxygen species in cell signaling. *American Journal of Physiology: Lung Cellular and Molecular Physiology*, Vol.279, No.6, pp. L1005-1028

Tokube, K.; Kiyosue, T. & Arita, M. (1996). Openings of cardiac K_{ATP} channel by oxygen free radicals produced by xanthine oxidase reaction. *American Journal of Physiology*, Vol.271, No.2, pp. H478-489

Vacher, H.; Mohapatra, D.P. & Trimmer, J.S. (2008). Localization and targeting of voltage-dependent ion channels in mammalian central neurons. *Physiological Reviews*, Vol.88, No.4, pp. 1407-1447

Van der Vliet, A. & Bast, A. (1992). Effect of oxidative stress on receptors and signal transmission. *Chemico-Biological Interactions*, Vol.85, No.2-3, pp. 95-116

Vega-Saenz de Miera, E. & Rudy, B. (1992). Modulation of K⁺ channels by hydrogen peroxide. *Biochemical and Biophysical Research Communications*, Vol.186, No.3, pp. 1681-1687

Vimard, F.; Saucet, M.; Nicole, O.; Feuilloley, M. & Duval, D. (2010). Toxicity induced by cumene hydroperoxide in PC12 cells: Protective role of thiol donors. *Journal of Biochemical and Molecular Toxicology*, Vol.25, No.4, pp.205-215

Ward, C.A. & Giles, W.R. (1997). Ionic mechanism of the effects of hydrogen peroxide in rat ventricular myocytes. *Journal of Physiology*, Vol.500, No.3, pp. 631-642

Wilson, J.X. (1997). Antioxidant defense of the brain: a role for astrocytes. *Canadian Journal of Physiology and Pharmacology*, Vol.7, No.10-11, pp. 1149-1163

Whittemore, E.R.; Loo, D.T.; Watt, J.A. & Cotman, C.W. (1995). A detailed analysis of hydrogen peroxide-induced cell death in primary neuronal culture. *Neuroscience*, Vol.67, No.4, pp. 921-993

Whyte, K.A.; Hogg, R.C.; Dyavanapalli, J.; Harper, A.A. & Adams, D.J. (2009). Reactive oxygen species modulate neuronal excitability in rat intrinsic cardiac ganglia. *Autonomic Neuroscience*, Vol.150, pp. 45–52

Yang, J.L.; Sykora, P.; Wilson, D.M.; Mattson, M.P. & Bohr, V.A. (2011). The excitatory neurotransmitter glutamate stimulates DNA repair to increase neuronal resiliency. *Mechanisms of Ageing and Development*, Vol.132, No.8-9, pp. 405-411

Zou, X.; Feng, Z.; Li, Y.; Wang, Y.; Wertz, K.; Weber, P.; Fu, Y. & Liu, J. (2011). Stimulation of GSH synthesis to prevent oxidative stress-induced apoptosis by hydroxytyrosol in human retinal pigment epithelial cells: activation of Nrf2 and JNK-p62/SQSTM1 pathways. *Journal of Nutritional Biochemistry*, [Epub ahead of print]

2

Oxidative Stress in Parkinson's Disease; Parallels Between Current Animal Models, Human Studies and Cells

Anwar Norazit[1,2], George Mellick[1] and Adrian C. B. Meedeniya[1,3*]
[1]Eskitis Institute for Cell and Molecular Therapies,
Griffith University, Nathan, Queensland
[2]Department of Molecular Medicine, Faculty of Medicine,
University of Kuala Lumpur,
[3]Griffith Health Institute, Griffith University, Gold Coast, Queensland,
[1,3]Australia
[2]Malaysia

1. Introduction

Reactive oxygen species (ROS) are byproducts generated primarily from the breakdown of oxygen during aerobic metabolism. ROS can be found as free radicals containing highly reactive unpaired electrons (super oxide, nitric oxide, hydroxyl radical) or as other molecules (hydrogen peroxide, peroxynitrate). Under normal physiological conditions, ROS are neutralized by antioxidants. However, an increased production of ROS, namely, oxidative stress, can occur under pathophysiological conditions. Oxidative stress can be defined as an imbalance between oxidants and antioxidants where the oxidants are favoured, potentially leading to cellular damage (Sies 1985, Sies 1986, Sies 1991).

The brain is highly susceptible to oxidative stress due to its high metabolic rate and limited regeneration capability (Andersen 2004). Oxidative stress has been implicated in a variety of neurodegenerative disease, including Parkinson's disease (PD), Alzheimer's disease, and amyotrophic lateral sclerosis. However, presence of oxidative stress as a cause or consequence of neurodegeneration, remains to be determined.

This review focuses on the pathogenesis of PD in relation to oxidative stress and the current animal models used to mimic the pathophysiology of human PD. We will also compare the animal and human data for PD like neurodegeneration with cell models of PD. This will also include a review on current experimentation and antioxidant therapies for counteracting oxidative stress.

2. Parkinson's disease

PD was first described in the early 19th century in the monologue "An essay on the shaking palsy" (Parkinson 2002). The three cardinal symptoms of PD are bradykinesia, rigidity, and

---

* Corresponding Author

postural instability (Gash et al. 1996). The clinical diagnosis of PD is usually based on the United Kingdom Parkinson's Disease Society Brain Research Centre criteria (Gibb and Lees 1988), with the accuracy of this diagnosis being high (90%) (Hughes, Daniel and Lees 2001). During the first 10 years of the disease, patients usually exhibit slowness of movement, mild gait hypokinesia, resting tremor, micrographic handwriting and reduced speech volume (Morris 2000). During the latter stages of the disease, festination, dyskinesia, akinesia, marked hypokinesia, postural instability and falls are usually more pronounced (Morris 2000). The symptoms of this condition only show after 80% of the striatal innervations and 60% of the substantia nigra par compacta dopaminergic neurons are lost, suggestive of a pathological process initiated at the synaptic end of the nigral neurons, with neuronal death as a result from a 'dying back' process (Dauer and Przedborski 2003)

2.1 Parkinson's disease pathology

PD can be defined as a progressive neurodegenerative disorder characterised histopathologically by the degeneration of the dopaminergic nigrostriatal pathway (Watts et al. 1997). Post-mortem analyses of the substantia nigra in PD patients have shown effects of oxidative stress with a decrease in glutathione (GSH) levels, increased levels of iron, neuromelanin associated redox-active iron, lipid peroxidation, protein oxidation and DNA damage (Jenner and Olanow 1998, Faucheux et al. 2003, Dexter et al. 1989). These changes may directly induce nigral cell degeneration via oxidative stress or render neurons susceptible to the actions of toxins.

The neurotrophic factor milieu of the brain is affected in PD patients. A decrease in brain derived neurotrophic factor (BDNF) in the post-mortem brains of clinically and pathologically diagnosed PD patients, compared to normal controls (Mogi et al. 1999, Parain et al. 1999). A decrease in BDNF mRNA expression in the substantia nigra of PD patients has also been demonstrated (Howells et al. 2000). A decrease in glial cell derived neurotrophic factor (GDNF) and ciliary neurotrophic factor (CNTF) also occurs in the brains of PD patients (Chauhan, Siegel and Lee 2001). The resultant inability to up-regulate neurotrophic factors in response to injury or stress may compromised defence mechanisms of the brain, thus contributing to cell degeneration (Olanow and Tatton 1999).

2.2 Pathogenic role of host/exogenous factors

There is increasing evidence of host and /or exogenous factors playing a role in the pathogenesis of PD. Many of these factors negatively impact mitochondrial function.

2.2.1 Age

The incidence of PD is related to increasing age (de Rijk et al. 1997, Mayeux et al. 1992, Van Den Eeden et al. 2003). With age, high levels of mitochondrial DNA deletion (Bender et al. 2006); increase in α-synuclein (Chu and Kordower 2007); decrease in dopamine transporter mRNA (Bannon et al. 1992); decrease in neurotrophic factor gene expression (Lee, Weindruch and Prolla 2000); reduced response to growth factors (Smith 1996); decrease in brain peroxidase and catalase activity (Ambani, Van Woert and Murphy 1975); and a decrease in dopamine binding sites (Severson et al. 1982) are apparent in the neuronal population of the substantia nigra. The relationship between PD and aging includes a superposition of a topographic gradient of neuronal loss in brainstem and basal forebrain structures related to the disease process and an age-related temporal gradient. Clinical progression of PD is determined by advancing age and not by disease duration; and a biological interaction is involved in the effects of the disease process and aging on non dopaminergic structures (Levy 2007).

2.2.2 Environmental toxins

Pesticide exposure is implicated as an environmental risk factor for PD (Herishanu et al. 2001, Le Couteur et al. 1999, Tanner 1989, Ho, Woo and Lee 1989). The susceptibility of humans to these pesticides has been reported to be linked to genetic factors (Menegon et al. 1998, Drozdzik et al. 2003). Many of these pesticides have a major site of action along the mitochondrial electron transport chain (Degli Esposti 1998), which results in increasing oxidative stress.

2.2.3 Genetic determinants

The role of genetic factors in PD has been the subject of intense scrutiny. The first clue of familial PD was provided in 1907 when Gowers reported approximately 15% of his PD patients reported an affected family member (Gowers 1902) Since then many genes have been identified as causing familial PD (Schapira 2008). The products associated with the gene mutations are either mitochondrial proteins or are associated with mitochondria. Namely, proteins that interface with the pathways of oxidative stress and free radical damage.

Gene	Locus	Inheritance	Gene product
PARK1/4	4q21	Autosomal dominant	α-synuclein
PARK2	6q25	Autosomal recessive	Parkin
PARK3	2p13	Autosomal dominant	-
UCH-L1	4p15	Autosomal dominant	Ubiquitin thiolesterase
PINK1	1p35	Autosomal recessive	PTEN-induced putative kinase 1
PARK7	1p36	Autosomal recessive	DJ-1
LRRK2	12p	Autosomal dominant	Leucine-rich repeat kinase 2
ATP13A2	1p36	Autosomal recessive	ATPase type 13A2
PARK10	1p32	Autosomal recessive	-
PARK11	2q36-q37	-	-

Table 1. Genetic causes of PD (modified from Schapira, 2008)

2.2.4 Sporadic Parkinson's disease

The Braak hypothesis suggests that the initial event in sporadic PD may be an infectious assault on susceptible neuronal types in the olfactory or enteric nervous system (Braak et al. 2003a, Hawkes, Del Tredici and Braak 2007, Braak et al. 2003b). The Lewy neuritis and Lewy bodies' progress rostrally in stages into the lower brainstem region (medulla oblongata and pontine tegmentum; stages 1 and 2), followed by the midbrain (substantia nigra; stage 3) and the basal prosencephalon and mesocortex (stage 4), and eventually the neocortex (stage 5 and 6) (Braak et al. 2003a, Braak et al. 2003b). Stages 1 to 3 have been characterised as the pre-symptomatic phase while stage 4 to 6 has been characterised as the symptomatic phase. The sequential ascending topography reported by Braak (2003b) has been reported to be only partially in line with the latest imaging of PD (Brooks 2010). It may instead reflect the more primitive regions of the nervous system (and perhaps the more active) showing a greater susceptibility.

3. Parkinson's disease animal models

There are both toxin and genetic animal models popularly used to represent PD. Both of these models increase ROS production directly or indirectly to induce the degeneration of the nigrostriatal pathway in laboratory animals.

3.1 Neurotoxins

There are four main toxin induced models popularly used to produce PD like symptoms in rodents. The neurotoxins used are 6-OHDA, MPTP, paraquat in combination with Maneb, and rotenone.

3.1.1 6-Hydroxydopamine (6-OHDA)

The neurotoxin 6-OHDA destroys catecholaminergic neurons by the combined effect of reactive oxygen species (ROS) and quinines (Cohen and Heikkila 1984). To specifically induce PD in an animal model, 6-OHDA is injected stereotaxically into the substantia nigra, the medial forebrain bundle, or the striatum (Javoy et al. 1976). 6-OHDA is delivered directly into the brain by stereotaxic means as 6-OHDA does not cross the blood-brain barrier, thus ruling out systemic injections (Bove et al. 2005). The administration of 6-OHDA into the substantia nigra or the medial forebrain bundle mediates its uptake anterogradely, while administration into the striatum causes the uptake of the chemical retrogradely. 6-OHDA is transported into dopaminergic neurons via their high-affinity catecholaminergic uptake system (Zigmond, Hastings and Abercrombie 1992). 6-OHDA is usually administered unilaterally to produce a unilateral lesion, allowing the unlesioned side to act as an internal control and to minimise morbidity and mortality (Betarbet, Sherer and Greenamyre 2002). The 6-OHDA lesioned rat has been extensively characterised behaviourally and pathologically, making it one of the models of choice when investigating PD (Schwarting and Huston 1996). However, the effects of 6-OHDA are acute and do not show the same cellular pathology (Lewy bodies) as seen in PD (Dawson and Dawson 2002).

3.1.2 1-Methyl-4-Phenyl-1,2,5,6-Tetrahydropyridine (MPTP)

MPTP was accidentally produced during the illegal production of 1-methyl-4-phenyl-4-propionoxypiperidine (MPPP), a synthetic opioid drug, causing heroin addicts to display Parkinson-like symptoms (Langston 1996). MPTP readily crosses the blood-brain-barrier and is converted by the enzyme monoamine oxidase B (MAO-B) to 1-methyl-4-phenyl-2, 3-dihydropyridium (MPDP+) that then deprotonates to generate the corresponding pyridium species, MPP+ (Smeyne and Jackson-Lewis 2005). MPP+ has a high affinity to the dopamine transporter, thus it is highly selective to dopaminergic neurons (Javitch et al. 1985). Its selective uptake leads to severe damage to the nigrostriatal dopaminergic system, acting as a neurotoxin that inhibits mitochondrial complex I, producing oxidative stress and disturbing intracellular calcium homeostasis (Sedelis, Schwarting and Huston 2001). When MPTP is delivered systemically, MPTP produces bilateral lesions of the dopamine neurons (Sedelis et al. 2001). Following systemic administration and collateral damage to the ventral tegmental area, an external source of dopamine is needed to stimulate adequate food and water uptake (Petzinger and Langston 1998). Bilateral lesions also have a high morbidity and mortality rate. The other difficulty with using MPTP lesioned animal models is that MPTP has varying effects on different animal models and strains due to differences in visceral functions (Betarbet et al. 2002). These drawbacks in mice are being looked at with the behavioural phenotyping of a MPTP mouse animal model for PD (Sedelis et al. 2001).

3.1.3 Paraquat and Maneb

1,1-dimethyl-4,4-bipyridinium, better known as paraquat is a herbicide that has a similar structure to MPP+, making it a putative risk factor for PD (Dawson and Dawson 2002). Paraquat when delivered systemically can pass the blood brain barrier (Brooks et al. 1999). Paraquat has a high affinity to the nigrostriatal dopaminergic system (Thiruchelvam et al. 2000) and exposure in mice can cause up-regulation and aggregation of α-synuclein, a pathological sign of PD in humans (Manning-Bog et al. 2002). Due to its structural similarity to MPP+, paraquat's mechanism of action is believed to involve oxidative stress and its toxic effect via the mitochondria (Betarbet et al. 2002). A link between paraquat and other types of herbicide/pesticides with a increased incidence of PD has been demonstrated (Liou et al. 1997). The effects of paraquat on the dopaminergic system can be increased when mixed with Maneb (manganese ethylenebisdithiocarbamate) (Thiruchelvam et al. 2000). Maneb is a fungicide that has been implicated in an increased incidence of PD in humans (Ferraz et al. 1988, Meco et al. 1994). This animal model is of use when examining PD like syndromes due to environmental factors.

3.1.4 Rotenone

Rotenone is a naturally occurring root extract of *Lonchocarpus utilis* and *Lonchocarpus urucu* used as an insecticide as well as a piscicide (Caboni et al. 2004). Rotenone is a high-affinity inhibitor of complex 1 of the mitochondrial electron transport chain (Sherer et al. 2003b). Rotenone cytotoxicity is not dependent on the dopamine transporter (Hirata et al. 2008). Complex 1 inhibition by rotenone can cause the production of ROS that causes oxidative damage in dopaminergic neurons (Testa, Sherer and Greenamyre 2005).

When neurons are exposed to rotenone in cell culture, they produce ROS and superoxides, with dopaminergic neurons showing higher susceptibility to rotenone compared to other neurons (Radad, Rausch and Gille 2006, Ahmadi et al. 2003, Moon et al. 2005). Over time, the increase in ROS and superoxides produced by rotenone exposure leads to cell death (Ahmadi et al. 2003, Moon et al. 2005). Thus rotenone is a relevant toxin for developing a rat model of PD.

In general, rotenone blocks the electron transfer between the Complex I-associated iron-sulfur clusters and ubiquinone binding site (Grivennikova et al. 1997) (Figure 1). Specifically, rotenone acts as a semiquinone antagonist and displaces the ubisemiquinone intermediate at the ubiquinone binding site (Degli Esposti 1998). By inhibiting the ubiquinone binding site, rotenone alters the state of complex I, leading to higher superoxide

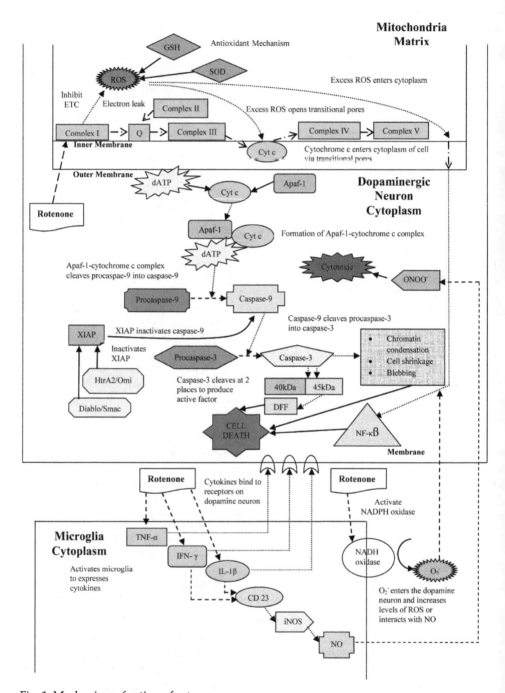

Fig. 1. Mechanism of action of rotenone

production at the same site (Lambert and Brand 2004). The inhibition of Complex I may overwhelm the mitochondria antioxidant system that consist of superoxide dismutase production with manganese (SOD) at the active site and GSH (Fridovich 1995, Betarbet et al. 2002), increasing the ROS concentration in the mitochondrial matrix. The increase of ROS activates the apoptotic intrinsic pathway, which increases the permeability of the outer membrane via the opening of transition pores (Turrens 2003). This allows cytochrome c to move from the intermembrane space into the cells cytoplasm, allowing it to bind with Apaf-1 (apoptopic protease activating factor) (Zou et al. 1997). In the presence of ATP or dATP, the Apaf-1 – cytochrome c complex changes its configuration, exposing the Apaf-1's CARD (caspase recruitment domain) to allow for the recruitment of procaspase-9 (Li et al. 1997). This interaction changes procaspase-9 into caspase-9 that in turn cleaves procaspase-3 into caspase-3. Cells contain an inhibitor of apoptosis (IAP) to prevent accidental caspase-9 activation. XIAP binds to the activated N terminus of caspase-9, making it inactive (Shiozaki et al. 2003). This process can be reversed by the mitochondrial proteins Diablo/Smac and HtrA2/Omi that are released during apoptosis (Vaux and Silke 2003), leading to caspase-3 activation. Caspase-3 cleaves the 45 kDa subunit of a two unit protein in two places producing a DNA Fragmentation Factor (DFF) (Liu et al. 1997). The increase in oxidative stress can also activate nuclear factor kappa β (NF-$\kappa\beta$) (Panet et al. 2001), which has been implicated in the beginning of a pro-apoptopic gene expression programme which may play a role in neurodegenerative disease (Qin et al. 1998, Panet et al. 2001).

Rotenone has the ability to activate microglia *in vivo* (Sherer et al. 2003a). This activation causes the release of tumour necrosis factor α (TNF- α), interleukin 1β (IL-1β), and interferon γ (IFN- γ). These cytokines can bind to their respective receptors and activate transduction pathways, leading to NF-$\kappa\beta$ activation and ultimately, apoptosis (Gao, Liu and Hong 2003). It was reported that NF-$\kappa\beta$ is present before caspase-3 (Wang et al. 2002), which suggests that superoxides and cytokines produced by the activation of microglia by rotenone play a role in the activation of NF-$\kappa\beta$ before complex I inhibition initiates the activation of caspase-3 apoptosis. However, excess ROS from the cytoplasm due to the Complex I inhibition could result in ROS crossing over into the cytoplasm via voltage-dependent anion channels (Han et al. 2003), which could also help activate NF-$\kappa\beta$. Excess ROS in the cytoplasm could also be complemented by the superoxides produced by the activated microglia via NADPH oxidase (Gao et al. 2002). The superoxides produced by NADPH oxidase can also interact with nitric oxide (NO) to produce peroxinitrate (ONOO-), a potent oxidant (Gao et al. 2003) implicated with being cytotoxic and able to damage the neuron's cell membrane (Hirsch 2000, Imao et al. 1998). NO is produced via the inducible nitric oxide synthase that is activated by CD23 which in turn is induced by the presence of TNF- α and IFN- γ (Munoz-Fernandez and Fresno 1998, Hirsch 2000).

Initial studies of the effect of rotenone used systemically in the rat showed a decrease in brain dopamine levels, decreases in tyrosine hydroxylase immunoreactivity in the substantia nigra, and motor deficits similar to those seen in PD (Betarbet et al. 2000, Sherer et al. 2003c, Alam and Schmidt 2002). Rotenone treatment also leads to intracytoplasmic inclusions within dopaminergic neurons thereby mimicking some aspects of PD histopathology (Betarbet et al. 2002). Unfortunately, systemic delivery of rotenone results in high mortality due to the systemic toxicity resulting in liver failure and inconsistent lesions of the substantia nigra (Dawson and Dawson 2002). These problems were alleviated with intra-peritoneal administration of rotenone in a specialised vehicle of a medium-chain triglyceride, although this model developed a debilitating behavioural phenotype in a relatively short time and is not appropriate as a slow onset chronic model (Cannon et al. 2009).

The side effects of peripheral rotenone treatment can be reduced or avoided by the infusion of a lower dose directly into the striatum, medial forebrain bundle or substantia nigra (Ravenstijn et al. 2008, Xiong et al. 2009). Bilateral infusion of rotenone into the medial forebrain bundle causes reductions in striatal dopamine and disruptions in motor behaviours (Alam, Mayerhofer and Schmidt 2004). However, bilateral infusions lead to weight loss and require specialised diets to maintain the animals (Betarbet et al. 2000, Sherer et al. 2003c), possibly due to bilateral lesioning of the ventral tegmental area. These problems are ameliorated by unilateral lesions which reduce dopamine signalling, dopamine level, DOPAC (3,4-dihydroxyphenylacetic acid) level and dopaminergic innervation, while increasing oxidative stress (hydroxyl radicals, GSH level, superoxide dismutase levels), as well as up-regulating α-synuclein expression in the ipsilateral substantia nigra (Sindhu et al. 2006, Antkiewicz-Michaluk et al. 2004, Saravanan, Sindhu and Mohanakumar 2005, Sindhu, Saravanan and Mohanakumar 2005, Ravenstijn et al. 2008, Xiong et al. 2009). Unilateral rotenone lesioned animals have shown differences in several behavioural indices, including rotarod and amphetamine or apomorphine-induced rotation, demonstrating the unilateral functional motor deficits associated with substantia nigra dopamine neuron loss (Sindhu et al. 2006, Sindhu et al. 2005, Ravenstijn et al. 2008, Xiong et al. 2009). Behavioural indices are also influenced by the lesion of the ventral tegmental area, as can occur with larger doses of rotenone in the medial forebrain bundle (Sindhu et al. 2005, Xiong et al. 2009, Thomas et al. 1994).

Our laboratory has recently reported that a low dose of rotenone injected into the medial forebrain pathway in adult rats caused progressive loss of dopaminergic neurons with the remaining neurons displaying pathophysiological hallmark of human PD (Norazit et al. 2010)(Figure 2). Unlike the complete lesion of dopaminergic neurons induced by focal 6-OHDA injection, rotenone injection into the medial forebrain bundle induced the up-regulation of markers of oxidative stress and markers of cell stress in dopaminergic neurons (Sindhu et al. 2005, Norazit et al. 2010). 0.5 µg of rotenone caused negligible necrosis, inflammation and a diffused glial response. A progressive loss of dopaminergic neurons in the substantia nigra and loss of striatal innervation was shown. The low dose of rotenone mediated dopaminergic cell death by oxidative stress as previously demonstrated (Rodrigues, Gomide and Chadi 2004). An increase in astrocytes and microglia occurs in human PD, where they have been ascribed both a neuroprotective and deleterious role (Imamura et al. 2003, Ishida et al. 2006, Vila et al. 2001). The study presented direct support for the hypothesis that rotenone induces a chronic state of oxidative stress in dopaminergic neurons. Rotenone exposure increased SOD2 immunoreactivity within surviving tyrosine hydroxylase positive neurons. The toxin leads to high super-oxide production, activating the apoptotic pathway as shown in the study with increased caspase-3 immunoreactivity (Esposti 1998, Lambert and Brand 2004, Turrens 2003). Under oxidative stress, cells generate G3BP positive cytoplasmic stress granules (Cande et al. 2004, Kedersha and Anderson 2002). This proposal was further validated by the presence of α-synuclein in tyrosine hydroxylase positive neurons 60 days following rotenone exposure (Norazit et al. 2010). α-synuclein is up-regulated in neurons subject to chronic oxidative stress, and plays a neuroprotective role and is expressed sporadically in the substantia nigra 28 days after rotenone infusion into the medial forebrain bundle (Hashimoto et al. 2002, Quilty et al. 2006). This model has now been successfully used to test the neuroprotective properties of chronic exposure to neurotrophic factors (unpublished data).

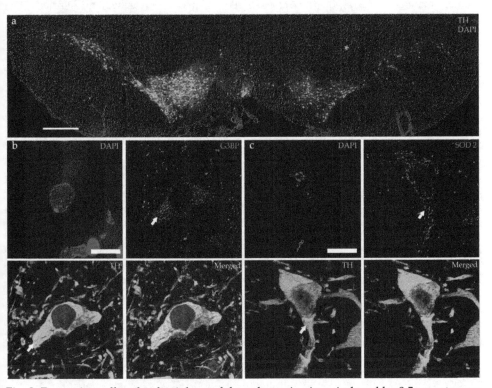

Fig. 2. Dopamine cell pathophysiology of the substantia nigra, induced by 0.5-μg rotenone delivered focally into the medial forebrain bundle and harvested after 14 days. a) The unilateral lesion is manifest as a subtle yet significant reduction in the number of dopaminergic neurons and dendritic arbour within the substantia nigra (asterisk) and a reduction in the dopaminergic innervation of the striatum (data not shown). The nigrostriatal circuitry contralateral to the lesion is uneffected by the rotenone treatment. b) Ras-GAP SH3 domain binding protein (G3BP) immunoreactivity (red, filled arrowhead) was expressed in tyrosine hydroxylase immunoreactive (green, filled arrowhead) dopaminergic neurons located in the rotenone exposed substantia nigra. c) Superoxide dismutase (SOD2) immunoreactivity (red, filled arrowhead) was also expressed in the rotenone exposed tyrosine hydroxylase immunoreactive (green, filled arrowhead) dopaminergic neurons. Nuclei for all sections are counterstained with the nuclear marker DAPI (blue) (scale bar=10 μm)

3.2 Cellular based models

Complete human disease phenotype is rarely observed in animal models introduced with human gene mutations. Thus, relevant human tissue is required to study the disease process and potential therapies for neurodegenerative disorders such as PD. Pathological human samples are often confounded by difficult-to-control artefacts resulting from the disease process itself (Sutherland et al. 2009), biased sampling, and the necessity to process tissue in a timely manner following death (Atz et al. 2007, Marcotte, Srivastava and Quirion 2003, Preece and Cairns 2003). Together with rare foetal derived tissue, patient derived olfactory stem cells and induced pluripotent stem cells are currently used as models for PD.

3.2.1 Patient derived olfactory stem cells

Procuring relevant neural cells from patients with central nervous system disorders is difficult. Therefore, we have developed olfactory stem cells as a model for PD. Neural stem cells from adult human olfactory mucosa may be harvested and expanded to enrich for the stem cells, which are then frozen, banked, thawed, and regrown in quantity for gene and protein expression analyses and functional investigations. Assays on olfactory stem cell function have shown a reduction in gluthathione while pathway analysis has demonstrated significantly dysregulated pathways associated with mitochondrial function and oxidative stress (Matigian et al. 2010). The ease of patient derived olfactory stem cells propagation and banking allows them to be used for extended genomic, proteomic, and functional studies, including drug and biomarker discovery.

3.2.2 Patient derived induced pluripotent stem cells

Induced pluripotent stem (iPS) cells are pluripotent cells derived from differentiated cells, for example by introducing key transcription factor genes, as demonstrated in adult mouse fibroblasts (Takahashi and Yamanaka 2006). iPS cells have successfully been used to generate neurons from patients with sporadic PD (Soldner et al. 2009). Notably, despite these iPS cells being derived from LRRK2 mutation carriers, no phenotypic differences between sporadic PD iPS and control iPS cells were demonstrated (Nguyen et al. 2011). Patient-derived iPS cells have the potential to be used to identify changes in neural cell biology associated with the identified mutations.

3.3 Genetic animal models

Transgenic models have been developed as genetic factors linked to PD have been identified. Several autosomal dominant and recessive genes linked to mitochondrial dysfunction have been identified in humans as reviewed by Schapira (Schapira 2008). This created an opportunity for transgenic animals to mimic familial forms of PD (Fleming, Fernagut and Chesselet 2005).

3.3.1 PARK2 (PARKIN)

PARK 2 (Parkin) is transcribed in the mitochondria (Schapira 2008). The function of parkin remains to be elucidated; however the direct association between parkin and the mitochondria is of interest and warrants further study. In the knockout transgenic mouse model, animals had a decrease in mitochondrial respiratory chain function in the striatum and reductions in specific respiratory chain and antioxidant proteins (Palacino et al. 2004). Midbrain neuronal cultures obtained from PARK2 knockout mice had an increased sensitivity to rotenone, suggesting an effect on the mitochondrial respiratory chain

(Casarejos et al. 2006). In some of the mouse knockout models, subtle abnormalities of the nigrostriatal pathway or the locus coeruleus noradrenergic system have been observed (Goldberg et al. 2003, von Coelln et al. 2006). Conversely, over-expression of mutant parkin produces a progressive loss of dopaminergic neurons in the nigrostriatal pathway in both mice and drosophila (Lu et al. 2009, Sang et al. 2007, Wang et al. 2007). This suggests that some parkin mutants may act in a dominant negative fashion. The parkin animal model exhibits several movement indices in both drosophila and mice (Greene et al. 2003, Whitworth et al. 2005).

3.3.2 PINK1

Similar to parkin knockouts, PINK1 knockouts also have mild mitochondrial defects (Gautier, Kitada and Shen 2008, Palacino et al. 2004).The PINK1 product is transcribed in the nucleus, translated in the cytoplasm, and imported intact into the mitochondria, with subsequent processing and intra mitochondrial sorting (Schapira 2008). The lack of PINK1 in transgenic mice causes enlargement of mitochondria as well as a decrease in mitochondrial numbers in dopaminergic neurons of the nigrostriatal pathway (Gautier et al. 2008, Gispert et al. 2009, Kitada et al. 2007). Although these animals do not exhibit any changes in the nigrostriatal pathway, a deficit in dopamine neurotransmission has been observed (Kitada et al. 2007). Changes in several behavioral indices have also been shown in the PINK1 knockout drosophila animal model (Clark et al. 2006, Park et al. 2006).

It is interesting to point out that while both the knockdown of parkin and PINK1 have been linked to mitochondrial dysfunction, the expression of parkin ameliorates PINK1-related abnormalities but not vice versa (Clark et al. 2006, Park et al. 2006). This suggests that parkin and PINK1 are part of a common pathway with PINK1 functioning upstream from parkin.

3.3.3 α -Synuclein

α -synuclein is a protein aggregate that is the main part of Lewy bodies in human PD. The function of α-synuclein is as yet unclear, however, there appears to be a reciprocal relationship between this protein and oxidative stress (Henchcliffe and Beal 2008). α-synuclein is up-regulated in neurons subject to chronic oxidative stress and expressed sporadically in the substantia nigra (Hashimoto et al. 2002, Quilty et al. 2006, Norazit et al. 2010). The association between the presence of α-synuclein and PD has led to the development of a variety of animal models (Table 1). The over-expression of α-synuclein increased the loss in dopaminergic neurons in both drosophila and C. elegans models. (Feany and Bender 2000, Kuwahara et al. 2006, Lakso et al. 2003). However, only the dopaminergic loss in the drosophila model is progressive. A loss of dopaminergic neurons with the over-expression of α-synuclein has been demonstrated in mice; however the phenotypic outcome depends on the promoters used to drive transgene expression (Chesselet 2008). Transgenic mice presented with several functional abnormalities in the nigrostriatal system, some of which are dopamine responsive (Chesselet 2008). However, the loss of dopaminergic neurons is not progressive. α-synuclein toxicity is induced though mitochondrial dysfunction, proteasomal and lysosomal impairments, and disruption of ER-Golgi trafficking (Cooper et al. 2006, Cuervo et al. 2004, Martin et al. 2006, Tanaka et al. 2001). The link between mitochondrial dysfunction and α-synuclein aggregation suggests a feed-forward loop that has the potential to initiate the progressive loss of dopaminergic neurons in the nigrostriatal system due to oxidative stress.

3.3.4 LRRK2

The LRRK2 protein functions as a serine-thereonine kinase, a known affector of mitochondrial function. A small percentage (10%) of LRRK2 is located in the outer mitochondrial membrane (West et al. 2005). Although the precise function of mitochondrial-located LRRK2 is not known, it has been suggested that it interacts with parkin. The current transgenic LRRK2 animal models are not robust enough to be used as a PD animal model due to the lack of loss of dopaminergic neurons of the nigrostriatal pathway (Li et al. 2010, Li et al. 2009, Lin et al. 2009, Wang et al. 2008, Tong et al. 2009). However, these animals do show several abnormalities in DA neurotransmission or in dopamine responsive behavior. It has been suggested that the mouse LRRK2 transgenic models do not exhibit more substantial pathology as LRRK2 mutations in humans are only partially penetrant, with further genetic and/or environmental insult required to induce the degeneration of dopaminergic neurons (Dawson, Ko and Dawson 2010).

4. Experimental therapies

4.1 Metal chelation therapy

Increased iron levels have been detected in the midbrain of PD patients suggesting that the increased levels of iron may be part of the disease pathology (Andersen 2004). Notably, iron participates as a catalyst to produce ROS.

4.1.1 8-Hydroxyquinolines

Previously the hydroxyquinoline clioquinol had been examined as a metal chelator in a clinical trial for Alzeimer's disease (Ritchie et al. 2003). The use of a variety of hydroxyquinolines to chelate iron into a form that does not catalyze ROS has shown promise in both *in vivo* and *in vitro* models (Table 2).

Hydroxyquinoline	Model	Reference
Clioquinol	MPTP mouse model	(Kaur et al. 2003)
HLA20	P19 cells exposuded to 6-OHDA	(Zheng et al. 2005)
VK-28	6-OHDA rat model	(Tsubaki, Honma and Hoshi 1971)
M30	MPTP mouse model In vitro	(Gal et al. 2005, Gal et al. 2006)
M98 and M99	SH-SY5Y and PC12 exposed to 6-OHDA	(Gal et al. 2006, Zheng, Blat and Fridkin 2006)
M10	PC12 cells exposed to 6-OHDA	(Ritchie et al. 2003)

Table 2. Hydroxyquinolines shown to chelate iron in both *in vivo* and *in vitro* models

Hydroxyquinolines are able to cross the blood brain barrier, allowing for oral administration. Hydroxyquinolines have also shown inhibitory effect on the activity of enzyme MAO-B (Yassin et al. 2000, Youdim, Fridkin and Zheng 2005).

4.1.2 Desferrioxamine

Previously, desferrioxamine was used to treat iron overload disease (Mandel et al. 2007). The use of Desferrioxamine has been reported to be neuroprotective *in vivo* and *in vitro* (Ben-Shachar et al. 1991, Jiang et al. 2006, Sangchot et al. 2002, Youdim, Stephenson and Ben Shachar 2004). Unlike hydroxyquinolines, desferrioxamine does not cross the blood brain barrier, thus removing the option of oral administration (Aouad et al. 2002).

4.2 Plant polyphenols

Polyphenols have been reported to have antioxidant properties, thus making them a candidate for antioxidant therapies (Malesev and Kuntic 2007). Green tea, cranberry, traditional chinese tea, and tumeric are sources of a variety of polyphenols (Reto et al. 2007, Ramassamy 2006, Perez, Wei and Guo 2009, Tan, Meng and Hostettmann 2000). Neuroprotective and neurorescue properties of plant polyphenols have been demonstrated *in vivo* and *in vitro* (Mandel, Maor and Youdim 2004, Mercer et al. 2005, Chen et al. 2006, Zbarsky et al. 2005, Mandel et al. 2006). Polyphenols exert their antioxidant effect via scavenging of free radicals and inhibition of the Fenton reaction (Perez et al. 2009, Pan, Jankovic and Le 2003).

4.3 Antioxidant therapy

The effects of oxidative stress are demonstrated in PD patients who have decreased GSH levels, increased levels of iron, neuromelanin associated redox-active iron, lipid peroxidation, protein oxidation and DNA damage in the substantia nigra (Jenner and Olanow 1998, Faucheux et al. 2003, Dexter et al. 1989). Antioxidant therapy has been suggested to ameliorate these effects.

4.4 Coenzyme Q_{10} (CoQ_{10})

CoQ_{10}, also known as ubiquinone is a cofactor that accepts electrons from Complex I and II of the electron transport chain in the mitochiondria (Beyer 1992, Dallner and Sindelar 2000). CoQ_{10} mediates its antioxidant effect via its interaction with a-tocopherol (Beyer 1992, Noack, Kube and Augustin 1994), inhibiting the activation of mitochondrial permeability, and acting independently of its free radical scavenging activity. Thus it blocks apoptosis (Papucci *et al.*, 2003), acting as a co-factor of mitochondrial uncoupling proteins which reduces mitochondrial-free radical generation (Echtay, Winkler and Klingenberg 2000, Echtay et al. 2002). Neuroprotection has been associated with the ability of CoQ_{10} to induce mitochondrial uncoupling in the substantia nigra of primates, after MPTP toxicity (Horvath et al. 2003). The use of CoQ_{10} as an antioxidant has been translated into phase II clinical trials (Shults et al. 2002) which showed a slowing of disease progression following 16 months of treatment.

4.5 Deprenyl and Tocopherol Antioxidative Therapy of Parkinsonism (DATATOP)

Deprenyl and tocopherol antioxidative therapy have been clinically trialed to explore its potential as therapy for PD (Shoulson et al. 2002). Deprenyl treatment delayed the initiation of levodopa therapy. Continued deprenyl treated subjects exhibited slower motor decline and lower likelihood of developing freezing of gait. However, this treatment increased the likelihood of developing dyskinesia.

5. Conclusion

There is increasing evidence that many factors including age, environmental toxins, genetic determinants, and lifestyle factors influence the risk for PD. Many of these impact on oxidative stress related pathways. Animal and cellular models have been developed to mimic the disease pathology in humans. Currently, antioxidant therapies are being investigated in both animals and human clinical trials, with promising outcomes. Notably, antioxidant therapies appear to delay the onset of disease and disease progression, although they do not prevent or reverse disease progression.

6. Acknowledgment

The authors would like to thank Suzita Mohd Noor, Maria Nguyen, Charlotte Dickson, and Brenton Cavanagh for their technical and editing support.

7. References

Ahmadi, F. A., D. A. Linseman, T. N. Grammatopoulos, S. M. Jones, R. J. Bouchard, C. R. Freed, K. A. Heidenreich & W. M. Zawada (2003) The pesticide rotenone induces caspase-3-mediated apoptosis in ventral mesencephalic dopaminergic neurons. *J Neurochem*, 87, 914-21.

Alam, M., A. Mayerhofer & W. J. Schmidt (2004) The neurobehavioral changes induced by bilateral rotenone lesion in medial forebrain bundle of rats are reversed by L-DOPA. *Behav Brain Res*, 151, 117-24.

Alam, M. & W. J. Schmidt (2002) Rotenone destroys dopaminergic neurons and induces parkinsonian symptoms in rats. *Behav Brain Res*, 136, 317-24.

Ambani, L. M., M. H. Van Woert & S. Murphy (1975) Brain peroxidase and catalase in Parkinson disease. *Arch Neurol*, 32, 114-8.

Andersen, J. K. (2004) Oxidative stress in neurodegeneration: cause or consequence? *Nature medicine*, 10 Suppl, S18-25.

Antkiewicz-Michaluk, L., J. Wardas, J. Michaluk, I. Romaska, A. Bojarski & J. Vetulani (2004) Protective effect of 1-methyl-1,2,3,4-tetrahydroisoquinoline against dopaminergic neurodegeneration in the extrapyramidal structures produced by intracerebral injection of rotenone. *Int J Neuropsychopharmacol*, 7, 155-63.

Aouad, F., A. Florence, Y. Zhang, F. Collins, C. Henry, R. J. Ward & R. Crichton (2002) Evaluation of new iron chelators and their therapeutic potential. *Inorganica Chimica Acta*, 339, 470-480.

Atz, M., D. Walsh, P. Cartagena, J. Li, S. Evans, P. Choudary, K. Overman, R. Stein, H. Tomita, S. Potkin, R. Myers, S. J. Watson, E. G. Jones, H. Akil, W. E. Bunney, Jr. & M. P. Vawter (2007) Methodological considerations for gene expression profiling of human brain. *Journal of neuroscience methods*, 163, 295-309.

Bannon, M. J., M. S. Poosch, Y. Xia, D. J. Goebel, B. Cassin & G. Kapatos (1992) Dopamine transporter mRNA content in human substantia nigra decreases precipitously with age. *Proc Natl Acad Sci U S A*, 89, 7095-9.

Ben-Shachar, D., G. Eshel, J. P. Finberg & M. B. Youdim (1991) The iron chelator desferrioxamine (Desferal) retards 6-hydroxydopamine-induced degeneration of nigrostriatal dopamine neurons. *Journal of neurochemistry*, 56, 1441-4.

Bender, A., K. J. Krishnan, C. M. Morris, G. A. Taylor, A. K. Reeve, R. H. Perry, E. Jaros, J. S. Hersheson, J. Betts, T. Klopstock, R. W. Taylor & D. M. Turnbull (2006) High levels of mitochondrial DNA deletions in substantia nigra neurons in aging and Parkinson disease. *Nat Genet*, 38, 515-7.

Betarbet, R., T. B. Sherer & J. T. Greenamyre (2002) Animal models of Parkinson's disease. *Bioessays*, 24, 308-18.

Betarbet, R., T. B. Sherer, G. MacKenzie, M. Garcia-Osuna, A. V. Panov & J. T. Greenamyre (2000) Chronic systemic pesticide exposure reproduces features of Parkinson's disease. *Nat Neurosci*, 3, 1301-6.

Beyer, R. E. (1992) An analysis of the role of coenzyme Q in free radical generation and as an antioxidant. *Biochemistry and cell biology = Biochimie et biologie cellulaire*, 70, 390-403.

Bove, J., D. Prou, C. Perier & S. Przedborski (2005) Toxin-induced models of Parkinson's disease. *NeuroRx*, 2, 484-94.

Braak, H., K. Del Tredici, U. Rub, R. A. de Vos, E. N. Jansen Steur & E. Braak (2003a) Staging of brain pathology related to sporadic Parkinson's disease. *Neurobiol Aging*, 24, 197-211.

Braak, H., U. Rub, W. P. Gai & K. Del Tredici (2003b) Idiopathic Parkinson's disease: possible routes by which vulnerable neuronal types may be subject to neuroinvasion by an unknown pathogen. *J Neural Transm*, 110, 517-36.

Brooks, A. I., C. A. Chadwick, H. A. Gelbard, D. A. Cory-Slechta & H. J. Federoff (1999) Paraquat elicited neurobehavioral syndrome caused by dopaminergic neuron loss. *Brain Res*, 823, 1-10.

Brooks, D. J. (2010) Examining Braak's hypothesis by imaging Parkinson's disease. *Mov Disord*, 25 Suppl 1, S83-8.

Caboni, P., T. B. Sherer, N. Zhang, G. Taylor, H. M. Na, J. T. Greenamyre & J. E. Casida (2004) Rotenone, deguelin, their metabolites, and the rat model of Parkinson's disease. *Chem Res Toxicol*, 17, 1540-8.

Cande, C., N. Vahsen, D. Metivier, H. Tourriere, K. Chebli, C. Garrido, J. Tazi & G. Kroemer (2004) Regulation of cytoplasmic stress granules by apoptosis-inducing factor. *J Cell Sci*, 117, 4461-8.

Cannon, J. R., V. Tapias, H. M. Na, A. S. Honick, R. E. Drolet & J. T. Greenamyre (2009) A highly reproducible rotenone model of Parkinson's disease. *Neurobiol Dis*, 34, 279-90.

Casarejos, M. J., J. Menendez, R. M. Solano, J. A. Rodriguez-Navarro, J. Garcia de Yebenes & M. A. Mena (2006) Susceptibility to rotenone is increased in neurons from parkin null mice and is reduced by minocycline. *Journal of neurochemistry*, 97, 934-46.

Chauhan, N. B., G. J. Siegel & J. M. Lee (2001) Depletion of glial cell line-derived neurotrophic factor in substantia nigra neurons of Parkinson's disease brain. *J Chem Neuroanat*, 21, 277-88.

Chen, J., X. Q. Tang, J. L. Zhi, Y. Cui, H. M. Yu, E. H. Tang, S. N. Sun, J. Q. Feng & P. X. Chen (2006) Curcumin protects PC12 cells against 1-methyl-4-phenylpyridinium ion-induced apoptosis by bcl-2-mitochondria-ROS-iNOS pathway. *Apoptosis : an international journal on programmed cell death*, 11, 943-53.

Chesselet, M. F. (2008) In vivo alpha-synuclein overexpression in rodents: a useful model of Parkinson's disease? *Experimental neurology*, 209, 22-7.

Chu, Y. & J. H. Kordower (2007) Age-associated increases of alpha-synuclein in monkeys and humans are associated with nigrostriatal dopamine depletion: Is this the target for Parkinson's disease? *Neurobiol Dis,* 25, 134-49.

Clark, I. E., M. W. Dodson, C. Jiang, J. H. Cao, J. R. Huh, J. H. Seol, S. J. Yoo, B. A. Hay & M. Guo (2006) Drosophila pink1 is required for mitochondrial function and interacts genetically with parkin. *Nature,* 441, 1162-6.

Cohen, G. & R. E. Heikkila (1984) Alloxan and 6-hydroxydopamine: cellular toxins. *Methods Enzymol,* 105, 510-6.

Cooper, A. A., A. D. Gitler, A. Cashikar, C. M. Haynes, K. J. Hill, B. Bhullar, K. Liu, K. Xu, K. E. Strathearn, F. Liu, S. Cao, K. A. Caldwell, G. A. Caldwell, G. Marsischky, R. D. Kolodner, J. Labaer, J. C. Rochet, N. M. Bonini & S. Lindquist (2006) Alpha synuclein blocks ER-Golgi traffic and Rab1 rescues neuron loss in Parkinson's models. *Science,* 313, 324-8.

Cuervo, A. M., L. Stefanis, R. Fredenburg, P. T. Lansbury & D. Sulzer (2004) Impaired degradation of mutant alpha-synuclein by chaperone-mediated autophagy. *Science,* 305, 1292-5.

Dallner, G. & P. J. Sindelar (2000) Regulation of ubiquinone metabolism. *Free radical biology & medicine,* 29, 285-94.

Dauer, W. & S. Przedborski (2003) Parkinson's disease: mechanisms and models. *Neuron,* 39, 889-909.

Dawson, T. M. & V. L. Dawson (2002) Neuroprotective and neurorestorative strategies for Parkinson's disease. *Nat Neurosci,* 5 Suppl, 1058-61.

Dawson, T. M., H. S. Ko & V. L. Dawson (2010) Genetic animal models of Parkinson's disease. *Neuron,* 66, 646-61.

de Rijk, M. C., C. Tzourio, M. M. Breteler, J. F. Dartigues, L. Amaducci, S. Lopez-Pousa, J. M. Manubens-Bertran, A. Alperovitch & W. A. Rocca (1997) Prevalence of parkinsonism and Parkinson's disease in Europe: the EUROPARKINSON Collaborative Study. European Community Concerted Action on the Epidemiology of Parkinson's disease. *J Neurol Neurosurg Psychiatry,* 62, 10-5.

Degli Esposti, M. (1998) Inhibitors of NADH-ubiquinone reductase: an overview. *Biochim Biophys Acta,* 1364, 222-35.

Dexter, D. T., C. J. Carter, F. R. Wells, F. Javoy-Agid, Y. Agid, A. Lees, P. Jenner & C. D. Marsden (1989) Basal lipid peroxidation in substantia nigra is increased in Parkinson's disease. *Journal of neurochemistry,* 52, 381-9.

Drozdzik, M., M. Bialecka, K. Mysliwiec, K. Honczarenko, J. Stankiewicz & Z. Sych (2003) Polymorphism in the P-glycoprotein drug transporter MDR1 gene: a possible link between environmental and genetic factors in Parkinson's disease. *Pharmacogenetics,* 13, 259-63.

Echtay, K. S., D. Roussel, J. St-Pierre, M. B. Jekabsons, S. Cadenas, J. A. Stuart, J. A. Harper, S. J. Roebuck, A. Morrison, S. Pickering, J. C. Clapham & M. D. Brand (2002) Superoxide activates mitochondrial uncoupling proteins. *Nature,* 415, 96-9.

Echtay, K. S., E. Winkler & M. Klingenberg (2000) Coenzyme Q is an obligatory cofactor for uncoupling protein function. *Nature,* 408, 609-13.

Esposti, M. D. (1998) Apoptosis: who was first? *Cell Death Differ,* 5, 719.

Faucheux, B. A., M. E. Martin, C. Beaumont, J. J. Hauw, Y. Agid & E. C. Hirsch (2003) Neuromelanin associated redox-active iron is increased in the substantia nigra of patients with Parkinson's disease. *J Neurochem,* 86, 1142-8.

Feany, M. B. & W. W. Bender (2000) A Drosophila model of Parkinson's disease. *Nature,* 404, 394-8.

Ferraz, H. B., P. H. Bertolucci, J. S. Pereira, J. G. Lima & L. A. Andrade (1988) Chronic exposure to the fungicide maneb may produce symptoms and signs of CNS manganese intoxication. *Neurology,* 38, 550-3.

Fleming, S. M., P. O. Fernagut & M. F. Chesselet (2005) Genetic mouse models of parkinsonism: strengths and limitations. *NeuroRx* 2, 495-503.

Fridovich, I. (1995) Superoxide radical and superoxide dismutases. *Annu Rev Biochem,* 64, 97-112.

Gal, S., M. Fridkin, T. Amit, H. Zheng & M. B. Youdim (2006) M30, a novel multifunctional neuroprotective drug with potent iron chelating and brain selective monoamine oxidase-ab inhibitory activity for Parkinson's disease. *Journal of neural transmission. Supplementum,* 447-56.

Gal, S., H. Zheng, M. Fridkin & M. B. Youdim (2005) Novel multifunctional neuroprotective iron chelator-monoamine oxidase inhibitor drugs for neurodegenerative diseases. In vivo selective brain monoamine oxidase inhibition and prevention of MPTP-induced striatal dopamine depletion. *Journal of neurochemistry,* 95, 79-88.

Gao, H. M., J. S. Hong, W. Zhang & B. Liu (2002) Distinct role for microglia in rotenone-induced degeneration of dopaminergic neurons. *J Neurosci,* 22, 782-90.

Gao, H. M., B. Liu & J. S. Hong (2003) Critical role for microglial NADPH oxidase in rotenone-induced degeneration of dopaminergic neurons. *J Neurosci,* 23, 6181-7.

Gash, D. M., Z. Zhang, A. Ovadia, W. A. Cass, A. Yi, L. Simmerman, D. Russell, D. Martin, P. A. Lapchak, F. Collins, B. J. Hoffer & G. A. Gerhardt (1996) Functional recovery in parkinsonian monkeys treated with GDNF. *Nature,* 380, 252-5.

Gautier, C. A., T. Kitada & J. Shen (2008) Loss of PINK1 causes mitochondrial functional defects and increased sensitivity to oxidative stress. *Proceedings of the National Academy of Sciences of the United States of America,* 105, 11364-9.

Gibb, W. R. & A. J. Lees (1988) The relevance of the Lewy body to the pathogenesis of idiopathic Parkinson's disease. *J Neurol Neurosurg Psychiatry,* 51, 745-52.

Gispert, S., F. Ricciardi, A. Kurz, M. Azizov, H. H. Hoepken, D. Becker, W. Voos, K. Leuner, W. E. Muller, A. P. Kudin, W. S. Kunz, A. Zimmermann, J. Roeper, D. Wenzel, M. Jendrach, M. Garcia-Arencibia, J. Fernandez-Ruiz, L. Huber, H. Rohrer, M. Barrera, A. S. Reichert, U. Rub, A. Chen, R. L. Nussbaum & G. Auburger (2009) Parkinson phenotype in aged PINK1-deficient mice is accompanied by progressive mitochondrial dysfunction in absence of neurodegeneration. *PloS one,* 4, e5777.

Goldberg, M. S., S. M. Fleming, J. J. Palacino, C. Cepeda, H. A. Lam, A. Bhatnagar, E. G. Meloni, N. Wu, L. C. Ackerson, G. J. Klapstein, M. Gajendiran, B. L. Roth, M. F. Chesselet, N. T. Maidment, M. S. Levine & J. Shen (2003) Parkin-deficient mice exhibit nigrostriatal deficits but not loss of dopaminergic neurons. *The Journal of biological chemistry,* 278, 43628-35.

Gowers, W. R. (1902) A lecture on myopathy and a distal form. *British Medical Journal,* 2, 89-92.

Greene, J. C., A. J. Whitworth, I. Kuo, L. A. Andrews, M. B. Feany & L. J. Pallanck (2003) Mitochondrial pathology and apoptotic muscle degeneration in Drosophila parkin mutants. *Proceedings of the National Academy of Sciences of the United States of America*, 100, 4078-83.

Grivennikova, V. G., E. O. Maklashina, E. V. Gavrikova & A. D. Vinogradov (1997) Interaction of the mitochondrial NADH-ubiquinone reductase with rotenone as related to the enzyme active/inactive transition. *Biochimica et biophysica acta*, 1319, 223-32.

Han, D., F. Antunes, R. Canali, D. Rettori & E. Cadenas (2003) Voltage-dependent anion channels control the release of the superoxide anion from mitochondria to cytosol. *The Journal of biological chemistry*, 278, 5557-63.

Hashimoto, M., L. J. Hsu, E. Rockenstein, T. Takenouchi, M. Mallory & E. Masliah (2002) alpha-Synuclein protects against oxidative stress via inactivation of the c-Jun N-terminal kinase stress-signaling pathway in neuronal cells. *J Biol Chem*, 277, 11465-72.

Hawkes, C. H., K. Del Tredici & H. Braak (2007) Parkinson's disease: a dual-hit hypothesis. *Neuropathol Appl Neurobiol*, 33, 599-614.

Henchcliffe, C. & M. F. Beal (2008) Mitochondrial biology and oxidative stress in Parkinson disease pathogenesis. *Nature clinical practice. Neurology*, 4, 600-9.

Herishanu, Y. O., M. Medvedovski, J. R. Goldsmith & E. Kordysh (2001) A case-control study of Parkinson's disease in urban population of southern Israel. *Can J Neurol Sci*, 28, 144-7.

Hirata, Y., H. Suzuno, T. Tsuruta, K. Oh-hashi & K. Kiuchi (2008) The role of dopamine transporter in selective toxicity of manganese and rotenone. *Toxicology*, 244, 249-56.

Hirsch, E. C. (2000) Glial cells and Parkinson's disease. *Journal of neurology*, 247 Suppl 2, II58-62.

Ho, S. C., J. Woo & C. M. Lee (1989) Epidemiologic study of Parkinson's disease in Hong Kong. *Neurology*, 39, 1314-8.

Horvath, T. L., S. Diano, C. Leranth, L. M. Garcia-Segura, M. A. Cowley, M. Shanabrough, J. D. Elsworth, P. Sotonyi, R. H. Roth, E. H. Dietrich, R. T. Matthews, C. J. Barnstable & D. E. Redmond, Jr. (2003) Coenzyme Q induces nigral mitochondrial uncoupling and prevents dopamine cell loss in a primate model of Parkinson's disease. *Endocrinology*, 144, 2757-60.

Howells, D. W., M. J. Porritt, J. Y. Wong, P. E. Batchelor, R. Kalnins, A. J. Hughes & G. A. Donnan (2000) Reduced BDNF mRNA expression in the Parkinson's disease substantia nigra. *Exp Neurol*, 166, 127-35.

Hughes, A. J., S. E. Daniel & A. J. Lees (2001) Improved accuracy of clinical diagnosis of Lewy body Parkinson's disease. *Neurology*, 57, 1497-9.

Imamura, K., N. Hishikawa, M. Sawada, T. Nagatsu, M. Yoshida & Y. Hashizume (2003) Distribution of major histocompatibility complex class II-positive microglia and cytokine profile of Parkinson's disease brains. *Acta Neuropathol*, 106, 518-26.

Imao, K., H. Wang, M. Komatsu & M. Hiramatsu (1998) Free radical scavenging activity of fermented papaya preparation and its effect on lipid peroxide level and superoxide dismutase activity in iron-induced epileptic foci of rats. *Biochemistry and molecular biology international*, 45, 11-23.

Ishida, Y., A. Nagai, S. Kobayashi & S. U. Kim (2006) Upregulation of protease-activated receptor-1 in astrocytes in Parkinson disease: astrocyte-mediated neuroprotection through increased levels of glutathione peroxidase. *J Neuropathol Exp Neurol*, 65, 66-77.

Javitch, J. A., R. J. D'Amato, S. M. Strittmatter & S. H. Snyder (1985) Parkinsonism-inducing neurotoxin, N-methyl-4-phenyl-1,2,3,6 -tetrahydropyridine: uptake of the metabolite N-methyl-4-phenylpyridine by dopamine neurons explains selective toxicity. *Proc Natl Acad Sci U S A*, 82, 2173-7.

Javoy, F., C. Sotelo, A. Herbet & Y. Agid (1976) Specificity of dopaminergic neuronal degeneration induced by intracerebral injection of 6-hydroxydopamine in the nigrostriatal dopamine system. *Brain Res*, 102, 201-15.

Jenner, P. & C. W. Olanow (1998) Understanding cell death in Parkinson's disease. *Ann Neurol*, 44, S72-84.

Jiang, H., Z. Luan, J. Wang & J. Xie (2006) Neuroprotective effects of iron chelator Desferal on dopaminergic neurons in the substantia nigra of rats with iron-overload. *Neurochemistry international*, 49, 605-9.

Kaur, D., F. Yantiri, S. Rajagopalan, J. Kumar, J. Q. Mo, R. Boonplueang, V. Viswanath, R. Jacobs, L. Yang, M. F. Beal, D. DiMonte, I. Volitaskis, L. Ellerby, R. A. Cherny, A. I. Bush & J. K. Andersen (2003) Genetic or pharmacological iron chelation prevents MPTP-induced neurotoxicity in vivo: a novel therapy for Parkinson's disease. *Neuron*, 37, 899-909.

Kedersha, N. & P. Anderson (2002) Stress granules: sites of mRNA triage that regulate mRNA stability and translatability. *Biochem Soc Trans*, 30, 963-9.

Kitada, T., A. Pisani, D. R. Porter, H. Yamaguchi, A. Tscherter, G. Martella, P. Bonsi, C. Zhang, E. N. Pothos & J. Shen (2007) Impaired dopamine release and synaptic plasticity in the striatum of PINK1-deficient mice. *Proceedings of the National Academy of Sciences of the United States of America*, 104, 11441-6.

Kuwahara, T., A. Koyama, K. Gengyo-Ando, M. Masuda, H. Kowa, M. Tsunoda, S. Mitani & T. Iwatsubo (2006) Familial Parkinson mutant alpha-synuclein causes dopamine neuron dysfunction in transgenic Caenorhabditis elegans. *The Journal of biological chemistry*, 281, 334-40.

Lakso, M., S. Vartiainen, A. M. Moilanen, J. Sirvio, J. H. Thomas, R. Nass, R. D. Blakely & G. Wong (2003) Dopaminergic neuronal loss and motor deficits in Caenorhabditis elegans overexpressing human alpha-synuclein. *Journal of neurochemistry*, 86, 165-72.

Lambert, A. J. & M. D. Brand (2004) Inhibitors of the quinone-binding site allow rapid superoxide production from mitochondrial NADH:ubiquinone oxidoreductase (complex I). *J Biol Chem*, 279, 39414-20.

Langston, J. W. (1996) The etiology of Parkinson's disease with emphasis on the MPTP story. *Neurology*, 47, S153-60.

Le Couteur, D. G., A. J. McLean, M. C. Taylor, B. L. Woodham & P. G. Board (1999) Pesticides and Parkinson's disease. *Biomed Pharmacother*, 53, 122-30.

Lee, C. K., R. Weindruch & T. A. Prolla (2000) Gene-expression profile of the ageing brain in mice. *Nat Genet*, 25, 294-7.

Levy, G. (2007) The relationship of Parkinson disease with aging. *Arch Neurol*, 64, 1242-6.

Li, P., D. Nijhawan, I. Budihardjo, S. M. Srinivasula, M. Ahmad, E. S. Alnemri & X. Wang (1997) Cytochrome c and dATP-dependent formation of Apaf-1/caspase-9 complex initiates an apoptotic protease cascade. *Cell*, 91, 479-89.

Li, X., J. C. Patel, J. Wang, M. V. Avshalumov, C. Nicholson, J. D. Buxbaum, G. A. Elder, M.
    E. Rice & Z. Yue (2010) Enhanced striatal dopamine transmission and motor
    performance with LRRK2 overexpression in mice is eliminated by familial
    Parkinson's disease mutation G2019S. *The Journal of neuroscience : the official journal
    of the Society for Neuroscience*, 30, 1788-97.
Li, Y., W. Liu, T. F. Oo, L. Wang, Y. Tang, V. Jackson-Lewis, C. Zhou, K. Geghman, M.
    Bogdanov, S. Przedborski, M. F. Beal, R. E. Burke & C. Li (2009) Mutant
    LRRK2(R1441G) BAC transgenic mice recapitulate cardinal features of Parkinson's
    disease. *Nature neuroscience*, 12, 826-8.
Lin, X., L. Parisiadou, X. L. Gu, L. Wang, H. Shim, L. Sun, C. Xie, C. X. Long, W. J. Yang, J.
    Ding, Z. Z. Chen, P. E. Gallant, J. H. Tao-Cheng, G. Rudow, J. C. Troncoso, Z. Liu,
    Z. Li & H. Cai (2009) Leucine-rich repeat kinase 2 regulates the progression of
    neuropathology induced by Parkinson's-disease-related mutant alpha-synuclein.
    Neuron, 64, 807-27.
Liou, H. H., M. C. Tsai, C. J. Chen, J. S. Jeng, Y. C. Chang, S. Y. Chen & R. C. Chen (1997)
    Environmental risk factors and Parkinson's disease: a case-control study in Taiwan.
    Neurology, 48, 1583-8.
Liu, X., H. Zou, C. Slaughter & X. Wang (1997) DFF, a heterodimeric protein that functions
    downstream of caspase-3 to trigger DNA fragmentation during apoptosis. *Cell*, 89,
    175-84.
Lu, X. H., S. M. Fleming, B. Meurers, L. C. Ackerson, F. Mortazavi, V. Lo, D. Hernandez, D.
    Sulzer, G. R. Jackson, N. T. Maidment, M. F. Chesselet & X. W. Yang (2009)
    Bacterial artificial chromosome transgenic mice expressing a truncated mutant
    parkin exhibit age-dependent hypokinetic motor deficits, dopaminergic neuron
    degeneration, and accumulation of proteinase K-resistant alpha-synuclein. *The
    Journal of neuroscience : the official journal of the Society for Neuroscience*, 29, 1962-76.
Malesev, D. & V. Kuntic (2007) Investigation of metal-flavonoid chelates and the
    determination of flavonoids via metal-flavonoid complexing reactions. *Journal of the
    Serbia Chemical Society* 72, 921-939.
Mandel, S., T. Amit, O. Bar-Am & M. B. Youdim (2007) Iron dysregulation in Alzheimer's
    disease: multimodal brain permeable iron chelating drugs, possessing
    neuroprotective-neurorescue and amyloid precursor protein-processing regulatory
    activities as therapeutic agents. *Progress in neurobiology*, 82, 348-60.
Mandel, S., G. Maor & M. B. Youdim (2004) Iron and alpha-synuclein in the substantia nigra
    of MPTP-treated mice: effect of neuroprotective drugs R-apomorphine and green
    tea polyphenol (-)-epigallocatechin-3-gallate. *Journal of molecular neuroscience : MN*,
    24, 401-16.
Mandel, S., O. Weinreb, L. Reznichenko, L. Kalfon & T. Amit (2006) Green tea catechins as
    brain-permeable, non toxic iron chelators to "iron out iron" from the brain. *Journal of
    neural transmission. Supplementum*, 249-57.
Manning-Bog, A. B., A. L. McCormack, J. Li, V. N. Uversky, A. L. Fink & D. A. Di Monte
    (2002) The herbicide paraquat causes up-regulation and aggregation of alpha-
    synuclein in mice: paraquat and alpha-synuclein. *J Biol Chem*, 277, 1641-4.
Marcotte, E. R., L. K. Srivastava & R. Quirion (2003) cDNA microarray and proteomic
    approaches in the study of brain diseases: focus on schizophrenia and Alzheimer's
    disease. *Pharmacology & therapeutics*, 100, 63-74.

Martin, L. J., Y. Pan, A. C. Price, W. Sterling, N. G. Copeland, N. A. Jenkins, D. L. Price & M. K. Lee (2006) Parkinson's disease alpha-synuclein transgenic mice develop neuronal mitochondrial degeneration and cell death. *The Journal of neuroscience : the official journal of the Society for Neuroscience*, 26, 41-50.

Matigian, N., G. Abrahamsen, R. Sutharsan, A. L. Cook, A. M. Vitale, A. Nouwens, B. Bellette, J. An, M. Anderson, A. G. Beckhouse, M. Bennebroek, R. Cecil, A. M. Chalk, J. Cochrane, Y. Fan, F. Feron, R. McCurdy, J. J. McGrath, W. Murrell, C. Perry, J. Raju, S. Ravishankar, P. A. Silburn, G. T. Sutherland, S. Mahler, G. D. Mellick, S. A. Wood, C. M. Sue, C. A. Wells & A. Mackay-Sim (2010) Disease-specific, neurosphere-derived cells as models for brain disorders. *Disease models & mechanisms*, 3, 785-98.

Mayeux, R., J. Denaro, N. Hemenegildo, K. Marder, M. X. Tang, L. J. Cote & Y. Stern (1992) A population-based investigation of Parkinson's disease with and without dementia. Relationship to age and gender. *Arch Neurol*, 49, 492-7.

Meco, G., V. Bonifati, N. Vanacore & E. Fabrizio (1994) Parkinsonism after chronic exposure to the fungicide maneb (manganese ethylene-bis-dithiocarbamate). *Scand J Work Environ Health*, 20, 301-5.

Menegon, A., P. G. Board, A. C. Blackburn, G. D. Mellick & D. G. Le Couteur (1998) Parkinson's disease, pesticides, and glutathione transferase polymorphisms. *Lancet*, 352, 1344-6.

Mercer, L. D., B. L. Kelly, M. K. Horne & P. M. Beart (2005) Dietary polyphenols protect dopamine neurons from oxidative insults and apoptosis: investigations in primary rat mesencephalic cultures. *Biochemical pharmacology*, 69, 339-45.

Mogi, M., A. Togari, T. Kondo, Y. Mizuno, O. Komure, S. Kuno, H. Ichinose & T. Nagatsu (1999) Brain-derived growth factor and nerve growth factor concentrations are decreased in the substantia nigra in Parkinson's disease. *Neurosci Lett*, 270, 45-8.

Moon, Y., K. H. Lee, J. H. Park, D. Geum & K. Kim (2005) Mitochondrial membrane depolarization and the selective death of dopaminergic neurons by rotenone: protective effect of coenzyme Q10. *J Neurochem*, 93, 1199-208.

Morris, M. E. (2000) Movement disorders in people with Parkinson disease: a model for physical therapy. *Phys Ther*, 80, 578-97.

Munoz-Fernandez, M. A. & M. Fresno (1998) The role of tumour necrosis factor, interleukin 6, interferon-gamma and inducible nitric oxide synthase in the development and pathology of the nervous system. *Progress in neurobiology*, 56, 307-40.

Nguyen, H. N., B. Byers, B. Cord, A. Shcheglovitov, J. Byrne, P. Gujar, K. Kee, B. Schule, R. E. Dolmetsch, W. Langston, T. D. Palmer & R. R. Pera (2011) LRRK2 mutant iPSC-derived DA neurons demonstrate increased susceptibility to oxidative stress. *Cell stem cell*, 8, 267-80.

Noack, H., U. Kube & W. Augustin (1994) Relations between tocopherol depletion and coenzyme Q during lipid peroxidation in rat liver mitochondria. *Free radical research*, 20, 375-86.

Norazit, A., A. C. Meedeniya, M. N. Nguyen & A. Mackay-Sim (2010) Progressive loss of dopaminergic neurons induced by unilateral rotenone infusion into the medial forebrain bundle. *Brain Res*, 1360, 119-29.

Olanow, C. W. & W. G. Tatton (1999) Etiology and pathogenesis of Parkinson's disease. *Annu Rev Neurosci*, 22, 123-44.

Palacino, J. J., D. Sagi, M. S. Goldberg, S. Krauss, C. Motz, M. Wacker, J. Klose & J. Shen (2004) Mitochondrial dysfunction and oxidative damage in parkin-deficient mice. *The Journal of biological chemistry*, 279, 18614-22.

Pan, T., J. Jankovic & W. Le (2003) Potential therapeutic properties of green tea polyphenols in Parkinson's disease. *Drugs & aging*, 20, 711-21.

Panet, H., A. Barzilai, D. Daily, E. Melamed & D. Offen (2001) Activation of nuclear transcription factor kappa B (NF-kappaB) is essential for dopamine-induced apoptosis in PC12 cells. *Journal of neurochemistry*, 77, 391-8.

Parain, K., M. G. Murer, Q. Yan, B. Faucheux, Y. Agid, E. Hirsch & R. Raisman-Vozari (1999) Reduced expression of brain-derived neurotrophic factor protein in Parkinson's disease substantia nigra. *Neuroreport*, 10, 557-61.

Park, J., S. B. Lee, S. Lee, Y. Kim, S. Song, S. Kim, E. Bae, J. Kim, M. Shong, J. M. Kim & J. Chung (2006) Mitochondrial dysfunction in Drosophila PINK1 mutants is complemented by parkin. *Nature*, 441, 1157-61.

Parkinson, J. (2002) An essay on the shaking palsy. 1817. *J Neuropsychiatry Clin Neurosci*, 14, 223-36; discussion 222.

Perez, C. A., Y. Wei & M. Guo (2009) Iron-binding and anti-Fenton properties of baicalein and baicalin. *Journal of inorganic biochemistry*, 103, 326-32.

Petzinger, G. M. & J. W. Langston. 1998. The MPTP-lesioned non-human primate: a model for Parkinson's disease. In *Advances in Neurodegenerative Disorders. Parkinson's Disease*, eds. J. Marwah & H. Teitelbaum, 113-148. Scottsdale: Prominent Press.

Preece, P. & N. J. Cairns (2003) Quantifying mRNA in postmortem human brain: influence of gender, age at death, postmortem interval, brain pH, agonal state and inter-lobe mRNA variance. *Brain research. Molecular brain research*, 118, 60-71.

Qin, Z. H., Y. Wang, M. Nakai & T. N. Chase (1998) Nuclear factor-kappa B contributes to excitotoxin-induced apoptosis in rat striatum. *Molecular pharmacology*, 53, 33-42.

Quilty, M. C., A. E. King, W. P. Gai, D. L. Pountney, A. K. West, J. C. Vickers & T. C. Dickson (2006) Alpha-synuclein is upregulated in neurones in response to chronic oxidative stress and is associated with neuroprotection. *Exp Neurol*, 199, 249-56.

Radad, K., W. D. Rausch & G. Gille (2006) Rotenone induces cell death in primary dopaminergic culture by increasing ROS production and inhibiting mitochondrial respiration. *Neurochem Int*, 49, 379-86.

Ramassamy, C. (2006) Emerging role of polyphenolic compounds in the treatment of neurodegenerative diseases: a review of their intracellular targets. *European journal of pharmacology*, 545, 51-64.

Ravenstijn, P. G., M. Merlini, M. Hameetman, T. K. Murray, M. A. Ward, H. Lewis, G. Ball, C. Mottart, C. de Ville de Goyet, T. Lemarchand, K. van Belle, M. J. O'Neill, M. Danhof & E. C. de Lange (2008) The exploration of rotenone as a toxin for inducing Parkinson's disease in rats, for application in BBB transport and PK-PD experiments. *J Pharmacol Toxicol Methods*, 57, 114-30.

Reto, M., M. E. Figueira, H. M. Filipe & C. M. Almeida (2007) Chemical composition of green tea (Camellia sinensis) infusions commercialized in Portugal. *Plant foods for human nutrition*, 62, 139-44.

Ritchie, C. W., A. I. Bush, A. Mackinnon, S. Macfarlane, M. Mastwyk, L. MacGregor, L. Kiers, R. Cherny, Q. X. Li, A. Tammer, D. Carrington, C. Mavros, I. Volitakis, M. Xilinas, D. Ames, S. Davis, K. Beyreuther, R. E. Tanzi & C. L. Masters (2003) Metal-protein attenuation with iodochlorhydroxyquin (clioquinol) targeting Abeta amyloid deposition and toxicity in Alzheimer disease: a pilot phase 2 clinical trial. *Archives of neurology*, 60, 1685-91.

Rodrigues, R. W., V. C. Gomide & G. Chadi (2004) Astroglial and microglial activation in the wistar rat ventral tegmental area after a single striatal injection of 6-hydroxydopamine. *Int J Neurosci*, 114, 197-216.

Sang, T. K., H. Y. Chang, G. M. Lawless, A. Ratnaparkhi, L. Mee, L. C. Ackerson, N. T. Maidment, D. E. Krantz & G. R. Jackson (2007) A Drosophila model of mutant human parkin-induced toxicity demonstrates selective loss of dopaminergic neurons and dependence on cellular dopamine. *The Journal of neuroscience : the official journal of the Society for Neuroscience*, 27, 981-92.

Sangchot, P., S. Sharma, B. Chetsawang, J. Porter, P. Govitrapong & M. Ebadi (2002) Deferoxamine attenuates iron-induced oxidative stress and prevents mitochondrial aggregation and alpha-synuclein translocation in SK-N-SH cells in culture. *Developmental neuroscience*, 24, 143-53.

Saravanan, K. S., K. M. Sindhu & K. P. Mohanakumar (2005) Acute intranigral infusion of rotenone in rats causes progressive biochemical lesions in the striatum similar to Parkinson's disease. *Brain Res*, 1049, 147-55.

Schapira, A. H. (2008) Mitochondria in the aetiology and pathogenesis of Parkinson's disease. *Lancet Neurol*, 7, 97-109.

Schwarting, R. K. & J. P. Huston (1996) The unilateral 6-hydroxydopamine lesion model in behavioral brain research. Analysis of functional deficits, recovery and treatments. *Prog Neurobiol*, 50, 275-331.

Sedelis, M., R. K. Schwarting & J. P. Huston (2001) Behavioral phenotyping of the MPTP mouse model of Parkinson's disease. *Behav Brain Res*, 125, 109-25.

Severson, J. A., J. Marcusson, B. Winblad & C. E. Finch (1982) Age-correlated loss of dopaminergic binding sites in human basal ganglia. *J Neurochem*, 39, 1623-31.

Sherer, T. B., R. Betarbet, J. H. Kim & J. T. Greenamyre (2003a) Selective microglial activation in the rat rotenone model of Parkinson's disease. *Neurosci Lett*, 341, 87-90.

Sherer, T. B., R. Betarbet, C. M. Testa, B. B. Seo, J. R. Richardson, J. H. Kim, G. W. Miller, T. Yagi, A. Matsuno-Yagi & J. T. Greenamyre (2003b) Mechanism of toxicity in rotenone models of Parkinson's disease. *J Neurosci*, 23, 10756-64.

Sherer, T. B., J. H. Kim, R. Betarbet & J. T. Greenamyre (2003c) Subcutaneous rotenone exposure causes highly selective dopaminergic degeneration and alpha-synuclein aggregation. *Exp Neurol*, 179, 9-16.

Shiozaki, E. N., J. Chai, D. J. Rigotti, S. J. Riedl, P. Li, S. M. Srinivasula, E. S. Alnemri, R. Fairman & Y. Shi (2003) Mechanism of XIAP-mediated inhibition of caspase-9. *Molecular cell*, 11, 519-27.

Shoulson, I., D. Oakes, S. Fahn, A. Lang, J. W. Langston, P. LeWitt, C. W. Olanow, J. B. Penney, C. Tanner, K. Kieburtz & A. Rudolph (2002) Impact of sustained deprenyl (selegiline) in levodopa-treated Parkinson's disease: a randomized placebo-controlled extension of the deprenyl and tocopherol antioxidative therapy of parkinsonism trial. *Annals of neurology*, 51, 604-12.

Shults, C. W., D. Oakes, K. Kieburtz, M. F. Beal, R. Haas, S. Plumb, J. L. Juncos, J. Nutt, I. Shoulson, J. Carter, K. Kompoliti, J. S. Perlmutter, S. Reich, M. Stern, R. L. Watts, R. Kurlan, E. Molho, M. Harrison & M. Lew (2002) Effects of coenzyme Q10 in early

Parkinson disease: evidence of slowing of the functional decline. *Archives of neurology*, 59, 1541-50.

Sies, H. 1985. *Oxidative stress: introductory remarks*. London: Academic Press.--- (1986) Biochemistry of oxidative stress. *Angewaudte Chemie, International Edition in English*, 28, 1058-1071.---. 1991. *Oxidative stress: introduction*. London: Academic Press.

Sindhu, K. M., R. Banerjee, K. S. Senthilkumar, K. S. Saravanan, B. C. Raju, J. M. Rao & K. P. Mohanakumar (2006) Rats with unilateral median forebrain bundle, but not striatal or nigral, lesions by the neurotoxins MPP+ or rotenone display differential sensitivity to amphetamine and apomorphine. *Pharmacol Biochem Behav*, 84, 321-9.

Sindhu, K. M., K. S. Saravanan & K. P. Mohanakumar (2005) Behavioral differences in a rotenone-induced hemiparkinsonian rat model developed following intranigral or median forebrain bundle infusion. *Brain Res*, 1051, 25-34.

Smeyne, R. J. & V. Jackson-Lewis (2005) The MPTP model of Parkinson's disease. *Brain Res Mol Brain Res*, 134, 57-66.

Smith, M. A. (1996) Hippocampal vulnerability to stress and aging: possible role of neurotrophic factors. *Behav Brain Res*, 78, 25-36.

Soldner, F., D. Hockemeyer, C. Beard, Q. Gao, G. W. Bell, E. G. Cook, G. Hargus, A. Blak, O. Cooper, M. Mitalipova, O. Isacson & R. Jaenisch (2009) Parkinson's disease patient-derived induced pluripotent stem cells free of viral reprogramming factors. *Cell*, 136, 964-77.

Sutherland, G. T., N. A. Matigian, A. M. Chalk, M. J. Anderson, P. A. Silburn, A. Mackay-Sim, C. A. Wells & G. D. Mellick (2009) A cross-study transcriptional analysis of Parkinson's disease. *PloS one*, 4, e4955.

Takahashi, K. & S. Yamanaka (2006) Induction of pluripotent stem cells from mouse embryonic and adult fibroblast cultures by defined factors. *Cell*, 126, 663-76.

Tan, R. X., J. C. Meng & K. Hostettmann (2000) Phytochemical investigation of some traditional Chinese medicines and endophyte cultures. *Pharmaceutical Biology*, 38, 25-32.

Tanaka, K., T. Suzuki, T. Chiba, H. Shimura, N. Hattori & Y. Mizuno (2001) Parkin is linked to the ubiquitin pathway. *Journal of molecular medicine*, 79, 482-94.

Tanner, C. M. (1989) The role of environmental toxins in the etiology of Parkinson's disease. *Trends Neurosci*, 12, 49-54.

Testa, C. M., T. B. Sherer & J. T. Greenamyre (2005) Rotenone induces oxidative stress and dopaminergic neuron damage in organotypic substantia nigra cultures. *Brain Res Mol Brain Res*, 134, 109-18.

Thiruchelvam, M., E. K. Richfield, R. B. Baggs, A. W. Tank & D. A. Cory-Slechta (2000) The nigrostriatal dopaminergic system as a preferential target of repeated exposures to combined paraquat and maneb: implications for Parkinson's disease. *J Neurosci*, 20, 9207-14.

Thomas, J., J. Wang, H. Takubo, J. Sheng, S. de Jesus & K. S. Bankiewicz (1994) A 6-hydroxydopamine-induced selective parkinsonian rat model: further biochemical and behavioral characterization. *Exp Neurol*, 126, 159-67.

Tong, Y., A. Pisani, G. Martella, M. Karouani, H. Yamaguchi, E. N. Pothos & J. Shen (2009) R1441C mutation in LRRK2 impairs dopaminergic neurotransmission in mice. *Proceedings of the National Academy of Sciences of the United States of America*, 106, 14622-7.

Tsubaki, T., Y. Honma & M. Hoshi (1971) Neurological syndrome associated with clioquinol. *Lancet*, 1, 696-7.

Turrens, J. F. (2003) Mitochondrial formation of reactive oxygen species. *J Physiol*, 552, 335-44.

Van Den Eeden, S. K., C. M. Tanner, A. L. Bernstein, R. D. Fross, A. Leimpeter, D. A. Bloch & L. M. Nelson (2003) Incidence of Parkinson's disease: variation by age, gender, and race/ethnicity. *Am J Epidemiol*, 157, 1015-22.

Vaux, D. L. & J. Silke (2003) Mammalian mitochondrial IAP binding proteins. *Biochemical and biophysical research communications*, 304, 499-504.

Vila, M., V. Jackson-Lewis, C. Guegan, D. C. Wu, P. Teismann, D. K. Choi, K. Tieu & S. Przedborski (2001) The role of glial cells in Parkinson's disease. *Curr Opin Neurol*, 14, 483-9.

von Coelln, R., B. Thomas, S. A. Andrabi, K. L. Lim, J. M. Savitt, R. Saffary, W. Stirling, K. Bruno, E. J. Hess, M. K. Lee, V. L. Dawson & T. M. Dawson (2006) Inclusion body formation and neurodegeneration are parkin independent in a mouse model of alpha-synucleinopathy. *The Journal of neuroscience : the official journal of the Society for Neuroscience*, 26, 3685-96.

Wang, C., R. Lu, X. Ouyang, M. W. Ho, W. Chia, F. Yu & K. L. Lim (2007) Drosophila overexpressing parkin R275W mutant exhibits dopaminergic neuron degeneration and mitochondrial abnormalities. *The Journal of neuroscience : the official journal of the Society for Neuroscience*, 27, 8563-70.

Wang, L., C. Xie, E. Greggio, L. Parisiadou, H. Shim, L. Sun, J. Chandran, X. Lin, C. Lai, W. J. Yang, D. J. Moore, T. M. Dawson, V. L. Dawson, G. Chiosis, M. R. Cookson & H. Cai (2008) The chaperone activity of heat shock protein 90 is critical for maintaining the stability of leucine-rich repeat kinase 2. *The Journal of neuroscience : the official journal of the Society for Neuroscience*, 28, 3384-91.

Wang, X., Z. H. Qin, Y. Leng, Y. Wang, X. Jin, T. N. Chase & M. C. Bennett (2002) Prostaglandin A1 inhibits rotenone-induced apoptosis in SH-SY5Y cells. *Journal of neurochemistry*, 83, 1094-102.

Watts, R. L., T. Subramanian, A. Freeman, C. G. Goetz, R. D. Penn, G. T. Stebbins, J. H. Kordower & R. A. Bakay (1997) Effect of stereotaxic intrastriatal cografts of autologous adrenal medulla and peripheral nerve in Parkinson's disease: two-year follow-up study. *Exp Neurol*, 147, 510-7.

West, A. B., D. J. Moore, S. Biskup, A. Bugayenko, W. W. Smith, C. A. Ross, V. L. Dawson & T. M. Dawson (2005) Parkinson's disease-associated mutations in leucine-rich repeat kinase 2 augment kinase activity. *Proceedings of the National Academy of Sciences of the United States of America*, 102, 16842-7.

Whitworth, A. J., D. A. Theodore, J. C. Greene, H. Benes, P. D. Wes & L. J. Pallanck (2005) Increased glutathione S-transferase activity rescues dopaminergic neuron loss in a Drosophila model of Parkinson's disease. *Proceedings of the National Academy of Sciences of the United States of America*, 102, 8024-9.

Xiong, N., J. Huang, Z. Zhang, J. Xiong, X. Liu, M. Jia, F. Wang, C. Chen, X. Cao, Z. Liang, S. Sun, Z. Lin & T. Wang (2009) Stereotaxical infusion of rotenone: a reliable rodent model for Parkinson's disease. *PLoS One*, 4, e7878.

Yassin, M. S., J. Ekblom, M. Xilinas, C. G. Gottfries & L. Oreland (2000) Changes in uptake of vitamin B(12) and trace metals in brains of mice treated with clioquinol. *Journal of the neurological sciences,* 173, 40-4.

Youdim, M. B., M. Fridkin & H. Zheng (2005) Bifunctional drug derivatives of MAO-B inhibitor rasagiline and iron chelator VK-28 as a more effective approach to treatment of brain ageing and ageing neurodegenerative diseases. *Mechanisms of ageing and development,* 126, 317-26.

Youdim, M. B., G. Stephenson & D. Ben Shachar (2004) Ironing iron out in Parkinson's disease and other neurodegenerative diseases with iron chelators: a lesson from 6-hydroxydopamine and iron chelators, desferal and VK-28. *Annals of the New York Academy of Sciences,* 1012, 306-25.

Zbarsky, V., K. P. Datla, S. Parkar, D. K. Rai, O. I. Aruoma & D. T. Dexter (2005) Neuroprotective properties of the natural phenolic antioxidants curcumin and naringenin but not quercetin and fisetin in a 6-OHDA model of Parkinson's disease. *Free radical research,* 39, 1119-25.

Zheng, H., D. Blat & M. Fridkin (2006) Novel neuroprotective neurotrophic NAP analogs targeting metal toxicity and oxidative stress: potential candidates for the control of neurodegenerative diseases. *Journal of neural transmission. Supplementum,* 163-72.

Zheng, H., L. M. Weiner, O. Bar-Am, S. Epsztejn, Z. I. Cabantchik, A. Warshawsky, M. B. Youdim & M. Fridkin (2005) Design, synthesis, and evaluation of novel bifunctional iron-chelators as potential agents for neuroprotection in Alzheimer's, Parkinson's, and other neurodegenerative diseases. *Bioorganic & medicinal chemistry,* 13, 773-83.

Zigmond, M. J., T. G. Hastings & E. D. Abercrombie (1992) Neurochemical responses to 6-hydroxydopamine and L-dopa therapy: implications for Parkinson's disease. *Ann N Y Acad Sci,* 648, 71-86.

Zou, H., W. J. Henzel, X. Liu, A. Lutschg & X. Wang (1997) Apaf-1, a human protein homologous to C. elegans CED-4, participates in cytochrome c-dependent activation of caspase-3. *Cell,* 90, 405-13.

3

Oxidative Stress in Human Autoimmune Joint Diseases

Martina Škurlová

Department of Normal, Pathological, and Clinical Physiology,
Third Faculty of Medicine, Charles University in Prague,
Czech Republic

1. Introduction

Living with oxygen is basically unsafe, but vital. During evolution, oxygen originally a waste product of the metabolism in primitive unicellular organisms became normal product of the metabolism in higher animal species involving humans. Even when oxidative reactions are toxic, and destructive, they are tolerated by all organisms to some extent. The fact has opened the discussion about efficiency of antioxidant mechanisms. The classical enzyme antioxidant defence alone does not explain high tolerance of the organism for oxygen. Moreover, enzyme antioxidant mechanisms are not hundred percent effective in preventing oxidation what allows oxidative damage to continue.

The pathogenesis of autoimmune joint inflammatory diseases is related to activation of native immune system. At site of inflammation, activated neutrophils and macrophages consume large amounts of oxygen, whose corollary is the increase of reactive oxygen species (ROS) production. There are several mechanisms how oxidative stress is involved into the pathogenesis of autoimmune joint inflammatory diseases. Excess production of ROS in the joint area encourages process of re-oxygenation, which then promotes joint inflammation. ROS further inhibit connective tissue cell proliferation, in some cases ROS have been shown to induce cell death to these cells inducing apoptosis.

2. Oxidative stress

2.1 Biology of oxidative stress

Cellular responses to oxidative stress depend on the cellular redox status. When the oxidants' level does not exceed the redox capacities in a cell, oxidants are beneficial to the cell controlling cellular functions such as signal transduction. In contrast, when the cellular antioxidant capacity is insufficient, the production of oxidants exceeds the capacity to neutralize them (Hitchon & El- Gabalawy, 2004). Insufficient oxidative defence mechanisms shift the balance between oxidants and antioxidants in the direction of oxidants leading to oxidative stress. Insufficient oxidative defence mechanisms result from depletion of enzymatic (e.g., superoxide dismuthase, catalase, glutathione peroxidase), and non-enzymatic (e.g., glutathione, vitamins A, C, and E, and selenium) antioxidants (Hovatta et al., 2010). The pro-oxidant conditions of the *'internal milieu'*, due to low redox status of the cell led to a new definition of oxidative stress. Oxidative stress is defined as " disruption between ROS production and elimination leading to their enhanced steady- state in the body " (Lushchak, 2011).

Reactive oxygen species destroy not only intracellular components, but also cell membranes, and extracellular components. ROS modify proteins by oxidation, nitrosylation, nitration or chlorination of specific amino acids, leading to their impaired biological activity, changes in protein structure and accumulation of damaged proteins in the tissue. During lipid peroxidation, which is a marker of oxidative stress, polyunsaturated fatty lipids are oxidized and produce lipid peroxyl radicals that in turn up-regulate oxidation, and cell membrane damage (Hitchon & El- Gabalawy, 2004). Genotoxic effects of oxidative stress involve direct breakage of DNA and DNA repair mechanisms. Oxidative stress may also cause cell death. Cellular content containing oxidized molecules when released into the extracellular environment may contribute to the exacerbation of synovial inflammation as newly formed ROS and degradation products form a vicious inflammatory circle (Henrontin et al., 2003).

3. Reactive oxidant species

Free radicals are very reactive chemical species that have unpaired valence shell electrons in their outer orbitals (Afonso et al., 2007). Highly reactive and partly reduced oxygen metabolites are a by-product of oxidative phosphorylation process, which takes part in mitochondria. These metabolites called ROS include oxygen radicals [superoxide ($O_2^{\bullet-}$), hydroxyl ($HO^{\bullet}$), peroxyl ($O_2R^{\bullet}$), and alkoxyl ($OR^{\bullet}$)] and certain non radicals that are either oxidizing agents or are easily converted into radicals, such as hypochlorous acid (HOCl), ozone (O_3), singlet oxygen ($^1O_2$), and hydrogen peroxide (H_2O_2). Other oxidants, generated by interactions with these molecules, include reactive nitrogen species (RNS), as an example nitric oxide (NO), peroxynitrite (ONOO-).

3.1 Generation of reactive oxidant species

Generation of ROS is generally cascade of reactions that starts with the production of $O_2^{\bullet-}$. Superoxide rapidly dismutates to H_2O_2 either spontaneously, particularly at low pH, or catalyzed by superoxide dismutase (SOD) enzyme. Other steps in the cascade of ROS generation include the reaction of $O_2^{\bullet-}$ with $ON^{\bullet}$ to form ONOO ⋅ the peroxidase-catalyzed formation of HOCl from H_2O_2, and the iron-catalyzed Fenton reaction leading to the generation of $HO^{\bullet}$. Hydroxyl radical is one of the most reactive oxygen radicals. In biologic systems, the $HO^{\bullet}$ is formed by the reaction between H_2O_2 and iron or copper in a low valence state. The oxidized halogens are almost as diverse group of reactive oxidants as are the free radicals. They consist of HOCl and the vast number of chloramines. Chloramines can be produced from the reaction of HOCl with the many amines that are found in biological systems. Like HOCl, the chloramines are oxidizing species. Chloramine (NH_2Cl) is formed by the reaction of HOCl with ammonia (NH_3). The reaction of HOCl with amino acids leads through chloramines to aldehydes. Altogether, the oxidized halogens represent probably the most important microbicidal oxidants produced by neutrophils. Oxygen (O_2) alone is a diradical with two unpaired electrons. There is also a much more reactive form of oxygen known as $^1O_2$, in which those two electrons are paired. The $^1O_2$ is produced by neutrophils, which manufacture it by the reaction between H_2O_2 and an oxidized halogen.

Reactive nitrogen species are produced by phagocytes during the reaction of NO with $O_2^{\bullet-}$, and other oxidizing species. The first characterized of these species is $ONOO^-$, formed in a reaction between $ON^{\bullet}$ and $O_2^{\bullet-}$, both of which are free radicals. Peroxynitrite then undergoes a secondary reaction to produce an agent that is able to nitrate tyrosine. The precise composition of the agent is not known, but several candidates have been proposed. Among them are various derivatives of $ONOO^-$ Major cellular sites of ROS generation include the mitochondria, and non- mitochondrial membrane bound enzyme systems (Babior, 2000).

3.2 Reactive oxidant species in the joint tissue

Articular cartilage is a unique tissue for its constituent cells, the chondrocytes, which maintain the cartilage matrix through their continual synthesis and degradation. The environment is avascular, hyperosmotic and acidic. Cartilage cells are synoviocytes, and chondrocytes. Chondrocytes display a metabolism adapted to anaerobic conditions. Synoviocytes supply the avascular cartilage tissue with nutrients via synovial fluid. Because cartilage is an avascular environment, the oxygen tension in the area is usually low. In pathological conditions, like inflammation, oxygen tension is subject of fluctuations. These variations of oxygen tension force chondrocytes to produce reactive species. The main reactive species produced by chondrocytes are $O_2^{\bullet-}$ radical, and NO that generate other derivative radicals, including $ONOO^-$ and H_2O_2 (Hiran et al., 1997; Stefanovic- Racic et al., 1997). The effects of free radicals on articular cartilage are dual. "Chondrocyte- derived" free radical levels are important for the maintenance of ion homeostasis. Natrium hydrogen (Na^+ - H^+) - exchanger (NHE- activity) and free radical levels exhibit a significant positive correlation. How exactly highly reactive species alter ion transport is not known, although interference with protein phosphorylation is possible (Gibson et al., 2008).

Except beneficial effect free radicals may have on articular cartilage when exceed in a cell, free radicals damage both chondrocytes, and extracellular matrix (ECM) components of articular cartilage. Free radicals shift the redox balance in articular cartilage in direction of oxidants. Hypochlorous acid, singlet oxygen, and peroxynitrite radicals balance the ascorbate, an antioxidant vitamin, from cartilage (Hajdigogos et al., 2003).

ROS and RNS damage articular cartilage directly or indirectly by up- regulating the mediators of the ECM degradation. Reactive oxygen/ nitrogen species, e.g. $ONOO^-$, also have been shown to degrade aggrecan, a major component of ECM, and this degradation is one of the initial events in the process of cartilage destruction (Billinghurst *et al.*, 1997). The incidence of sulfated glycosaminoglycans (GAGs) reflects the ratio of aggrecan degradation in the cartilage.

In seeking for a role of oxidative radicals in cartilage metabolism, it has been noted that endogenously generated NO suppresses the biosynthesis of aggrecan, a major macromolecular component of the cartilaginous matrix (Cao et al., 1997). Furthermore, it was discovered that oxygen radicals fragment hyaluronan and chondroitin sulphate (Kennett & Davies, 2009).

Collagen, which provides tensile strength and forms a network that resists the swelling pressure of aggrecan- hyaluronate aggregates, can be altered directly by oxygen radicals. Free radicals prime collagen to proteolytic enzymes. Incubation of cartilage slices with xanthine-oxidase- generated $O_2^{\bullet-}$ anion degrades type I collagen and fibril formation by this collagen. Hydroxyl radical in the presence of oxygen fragments collagen into small peptides. The cleavage seems to be specific to proline or 4- hydroxyproline residues (Monboisse & Borel, 1992). Interestingly, free radicals may destruct collagen synthesis indirectly. NO inhibits collagen synthesis via interleukin-1 (IL-1) (Cao et al., 1997). H_2O_2 inhibits cartilage proteoglycan synthesis interfering with adenosine triphosphate (ATP) synthesis, in part by inhibiting the glycolytic enzyme glyceraldehyde-3-phosphate dehydrogenase in chondrocytes. ONOO- and HOCl may facilitate cartilage damage by inactivating tissue inhibitor of metalloproteinases (TIMPs). TIMP-1 inhibits stromelysins, collagenases and gelatinases (Henrontin et al., 2003). Moreover it was observed, that treatment of chondrocytes with the NO-producing agent, S-nitroso-N-acetylpenicillamine, up regulates the collagenase mRNA levels, which is one of MMPs (Lo et al., 1998).

Exposure of the chondrocytes to H_2O_2 inhibits proteoglycan synthesis (Henrontin et al., 2003). Chondrocytes in arthritic cartilage respond poorly to insulin- like growth factor 1 (IGF-1) what may lead to abrogation of cartilage repair. In this context, ROS may participate in reducing the capacity of chondrogenic precursor cells to migrate and proliferate within joint area. Nitric oxide radical was also demonstrated to inhibit chondrocyte migration and attachment to fibronectin via modification of the actin cytoskeleton. Chondrocytes produce high levels of NO, which is a mediator of anti- proliferative effects of IL-1 in these articular cells (Blanco & Lotz, 1994). In addition, it was discovered that IL-1 induces apoptosis to chondrocytes via NO. Combination of IL-1 with the radical scavengers like N-acetyl cysteine, dimethyl sulfoxide, or 5, 5'-dimetylpyrroline 1-oxide induced apoptosis, which was inhibited in a dose dependent manner by the NO synthase inhibitor N-monomethyl L-arginine (Blanco et al., 1995). Chondrocyte death, determined as the percentage of empty lacunae in articular cartilage, was completely blocked in p47phox-/- mice confirming the "Nicotinamide adenine dinucleotide phosphate" (NADPH)- oxidase" driven oxygen radical production in mediating these effects (Lem van Lent et al., 2005).

Synoviocytes are the second kind of articular cells. These cells consume larger amount of oxygen when compared to chondrocytes (Schneider et al., 2005). An indirect evidence of oxidative stress in synoviocytes is the incidence of antioxidant enzymes such as superoxide dismutase, glutathione peroxidase and catalase in these cells (Mattey et al., 1993). Oxidative stress makes synoviocytes to undergo a cell death of an apoptotic nature (Galleron et al., 1999). The ROS scavenger system of synoviocytes protects chondrocytes from toxic effects of free radicals. In a co-culture of these cells synoviocytes reduced toxic effects of H_2O_2 on chondrocyte cell damage (Kurz et al., 1999). NO is the primary inducer of apoptosis in human articular chondrocytes (Blanco et al., 1995). NO- mediated chondrocyte cell death requires generation of additional reactive species like ONOO- and O_2^{-} (Del Carlo Jr. & Loeser, 2002).

Joint fluid is produced as a transudate of plasma from synovial cells and provides nutrition to the articular cartilage by diffusion of oxygen and other molecules. The primary catalytic antioxidant of the joint fluid is the extracellular SOD (Regan et al., 2008). SOD type III accounts for 80% of the enzyme's activity in the joint fluid (Afonso et al., 2007). Glycosaminoglycans (long-chain polysaccharides) are major components of the extracellular matrix, glycocalyx, and synovial fluid. Modifications to these materials are linked to multiple human pathologies including autoimmune diseases. Hyaluronan and chondroitin sulfate are extensively depolymerized by hydroxyl and carbonyl radicals, which may be formed from ONOO.. Polymer fragmentation is shown to be dependent on the radical flux (Kennett & Davies, 2009). NO, and O_{2-} inhibit type II collagen and proteoglycans synthesis and the sulfation of newly synthesized GaGs (Hickery & Bayliss, 1998).

4. Inflammatory synovitis: May oxidative stress be a cause?

4.1 ROS in inflammatory synovium

In autoimmune joint diseases systemic inflammation exists long before it exerts local effects on synovial membrane. At one point, systemic inflammation is translocated into synovium where it initiates the inflammatory response often leading to oxidative burst. Oxidative burst in rheumatoid joints is a result of the activation of innate immune system cells. Activated phagocytic cells such as neutrophils, and macrophages both produce free radicals in the joint area. Activated phagocytes produce reactive oxidants by enzymes: the NADPH- oxidase, and the nitric oxide synthase (NOS). Mechanism of free radicals production differs between these cell groups. While macrophages are stimulated by the "NADPH- oxidase" system to produce free radicals, the presence of NOS accompanied by the NADPH- oxidase is necessary for neutrophils to secrete free radicals. RA neutrophils also generate enhanced amount of ONOO- by NOS (El Benna et al., 2002). Chemiluminescence assays demonstrated significant activation of the neutrophil myeloperoxidase H_2O_2 system in synovial fluids from patients with RA further suggesting that oxidative stress may contribute to the cyclic, self- perpetuating nature of rheumatoid inflammation. ROS, produced by activated phagocytes alters the antigenic behaviour of immunoglobulin G (IgG). Radical-exposed IgG is able to bind rheumatoid factor and results in the generation of C3alpha complement component. This reaction may be self- perpetuating within the rheumatoid joint, suggesting that free radicals play a role in the chronicity of rheumatoid inflammation (Newkirk et al., 2002).

Pro-inflammatory cytokines´ presence and activity undoubtedly subjects to rheumatoid synovitis governing a variety of pathological processes including cell activation, cell proliferation, tissue resorption and chemotaxis (Schett et al., 2000). Experimental evidence confirms cytokine- induced oxidative stress in rheumatoid synovium. Thioredoxin, a cellular catalyst induced by oxidative stress, is found in high amounts in RA synovial cells and tissue. Thioredoxin acts as a co- factor for tumor necrosis factor- alpha (TNF- α) - induced synthesis of interleukins (IL-6 and IL-8) in synovial fibroblastlike cells (Yoshida et al., 1999). Edaravone, which is a clinically available antioxidant, suppresses IL-1β- induced synovial cells proliferation and migration under in vitro conditions (Arii et al., 2006). N- acetylcytein, a known thiol antioxidant, abrogated L-6 - induced proliferation of RA patients synovial fibroblasts (Ji-Yeon et al., 2000).

ROS are documented as mediators of synovial inflammation. Excessive production of ROS at the site of inflammation contributes to the inflammatory process in general, by induction of the expression of adhesion molecules, pro-inflammatory cytokines, and chemoattractants. Furthermore, ROS can directly increase tissue destruction through inactivation of the major inhibitor of degrading enzymes, α1-antiproteinase, which subsequently leads to the activation of extracellular matrix-degrading metalloproteinases (MMPs) (Maurice et al., 1997). 3-nitrotyrosine (3- NT) has been identified as a stable end product and marker of inflammation and RNS production. Nitrated proteins are generated in inflamed tissues by inflammatory cells producing ONOO-, a naturally occurring nitrating agent. A study by Khan & Siddiqui (2006) further investigated the binding characteristics of naturally occurring antibodies to 3- NT present in synovial fluid. Antibodies to 3- NT were found higher in the synovial fluid of RA patients.

Ex-vivo cultured rheumatoid synovium produces a significant amount of nitrite, and the addition of N- methylarginine (L-NMMA) significantly inhibits NO production. *In vitro* study revealed that NO production from freshly isolated synovial cells was up- regulated by stimulation with a combination of IL-1β, TNF-α and lipopolysaccharide. Inducible NOS expression was induced when human chondrocytes were stimulated with IL-1β, TNF-α or endotoxin in a dose- and time-dependent manner. Inflammatory reaction in the synovium of RA patients could be augmented by the autocrine or other cytokine-induced production of IL-6 with subsequent generation of ROS in the synoviocytes. Fibroblast- like synoviocytes proliferation by IL-6 was inhibited by N- acetylcysteine. Oxidative stress of RA synovial tissue can cause DNA damage and suppress the DNA mismatch repair (MMR) system in cultured synoviocytes. DNA MMR enzyme expression is greatest in the synovial intimal lining layer, where maximal oxidative stress in RA occurs (Šimelyte et al., 2004).

4.2 Cycles of hypoxia/ reoxygenation in inflammatory synovium

Angiogenesis in the synovial membrane is an important early step in pathogenesis of RA, in the perpetuation of the disease and may precede other pathological features. Elevated levels of pro-angiogenic factor, the vascular endothelial growth factor VEGF, are expressed in synovium, synovial fluids, and serum of RA patients. An increased VEGF level in RA is responsible for subsequent joint destruction. The synovium of RA is hypoxic as a result of synovial tissue proliferation outpacing in angiogenesis. Hypoxia is a potent inducer of cytokines, matrix degrading enzymes, angiogenic factors that play a central role in the inflammatory response. Hypoxia- inducible factor- 1α (HIF-1α) is a key transcription factor which is highly inducible by hypoxia and expressed predominantly in synovium of RA. Expression of HIF-1α is critical for joint inflammation. RA synovial fluids are hypoxic, acidic with low glucose and elevated lactate concentrations. This biochemical profile is indicative of a chronically hypoxic microenvironment that compensates by anaerobic metabolism. The oxygen partial pressure (pO₂) is extremely low in RA synovial fluids, and correlates with elevated plasma levels of lactate. Inflamed synovitis is a hallmark of RA which is hypoxic in nature (Shankar et al., 2009). The biological basis of this process relates to hypoxia- inducible factor (HIF). The response of mammalian cells to hypoxia is mediated partly through stabilization of certain transcription factors including HIF-1 and HIF-2. These oxygen sensitive transcription factors are multifunctional. Firstly, they program the cells to anaerobic metabolism, secondly, they enhance cell survival by inhibiting apoptosis, and thirdly they improve the supply of oxygen by promoting angiogenesis and increase oxygen-carrying capacity. In view of the crucial role of HIF-1 in cellular adaptation to hypoxia, its regulation needs to be rapidly responsive to changes in the cellular oxygen supply. Inhibition of degradation is the primary mechanism by which hypoxia directly regulates HIF- 1α. In the absence of oxygen critical step of degradation process of HIF-1α, the hydroxylation of proline and asparagine residues, becomes rate limiting, thus preventing HIF-1α from being degraded and leaving it free to bind to its constitutively expressed partner HIF-1β. In rheumatic patients, joint movement in a ratio of normally functional joint increases the pO_2 leading to re-oxygenation. In RA cycling transient episodes of hypoxia/re-oxygenation increase levels of ROS. Increased mitochondrial ROS levels stabilize the transcription factor HIF- 1α (Chandel et al., 2000). Of particular relevance to RA was the marked attenuation of synovitis and articular damage in an adjuvant arthritis model when HIF- 1α was absent. In RA synovitis, HIF-1a protein accumulates and translocates to the nucleus and directly activates transcription of pro-angiogenic factor like VEGF. VEGF is highly inducible by hypoxia which occurs in the inflamed joints of RA. The HIF-1α translocates and binds to core DNA motif in the hypoxia responsive elements (HRE) which are associated with target genes such as VEGF and induces its gene expression and thereby angiogenesis. HIF-1α is expressed abundantly by macrophages in most rheumatoid synovia, predominantly close to the intimal layer but also in the subintimal zone (Hollander et al., 2001). The synovial expression of HIF-1α also showed a mixed nuclear and cytoplasmic pattern mostly seen in lining cells, stromal cells, mononuclear cells, and blood vessels (Giatromanolaki et al., 2002). The number of HIF-1α -positive cells correlated strongly with the number of blood vessels in RA synovial tissue and with inflammatory endothelial cell infiltration (blood vessels), cell proliferation (Ki67) and the synovitis score (Brouwer et al., 2009). Reduction in the degradation rate of HIFs (HIF-1α, and HIF-2α) as occurs under hypoxic stress throughout arthritis, results in increased steady- state level of HIFs proteins and up-regulation of the neo- angiogenic process. Neo-angiogenic VEGF/ KDR pathway was shown persistently increased in RA, as indeed was microvessel density and the expression of PD- ECGF, irrespective of the extent of HIF expression (Giatromanolaki et al., 2002). The direct link exists between accumulation of HIFs and overexpression of VEGF in RA. Neo-angiogenesis contributes to pathogenesis of RA encouraging synovitis, pannus formation and articular cartilage destruction. Concluding, the HIF seems to be a promising factor that targets both synovitis and angiogenesis in RA.

5. Oxidative stress in human autoimmune joint diseases

5.1 Rheumatoid arthritis

Rheumatoid arthritis (RA) is a systemic autoimmune inflammatory disease primarily affecting synovial membranes of joints. Pathogenesis of RA is a multistep process where cellular and humoral interactions mediated by lymphocytes (T and B cells) and non hematopoietic cells like fibroblasts, connective tissue cells, and bone cells play a role. At site of inflammation, activation of T cells and macrophages leads to a large increase in oxygen consumption, whose corollary is increased release of ROS (Afonso et al., 2007). Both, hematopoietic, and connective tissue cells are subject of oxidative stress process in arthritis.

5.1.1 Hematopoietic cells and tissue as oxidative stress targets in RA

Activity of NADPH- oxidase is enhanced in circulating neutrophils and monocytes of RA patients. These phagocytic cells synthesize two- to eight- fold higher amounts of $O_2^{\bullet-}$ when compared to healthy controls. Production rates of $O_2^{\bullet-}$ in neutrophils and monocytes from rheumatic patients positively correlate with the plasma levels of TNF- α (Miesel et al., 1996). Other parameters of oxidative stress like decrease in neutrophil SOD activity and an increase in the levels of "loose" iron in the plasmalemma of RA neutrophils and monocytes were also observed (Ostrakhovitch & Afanas´ev, 2001). A characteristic feature of RA ROS-producing neutrophils is their functional state. The production of $O_2{}^{\bullet-}$ by blood neutrophils from RA patients in response to N-formyl-methionylleucyl-phenylalanine, a stimulatory agent, was greater in arthritic than control blood neutrophils (Eggleton et al., 1995). The study suggested that stimulated sub-population of neutrophils are source of ROS in arthritis.

5.1.2 Connective cells and tissue as oxidative stress targets in RA

Studies of RA synovial fluid (SF) and tissue have demonstrated oxidative damage to the tissue. Signs of oxidative stress such as DNA oxidative damage, and lipid peroxidation are present in the inflammatory synovium of RA patients. Immunohistochemical analysis revealed increased staining of 8- oxo- 7,8- dihydro- 2´- deoxyguanine, a marker of DNA oxidative damage, and increased staining of 4-hydroxy-2-nonenal in the lining and sublining layers of the RA inflammatory synovium (Šimelyte et al., 2004). Furthermore, it was shown that increased lipid peroxidation damage in RA inflammatory synovium is proportional to the levels of hypoxia in the joint, disease activity and angiogenic marker expression (Biniecka et al., 2009). Synovial fluids macrophages produce increased amounts of $O_2{}^{\bullet-}$. Also neutrophils from synovial fluids of rheumatic patients generate increased amounts of $O_2{}^{\bullet-}$ possibly because of their exposure to cytokines present in synovial fluids (Hitchon & El-Gabalawy, 2004). Chronic oxidative stress contributes to functional hyporesponsiveness of synovial T lymphocytes. The impaired mitogenic responses of SF T lymphocytes correlated with a significant decrease in the levels of the intracellular redox-regulating agent glutathione (GSH) (Maurice et al., 1997).

Indirect evidence for ROS implication in cartilage degradation comes from the presence of lipid peroxidation products, nitrite, nitrotyrosine, a nitrated type II collagen peptide, modified low-density lipoprotein (LDL) and oxidized IgG in the biological fluids of patients with arthritis. Furthermore, nitrotyrosine, nitrated proteins and oxidized LDL (ox-LDL) have been found to be accumulated in cartilage of arthritic patients demonstrating the direct implication of ROS in some joint diseases. Rheumatoid arthritis is characterized by irreversible damage to the cartilage matrix caused by enzymatic degradation of the proteins, e.g., collagen type II (CII), and proteoglycans of cartilage (e.g., aggrecan) (Billinghurst et al., 1997). As a result of the breakdown of the proteins and proteoglycans, CII degradation products and sulfated glycosaminoglycans (GAGs) appear in SFs of the affected joints. The level of GAGs in SF indicates the extent of proteoglycan degradation (Lark et al., 1997). Hydrogene peroxide, and singlet oxygen accelerate bone resorption by osteoclasts. Osteoclasts generate ROS through "NADPH- oxidase dependent mechanisms". Studies involving assays of nitrotyrosine residues in synovial tissues from patients with RA or exposure of chondrocytes to synthetic peroxynitrite *in vitro* have established that combination of the $O_2{}^{\bullet-}$ anion to nitric oxide (NO) causes cartilage damage (Abramson et al., 2001).

The cytokine network is involved in the pathogenesis of RA. IL-1 is a key mediator of bone resorption and cartilage destruction in arthritis. The cytokine activates bone resorption through its effects on osteoclast differentiation and activation. IL-1 destructs cartilage by stimulating release of MMPs from fibroblasts and chondrocytes. Neutralizing ROS activity significantly attenuated IL-1- induced collagenase gene expression in bovine chondrocytes (Lo et al., 1998).

In RA oxidative stress also features by oxidation of low- density lipoproteins (LDL). Oxidized LDLs promote inflammatory changes including local up- regulation of adhesion molecules and chemokines.

RA tissue has evidence of microsatellite instability reflecting ongoing mutagenesis. Such mutagenesis is normally corrected by DNA repair systems including the mismatch repair system, which is defective in RA, probably due to oxidative stress.

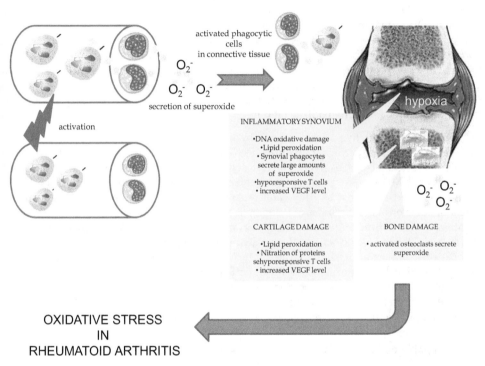

Fig. 1. Connective tissue and cells as targets of oxidative stress in rheumatoid arthritis.

6. Systemic lupus erythematosus

Systemic lupus erythematosus (SLE) is a prototype autoimmune, multisystem and multifactorial disease characterized by the presence of auto-antibodies to a variety of nuclear antigens such as DNA and histones, as well as protein antigens and protein–nucleic acid complexes. The initial immunizing antigen(s) that drive the development of SLE are unknown, but characteristics of the immune response in SLE suggest that it is an antigen-driven condition. Multiorgan inflammatory lesions also involve the joints. Immune complex deposits in the synovium are associated with mild inflammation and cartilage destruction is seldom severe. The arthritis of SLE is described as non-destructive and non-deforming. Immune complex deposits in the synovium trigger inflammatory reaction, which led continuously to cartilage destruction (Khan & Siddiqui, 2006). Free radicals synthesize autoantigens, which contribute to disease development. Oxidative stress and inflammation are interrelated in SLE. Malondialdehyd (MDA) levels positively correlated with levels of interferon- γ (INF-γ), and interleukin- 12 (IL- 12) in lupus disease (Shah et al., 2010).

6.1 Hematopoietic cells and tissue as oxidative stress targets in SLE

Oxidative damage to red blood cells and leukocytes are hallmarks of oxidative stress in SLE. Lipid peroxidation, in SLE erythrocytes, and generation of O_2^{-} and H_2O_2 in leukocytes were increased in SLE patients when compared to healthy controls (Tewthanom et al., 2008). Destroying effects of free radicals in lupus disease are apparent on the reactions with serum proteins. Oxidation makes the proteins more fragile. Modified proteins become autoantigens, and more, their presence enhances oxidative stress in serum 4-hydroxy-2-nonenal (4- HNE), which is a marker of lipid peroxidation in SLE, after binding protein forms dangerous protein adducts. The target of 4- HNE modification reported was catalase, a membrane protein of red blood cells (D'Souza et al., 2008). Hydroxyl radical binds human serum albumin. The increase in total serum protein carbonyl levels in the SLE patients was largely due to an increase in oxidized albumin (Shjekh et al., 2007).

In lupus disease, oxidative stress presents by lipid peroxidation mostly. Moreover it was discovered that lipid peroxidation influences pathogenesis of the disease. The disease activity score (SLEDAI) positively correlates with serum levels of MDA. T-lymphocyte apoptosis and MDA were positively associated with disease activity (Shah et al., 2011). MDA- modified proteins as the catalase, and SOD are targets of IgG circulated autoantibodies promoting disease development. On the other hand, SLEDAI score correlates negatively with serum antioxidant enzymes as SOD, and glutathione peroxidase (GPx) (Taysi et al., 2002 b). Auto- antibodies against plasma lipoproteins have been reported in SLE patients (Batuca et al., 2007). The study further documented incidence of antibodies toward high density lipoproteins (HDL) in patients with SLE, and identified Apo A-I as a target of oxidative damage. The incidence of auto-antibodies against lipid particles correlated with reduced activity of paraoxonase (PON), which is the most active antioxidant enzyme in lipids, and also prevents lipid peroxidation of low density lipoproteins (LDL).

6.2 Connective cells and tissue as oxidative stress targets in SLE

The arthritis of SLE is described as non- destructive and non- deforming. Immune complex deposits in the synovium are associated with mild inflammation and cartilage destruction is seldom severe. In SLE patients, NO and its intermediates may be mediators of inflammatory arthritis. Antibodies against 3- nitrotyrosine (3- NT) were found elevated in sera, and synovial fluid of SLE patients. Interestingly, the sera levels were much higher than in synovial fluid (Khan & Siddiqui, 2006).

7. Markers of oxidative stress in human autoimmune joint diseases

A short lifetime of free radicals in body fluids restricts their direct estimation instead effects of oxygen on lipids, proteins, and nucleic acids molecules are used. There are many chemical modifiers of protein, and lipid structures, whose activity is accelerated by oxidative stress. N-carboxymethyllysine (CML) represents a chemically modified amino acid and originates *in vivo* from carbohydrate as well as from lipid derived precursors. Oxidative damage to proteins is also reflected by increased levels of advanced oxidation protein products (AOPP), which form by the reaction between chlorinated oxidants (HOCl/OCl-) and proteins. AOPP are defined as dityrosine-containing cross-linked proteins. MDA is one of the end products of lipid peroxidation induced by ROS and marker of oxidative stress in lipids. 8-Hydroxy-2´-deoxyguanosine (8-OHdG) has been recognized as a biomarker of oxidative DNA damage by endogenously generated oxygen radicals. In addition, oxidative stress may be estimated by levels of enzymes like thioredoxin, a protein with reduction / oxidation active disulfide / dithiol groups in its active site. Signs of oxidative stress are apparent on cells and tissue affected by arthritis. Advanced glycation end products (AGEs), which are formed during the Maillard reaction by non-enzymatic glycation, and oxidation of proteins were detected in the synovial lining, sublining, and endothelium in RA patiens. CML showed positive immunostaining in some RA macrophages (CD68+) and T cells (CD45RO+) (Drinda et al., 2002). High levels of protein carbonyl (PCO), and AGE products were found in serum of collagen – induced arthritis mice (Choi, 2007). Plasma MDA concentrations are significantly higher in RA patients (Sarban et al., 2005). An excessive degree of oxidative stress in RA patients confirm decline of protein thiol levels and lower activity of antioxidant enzymes like glutathione (GSH), GPx, and CuZn SOD in blood of these patients (Seven et al., 2008). The study by Jikimoto (Jikimoto et al., 2001) discovered elevated levels of TRX in plasma, and synovial fluid of RA patients. Plasma TRX correlated with urinary excretion of 8-hydroxy-29-deoxyguanosine (8-OHdG). Serum MDA levels were increased in SLE patients, while serum antioxidant levels were decreased in these patients what confirms oxidative stress in the pathology of SLE (Taysi et al., 2002 a).

8. Oxidative stress in autoimmune joint diseases: Beneficial or harmful?

The recent research describes dual role of ROS in autoimmune- joint diseases. The ROS production by the activated NOX- family (NADPH) of oxidases has a principal function throughout the priming phase of the development of autoimmune joint diseases. The NADPH oxidase- deficient knockout mice developed a serious inflammatory arthritis with extensive bone erosions and a massive osteolysis (van de Loo et al., 2003). ROS might reduce arthritis development influencing circulating inflammatory cells before they reach the joint. ROS limit T cell responses to self- antigens inducing their apoptosis, what inhibits disease development (Olofsson et al., 2003). Arthritis development correlates with functionality of NADPH- oxidase complex, which is determined by the neutrophil cytosolic factor 1 (Ncf 1), a phagocytic oxidase. The naturally occurring polymorphism of Ncf1 allele limiting the NADPH functionality promotes activation of arthritogenic CD4+ cells. An expansion of arthritogenic T cells owing to the lower activity of the NADPH- oxidase complex was observed in the DA Ncf 1 allele (Olofsson et al., 2003). An increased severity of arthritis was observed in animals with either loss of function mutations or deletions in components of the phagocyte NADPH- oxidase like the p47phox (van de Loo et al., 2003). Furthermore, the effect of NOX-derived ROS might depend on the arthritis model: in IFN-gamma (INF- γ) enhanced imine complex arthritis, the p47phox-deficient animals showed a less severe joint destruction and decreased chondrocyte death (Lem van Lent et al., 2005).

The NADPH- oxidase- derived superoxide is not exclusive effector molecule in arthritis. Superoxide dismuthase (SOD) catalyses the dismutation of superoxide anion to oxygen and hydrogen peroxide. Three SOD enzyme isoforms have been characterized in humans. SOD I, and SOD III are copper, zinc, enzyme isoforms. SOD II is a manganese enzyme isoform.

Whereas type I SOD is found mostly in the cytoplasm, nucleus, and intermembrane space of mitochondria, type III of SOD is extracellular. Manganese is cofactor for SOD type II, which is a mitochondrial enzyme. The enzyme is involved in the pathogenesis of inflammatory joint disease. In adjuvant model of arthritis the enzyme suppressed swelling, and retarded bone destruction (Shingu et al., 1994). An intra-articular injection of SOD significantly reduced synovitis in streptococcal wall cell (SWC)- model of arthritis. Intra- aricular injection of native SOD (bovine orgotein) produced greater clinical improvements than did intra- articular aspirin in patients with RA involving the knee (Afonso et al., 2007). Available data suggest a protective role of SOD in inflammatory joint disease. In mice that are genetically deficient in SOD III, both the severity of collagen-induced arthritis and the production of pro-inflammatory cytokines are increased. SODIII gene transferred subcutaneously or intra- articularly decreased the severity of experimental arthritis in rodents.

9. Antioxidants as therapeutic possibilities in human autoimmune joint diseases

9.1 Antioxidant enzymes

SODs exert protective effects in animal models of inflammation. In mice, genetically deficient in SOD III, both the severity of collagen- induced arthritis and the production of pro- inflammatory cytokines are increased (Ross et al., 2004). SOD III gene transfer into the knee decreased the severity of experimental arthritis in rodents. In humans, serum SOD III levels correlated negatively with disease activity. Despite conflicting results of native SOD (orgotein) in RA patients, SOD mimetics have shown beneficial effects. Until now, the most promising are nitroxide (tempol) and Mn (II) pentaazamacrocyclic ligand (M40403). *In vitro*, tempol diminishes hydroxyl radical production, and decreases the cytotoxic effects of hydrogen peroxide and peroxynitrite. Tempol decreased collagen- induced arthritis in rats (Cuzzocrea et al., 2001). The anti-inflammatory effects of M40403 ligand are related to superoxide elimination, and restraint of nitration of tyrosine residues in proteins. M40403 ligand diminishes the expression of adhesion molecules such as integrins or selectins decreasing influx of neutrophils to inflammatory site. M40403 ligand inhibits TNF-α via blocking the nuclear factor kappa B (NF- κB). Its beneficial effects have been reported in rats with collagen- induced arthritis (Cuzzocrea et al., 2005).

9.2 Vitamins

Micronutrient supplementation has been recommended to arthritic patients based on findings that the antioxidant status is very low in these patients. It has been documented that the mean synovial fluid concentrations of α-tocopherol are significantly decreased in arthritic patients compared to healthy controls. Also, levels of vitamin E in peripheral blood cell were significantly decreased in patients with active arthritis than in healthy subjects (Vasanthi et al., 2009). Concentrations of other antioxidant vitamins such as A, C are of lower order in RA and SLE patients than in normal subjects too. Based on animal studies, there is evidence that antioxidant vitamins may prevent arthritis, and increase antioxidant status. Daily oral administration of vitamin E to male arthritic rats restored levels of thiol (- SH) groups to pre- arthritic levels (Kheir- Eldin et al., 1992). High doses of vitamin C (1-

2g/kg body weight) applied to arthritic Lewis rats, delayed incidence of paw oedema in these rats. Histological studies revealed a decreased inflammatory cells infiltration of superficial layer of synovium, and decreased synovial SOD activity in these rats (Sakai et al., 1999). In addition, administration of rutin (vitamin P) strongly inhibited spontaneous ROS production in RA neutrophils (Ostrakhovitch & Afanas´ev, 2001). Vitamin E exerts analgetic effects. The results from a double blind clinical trials suggested that a dose of 400–1200 mg of α-tocopherol daily was effective with respect to various pain parameters such as pain on pressure, pain at rest, and pain on movement. The postulated analgesic properties of vitamin E seem to be correlated to vitamin E plasma concentrations. Patients with active RA supplemented with a combination of conjugated fatty acids and vitamin E improved their disease activity score. The diet in these patients relieved from night and activity pain, and morning stiffness (Aryaeian et al., 2009). The analgesic effect of vitamin E is independent of a peripheral anti-inflammatory action, thereby suggesting a central rather than peripheral action. Dietary antioxidant micronutrients act as scavengers of reactive oxygen radicals and may protect against free radical mediated tissue damage in an inflamed joint. Vitamin E supplemented in soyabean oil reduced anti- double- stranded DNA IgG antibodies (Chia-Chien & Bi- Fong, 2005).

9.3 Fatty acids

Eicosapentaenoic acid and docosahexaenoic acid (EPA and DHA respectively) exert potent antioxidant properties. Supplementation with these acids to lupus patients restored levels of antioxidant enzymes as the GPx, and SOD to baseline levels (Mohan, & Das, 1997). Lupus autoimmune female mice fed with fish oil exhibited significantly higher liver of catalase, SOD, and GPx. The diet was shown to be beneficial against oxidative damage of hepatic tissue (Bhattacharya et al., 2003). Sera of lupus mice treated with conjugated linoleic acid (CLA) contained higher concentrations of total GSH which were negatively correlated with the levels of oxidative stress markers. Moreover, increased GSH, gammaGCL, glutathione S-transferase (GSTs), and NAD(P)H:quinone oxidoreductase (NQO1) activities were measured in liver and spleen of CLA-treated animals. The activation of detoxifying enzymes may be one of the mechanisms whereby dietary CLA down-regulates oxidative stress in lupus mice (Bergamo et al., 2007).

10. Conclusions and future perspectives

In all cell types, oxygen metabolism can lead to the production of reactive oxygen / nitrogen species. Reactive species are known for their dual role as both beneficial and harmful. Deleterious effects of oxidative stress on cell components (proteins, lipids, DNA) include changes in their structure, and function. On the other hand, beneficial effects of ROS/RNS involve physiological cellular responses to noxious agents, e.g., in defense against infectious agents, in a number of cellular signaling pathways and in the induction of a mitogenic response. Oxidants even protect cells against oxidative stress and re-establish or maintain redox homeostasis (Ortona et al., 2008).

Oxygen metabolism has an important role in the pathogenesis of autoimmune joint diseases. Reactive oxygen/nitrogen species are documented as mediators of synovial inflammation. Their excessive production at the site of inflammation contributes to inflammatory process in general, by induction of local production of chemoattractants, adhesion molecules, and pro- inflammatory cytokines. In addition, oxidative stress may contribute to the cyclic, self-perpetuating nature of autoimmune inflammation. Immune cells when "affected" by oxidants became auto- antigens intensifying auto- immune responses. Persisting auto-inflammation, and oxidative stress in joint area leads to damages of connective tissue and the ECM directly or indirectly. *In vitro*, lipid peroxidation, ONOO- formation is associated with decreased production of type II collagen and aggrecan, and with diminished chondrocyte responses to growth factors. Moreover, ONOO- interferes with metabolism of matrix enhancing the expression of matrix degrading enzymes inhibiting the production and activity of tissue inhibiting enzymes. Hydrogen peroxide and singlet oxygen accelerate bone resorption. Level of oxidative stress in rheumatoid synovium is proportional to the levels of hypoxia and angiogenesis in the joint area.

Oxidative stress generated within an inflammatory joint can produce autoimmune phenomena and connective tissue destruction in rheumatoid synovitis (Vasanthi et al., 2009). Swelling stiffness pain in rest immobility are common symptoms of autoimmune joint diseases. Rationale of antioxidant therapy should relieve from objective complications. Polyunsaturated fatty acids (PUFA) can modulate oxidative stress and may have a role as regulators in the synthesis of antioxidant enzymes (Mohan & Das, 1997). Arthritic patients taking these acids showed biochemical and chemical improvement. Dietary supplementation significantly decreased joint pain index. In combination with other dietary modifications modest improvement in morning stiffness and in the number of painful joints were reported. Clinical improvements were connected with anti- inflammatory effects as a decreased synthesis of leukotriens by neutrophils and lower synthesis of IL-1β by macrophages. Significant clinical benefit has been claimed in SLE patients given a low- fat diet with PUFA (reviewed in Darlington & Stone, 2001). The clinical work with diets containing PUFA has clearly demonstrated their anti-inflammatory effects, but it was also shown that these effects were attributed to omega- 3 (ω-3) rather than ω-6 PUFA. As there is no doubt about dietary fatty acids do decrease the generation of inflammatory agents, opposite results have been obtained on free radical formation. PUFA are especially potent at increasing levels of oxidative stress. On the other hand, EPA increases mitochondrial Mn-SOD mitochondrial activity.

The hypothesis about oxidative stress promotes arthritic process was challenged when oxidants were shown to decrease disease severity in mouse and rat arthritis models. Certain oils with an alkane structure such as phytol besides its oxidative effects protect against arthritis development. Its subcutaneous administration prevented development of pristane-induced arthritis (Hultqvist et al., 2006). Rats treated with phytol in acute phases of pristane arthritis showed no signs of inflammation. A decrease in COMP, a measurement of ongoing cartilage destruction, was prevented during chronic phases of the disease. The efficiency of phytol in preventing arthritis was compared to methotrexate and/or etanecerpt. Etanecerpt, TNF-α blocker, was highly effective in reducing collagen- induced arthritis. In pristane-induced arthritis, the preventive effects of phytol was more pronounced than that of etanecerpt. Also, in comparison to MTX, phytol was valid as a potential therapeutic agent. Concluded, ROS-promoting substances such as phytol represent a promising class of therapeutics for treatment of autoimmune joint inflammatory diseases what needs further research.

11. Acknowledgements

The work was supported by VZ 0021620816 and by 262708/SVV/2011 and by CN LC 554.

12. References

Abramson, S.B.; Amin, A.R.; Clancy, R.M. & Attur, M. (2001). The role of nitric oxide in tissue destruction. *Best Practice & Research Clinical Rheumatology*, Vol. 15, No.5, pp. 831-845.

Arii, K.; Kumon, Y.; Ikeda, Y; Suehiro, T. &Hashimoto, K. (2006). Edaravone inhibits rheumatoid synovial cell proliferation and migration. *Free Radic Res*, Vol. 40, No. 2, pp. 121-125.

Aryaeian, N.; Shahram, F.; Djalali, M.; Eshragian, M.R.; Djazayeri, A.; Sarrafnejad, A.; Salimzadeh, A.; Naderi, N.; Maryam, Ch. (2009). Effect of conjugated linoleic acids, vitamin E and thein combination on the clinical outcome of Iranian adults with active rheumatoid arthritis. *International Journal of Rheumatic Diseases*, Vol. 12, pp. 20–28.

Afonso, V.; Champy, R.; Mitrovic, D.; Collin, P. & Lomri, A. (2007). Reactive oxygen species and superoxide dismutases: Role in joint diseases. *Joint Bone Spine*, Vol. 74, No. 4 , pp. 324- 329.

Babior, B.M. (2000). Phagocytes and oxidative stress. *Am J Med*, Vol. 109, No. 1, pp. 33-44.

Batuca, J.R.; Ames, P.R.J.; Isenberg, D.A. & Delgado Aves, J. (2007). Antibodies toward high/densitz lipoprotein components inhibit paraoxonase activity in patients with systemic lupus erythematosus. *Ann. N.Y.Acad. Sci.* Vol. 1108, No. 1, pp. 137- 146.

Bergamo, P.; Maurano, F. & Rossi, M. (2007). Phase 2 enzyme induction by conjugated linoleic acid improves lupus- associated oxidative stress. *Free Radic Biol Med*, Vol. 43, No. 1, pp. 71-79.

Bhattacharya, A. & Lawrence, R.A. (2003). Effect of dietary n-3 and n-6 oils with and without food restriction on activity of antioxidant enzymes and lipid peroxidation in livers of cyclophosphamide treated autoimmune- prone NZB/W female mice. *J Am Coll Nutr*, Vol. 22, No. 5, pp. 388-399.

Billinghurst, R.C.;Dahlberg, L.; Ionescu, M., Reiner, A.; Bourne, R.; Rorabeck, C.; Mitchell, P.; Hambor, J.; Diekmann, O.; Tischesche, H.; Chen, J.; Van Wart, H.& Poole, A.R.. (1997). Enhanced cleavage of type ll collagen by collagenases in osteoarthritic articular cartilage. *J Clin Invest*, Vol. 99, No. 7, pp.1534–1545.

Biniecka, M. ; Kennedy, A. ; Fearon, U. ; Ng, Ch. T.. ; Veale, D.J. & O' Sullivan. (2009). Oxidative damage in synovial tissue is associated with in vivo hypoxic status in the arthritic joint. *Ann Rheum Dis*, Vol. 69, No.6 , pp. 1172- 1178.

Blanco, F. & Lotz, M. (1994). IL-1-induced nitric oxide inhibits chondrocyte proliferation via PGE2. *FASEB J*, Vol. 8, pp. A365.

Blanco, F.J. ; Ochs, R.L. ; Schwarz, H. & Lotz, M. (1995). Chondrocyte apoptosis induced by nitric oxide. *Am J Pathol*, Vol. 146, No. 1, pp.75-85.

Brouwer, E. ; Gouw, A.S. ; Posthumus, M.D. ; van Leeuwen, M.A. ; Boerboom, A.L. ; Bijzet, J. ; Bos, R. ; Limburg, P.C. ; Kallenberg, C.G. & Westra J. (2009). Hypoxia inducible factor- 1- alpha (HIF- 1alpha) is related to both angiogenesis and inflammation in rheumatoid arthritis. *Clin Exp Rheumatol*, Vol. 27, No. 6, pp. 945-951.

Cao, M. ; Westerhausen-Larson, A. ; Niyibizi, C. ; Kavalkovich, H.I. ; Georgescu, H.I. ; Riyyo, C. F. ; Hebda, P.A. ; Stefanovic-Racic, M. & Evans, C.H. (1997). Nitric oxide inhibits the synthesis of type II collagen without altering Col2A1 mRNA abundance : prolyl hydroxylase as a possible target. *Biochem. J*, Vol. 324, No. Pt1 , pp. 305- 310.

Chandel, N.S.; Maltepe, E.; Goldwasser, E.; Mathieu, C.E.; Simon, M.C. & Schumacker, P.T. (1998). Mitochondrial reactive oxygen species trigger hypoxia- induced transcription. *Cell Biology*, Vol. 95, No. 20, pp. 11715- 11720.

Chia- Chien, H. & Bi- Fong, L. (2005). The effects of vitamin E supplementation of autoimmune- prone New Zeland black 3 New Yeland white F1 mice fed an oxidized oil diet. *Br J Nutr*, Vol. 93, No. , pp. 655- 662.

Choi, E-M. (2007). Oxidative status of DBA/ 1J mice with type II collagen induced arthritis. *J Appl Toxicol*, Vol. 27, No.5, pp. 472- 481.

Cuzzocrea, S. ; Riley, D.P. ; Achille, P.C. & Salvemini, D. (2001). Antioxidant therapy : A new pharmacological approach in shock, inflammation, and ischemia / reperfusion injury. *Pharmacol Rev*, Vol. 53, No. 1, pp. 135- 159.

Cuzzocrea, S.; Mazzon, E.; di Paola, R.; Genovese, T.; Muia, C.; Caputi, A.P. (2005). Synergistic interaction between methotrexate and a superoxide dismutase mimetic: pharmacologic and potential clinical significance. *Arthritis Rheum*, Vol. 52, No. pp. 3755-3760.

Darlington, L.G & Stone, T.W. (2001). Antioxidants and fatty acids in the amelioration of rheumatoid arthritis and related disorders. *Br J Nutr*. Vol. 85, No. , pp. 251- 269.

Del Carlo Jr, M. & Loeser, R.F. (2002). Nitric oxide- mediated chondrocyte cell death requires the generation of additional reactive oxygen species. *Arthritis & Rheumatism*, Vol. 46, No. 2, pp/ 394- 403.

Drinda, S. ; Franke, S. ; Canet, C.C ; Petrow, P. ; Bräuer, R. ; Hüttich, C. ; Stein, G. & Hein, G. (2002). Identification of the advanced glycation end products N-carboxymethyllysine in the synovial tissue of patients with rheumatoid arthritis. *Ann Rheum Dis*, Vol. 61, No. 6, pp. 488-492.

D' Souza, A. ; Kurien, B. T. ; Ridgers, R. ; Shenoi, J. ; Kurono, S. ; Matsumo, H. ; Hensley. ; Nath, S.K. & Scofield, R.H. (2008). Detection of catalase as a major protein target of the lipid peroxidation product 4-HNE and the lack of its genetic association as a risk factor in SLE. *BMC Medical Genetics*, Vol. 9, pp.62.

Eggleton, P. ; Wang, L. ; Penhallow, J. ; Crawford, N. & Brown, K.A. (1995). Differences in oxidative response of subpopulations of neutrophils from healthy subjects and patients with rheumatoid arthritis. Ann Rheum Dis, Vol. 54, No. 11, pp. 916- 923.

El Benna, J. ; Hayem, G. ; My-Chan Dang, P. ; Fay, M. ; Chollet- Martin, S. ; Elbim, C. ; Meyer, O. & Gougerot- Pocidalo, M-A. (2002). NADPH oxidase priming and p47phox phosphorylation in neutrophils from synovial fluid of patients with rheumatoid arthritis and spondylarthropathy. *Inflammation*, Vol. 26, No. 6, pp. 273- 278.

Galleron, S.; Borderie, D.; Ponteziere, C.; Lemarechal, H.; Jambou, M.; Roch- Arveiller, M.; Ekindjan, O.G. & Calls, M.J. (1999). Reactive oxygen species induce apoptosis of synoviocytes in vitro. Alpha tocopherol provides no protection. *Cell Biol Int*, Vol. 23, No. 9, pp. 637- 642.

Giatromanolaki, A.; Sivridis, E.; Maltezos, E.; Athanassou, N.; Papazoglou, D.; Gatter, K.C.; Harris, A.L. & Koukourakis, M.I. (2003). Upregulated hypoxia inducible factor- 1α and -2α pathway in rheumatoid arthritis and osteoarthritis. *Arthritis Res Ther*, Vol. 5, No. 4, pp. R193- R201.

Gibson, J.S.; Milner, P.I.; White, R.; Fairfax, T.P.A. & Wilkins, R.J. (2008). Oxygen and reactive oxygen species in articular cartilage: modulators of ionic homeostasis. *Pflugers Arch - Eur J Physiol, Vol.* 455, No. 4, pp.563–573.

Hadjigogos, K. (2003). The role of free radicals in the pathogenesis of rheumatoid arthritis. *Panminerva Med,* Vol. 45, No. 1, pp. 7-13.

Henrotin,Y. E.; Bruckner, P.; Pujol, J.P.L. (2003). The role of reactive oxygen species in homeostasis and degradation of cartilage, *OsteoArthritis and Cartilage,* Vol. 11, No. 10, pp. 747–755.

Hickery, M.S. & Bayliss, M.T. (1998). Interleukin-1 induced nitric oxide inhibic sulphation of glycosaminoglycan chaos in human articular cartilage chondrocytes. *Biochimica et Biophysica Acta,* Vol. 1425, No.2 , pp. 282- 290.

Hiran, T. S.; Moulton, P.J. ; Hancoc, J.T. (1997). Detection of superoxide and NADPH oxidase in porcine articular chondrocytes. *Free Radical Biology & Medicine,* Vol. 23, No. 5, pp. 736–743.

Hitchon, C. A. & El- Gabalawy, H.S. (2004). Oxidation in rheumatoid arthritis. *Arthritis Res Ther,* Vol. 6, No. 6, pp. 265- 278.

Hollander, A.P.; Corke, K.P.; Freemont, A.J. & Lewis, C.E. (2001). Expression of hypoxia-inducible factor 1 a by macrophages in the rheumatoid synovium. Implications for targeting genes to the inflammaed joint. *Arthritis & Rheumatism,* Vol. 44, No. 7, pp. 1540-1544.

Hovatta, I.; Juhila, J.; Donner, J. (2010). Oxidative stress in anxiety and comorbid disorders. *Neurosci Res,* Vol. 68, No. 4, pp.261-275.

Hultqvist, M.; Olofsson, P.; Gelderman, K. A.; Holmberg, J.; Holmdahl, R. (2006). A New Arthritis Therapy with Oxidative Burst Inducers. *Plos Medicine,* Vol 3, No. 9, pp.1625-1636.

Jikimoto, T.; Nishikubo, Y.; Kanagawa, S.; Morinobu, S.; Saura, R.; Miyuno, K,; Kondo, S.; Toyokuni, S.; Nakamurad, H.; Yodoi, J. & Kumagai, S. (2001). Thioredoxin as a biomarker for oxidative stress in patients with rheumatoid arthritis. *Mol Immunol,* Vol. 38, No. 10 ,pp. 765- 772.

Ji-Yeon, S.; Jang- Hee, H.; Hyung-Sik, K.; Inpyo, Ch.; Sang-Deok, L.; June- Kyu, L.; Jeong-Ho, S.; Jae-Heun, L. & Gang- Min, H. (2000). Methotrexate suppresses the interleukin-6 induced generation of reactive oxygen species in the synoviocytes of rheumatoid arthritis. *Immunopharmacology,* Vol. 47, No. , pp. 35- 44.

Kennett, E.C. & Davies, M.J. (2009). Glycosaminoglycans are fragmented by hydroxy L carbonate and nitrogen dioxide radicals in a site- selective manner: implications for peroxynitrite- mediated damage at sites of inflammation. *Free Radical Biology & Medicine,* Vol. 47, No. 4 , pp. 389- 400.

Khan, F. & Siddiqui, A.A. (2006). Prevalence of anti- 3- nitrotyrosine antibodies in the joint synovial fluid of patients with rheumatoid arthritis, osteoarthritis and systemic lupus erythematosus. *Clin Chim Acta,* Vol. 370, No. 1-2, pp. 100-107.

Kheir- Eldin, A.A; Hamdy, M.A.; Motawi, T.K.; Shaheen, A.A. & Abd El Gawad, H.M. (1992). Biochemical changes in arthritic rats under the influence of vitamin E. *Agents Actions,* Vol. 36, No. 3-4, pp. 300-305.

Kurz, B.; Steinhagen, J. & Schünke, M. (1999). Articular chondrocytes and synoviocytes in a co-culture system: influence on reactive oxygen species- induced cytotoxicity and lipid peroxidation. *Cell Tissue Res,* Vol. 295, No. 3, pp. 555-563.

Lark, M.W.; Bayne, E.K.; Flanagan, J.; Harper, C.F.; Hoerner, L.A.; Hutchinson, N.I.; Singer, I.I.; Donatelli, S,A.; Weidner, J.R.; Williams,H.R.; Mumford, R.A.; Lohmander, L.S. (1997) Aggrecan degradation in human cartilage. Evidence for both matrix metalloproteinase and aggrecanase activity in normal, osteoarthritic, and rheumatoid joints. *J Clin Invest,* Vol. 100, No. , pp.93–106.

Lem van Lent, P.; Nabbe, K.C.; Blom, A.B.; Sloetjes, A.; Holthuysen, A.E.M.; Kolls, J.; Van De Loo, F. A.J.; Holland, S.M. & Van Den Berg, W.B. (2005). (NADPH)-oxidase- driven oxygen radical production determines chondrocyte death and partly regulates metalloproteinase- mediated cartilage matrix degradation during interferon- γ-stimulated immune complex arthritis. *Arth Res & Ther,* Vol. 7, No. 4, pp. 2005.

Lo, Y. C.L.; Wong, J.M.S. & Cruz, T.F. (1998). Reactive oxygen species mediate cytokine activation of c- Jun NH2- terminal kinases. *J Biol Chem,* Vol. 271, No. 26, pp. 15703- 15707.

Lushchak, V. (2011). Adaptive response to oxidative stress: Bacteria, fungi, plants and animals. *Comp Biochem Physiol,* Vol. , No. pp. 175- 190.

Mattey, D.L.; Nixon, N.; Alldersea, J.E.; Cotton, W.; Fryer, A.A.; Zhao, L.; Jones, P. & Strange, R.C. (1993). Alpha, mu and pi class glutathione S- transferases in human synovium and cultured synovial fibroblasts: effects of interleukin-1 alpha, hydrogen peroxide and inhibition of eicosanoid synthesis. *Free Radic Res Commun,* Vol. 19, No. 3, pp. 159- 617.

Maurice, M.M.; Nakamura, H.; van der Voort, E.A.; van Vliet, A.I.; Staal, F.J.; Tak, P.P; Breedveld, F.C.& Verweij, C.L. (1997). Evidence for the role of an altered redox state in hyporesponsiveness of synovial T cells in rheumatoid arthritis. *J Immunol,* Vol. 158, No.3, pp. 1458- 1465.

Miesel, R.; Kurpisz, M. & Kröger, H. (1996). Suppression of inflammatory arthritis by simultaneous inhibition of nitric oxide synthase and NADPH oxidase. *Free Radical Biology & Medicine,* Vol. 20, No. 1, pp. 75-81.

Mohan, I.K. & Das, U.N. (1997). Oxidant stress, anti-oxidants and essential fatty acids in systemic lupus erythematosus. *Prostag Leu Ess Fatty Acid,* Vol. 56, No.3, pp. 193- 198.

Monboisse, J.C. & Borel, J.P. (1992). Oxidative damage to collagen. *EXS,* Vol. 62, pp. 323- 327.

Newkirk, M.M.; Goldbach- Mansky, R.; Lee, J.; Hoxworth, J.; McCoy, A,; Yarboro, Ch.; Klippel, J. & El- Gabalawy, H.S. (2003). Advanced glycation end-product (AGE)-damaged IgG and IgM autoantibodies to IgG AGE in patients with early synovitis. *Arthritis Res Ther,* Vol. 5, No. 2 , pp. R82- R90.

Olofsson, P.; Holmberg, J.; Tordsson, J.; Lu, S.; Akerström, B. & Holmdahl, R. (2003). Positional identification of Ncf 1 as a gene that regulates arthritis severity in rats. *Nat Genet,* Vol. 33, No. 2 , pp. 25-32.

Ortona, E.; Margutti, P.; Matarrese, P.; Franconi, F.; Malorni, W. (2008). Redox state, cell death and autoimmune diseases: a gender perspective. *Autoimmun Rev,* Vol. 7, No. 7, pp. 579- 584.

Ostrakhovitch, E.A. & Afanas' ev, L.B. (2001). Oxidative stress in rheumatoid arthritis leukocytes: suppression by rutin and other antioxidants and chelators. *Biochemical Pharmacology,* Vol. 62, No. 6, pp. 743-746.

Regan, E.A.; Bowler, R.P. & Crapo, J.D. (2008). Joint fluid antioxidants are decreased in osteoarthritic joints compared to joints with macroscopically intact cartilage and subacute injury. *Osteoarthritis & Cartilage*, Vol. 16, No. 4 , pp. 515- 521.

Ross, A.D.; Banda, N.K.; Muggli, M.; Arend, W.P.(2004). Enhancement of collagen induced arthritis in mice genetically deficient in extracellular superoxide dismuthase. *Arthritis Rheum*, Vol. 50, No. , pp. 3702- 3711.

Sakai, A.; Hirano, T.; Okayaki, R.; Okimoto, N.; Tanaka, K. & Nakamura, T. (1999). Large-dose ascorbic acid administration suppresses the development of arthritis in adjuvant- injected rats. Arch Orthop Trauma Surg, Vol. 119, No. 3-4, pp. 121- 126.

Sarban, S.; Kocyigit, A.; Yazar, M. & Isikan, U.E. (2005). Plasma total antioxidant capacity, lipid peroxidation, and erythrocyte antioxidant enzyme activities in patients with rheumatoid arthritis and osteoarthritis. Clin Bioch, Vol. 38, No.11 , pp. 981- 986.

Seven, A.; Güzel, S.; Aslan, M. & Hamuryudan, V. (2008). Lipid, protein, DNA oxidation and antioxidant status in rheumatoid arthritis. Clin Bioch, Vol. 41, No. 7-8, pp. 538- 543.

Shah, D.; Kiran, R.; Wanchu, A. & Bhatnagar, A. (2010). Oxidative stress in systemic lupuys erythematosus: relationship to Th1 cytokine and disease activity. *Immunol Lett*, Vol. 129, pp. 7-12.

Shah, D.; Wanchu, A. & Bhatnagar, A. (2011). Interaction between oxidative stress and chemokines: Possible pathogenic role in systemic lupus erythematosus and rheumatoid arthritis. doi: 10. 1016/ j.imbio. 2001. 04. 001.

Shankar, J.; Thippegowda, P.B. & Kanum, S.A. (2009). Inhibition of HIF-1a activity by BP-1 ameliorates adjuvant induced arthritis in rats. *Biochemical and Biophysical Research Communications*, Vol. 387, No. 2 , pp. 223- 228.

Shingu, M.; Takahashi, S.; Ito, M.; Hamamatu, N.; Suenaga, Y.; Ichibangase, Y.; Nobunaga, M. (1994). Anti- inflammatory effects of recombinant human manganese superoxide dismuthase on adjuvant arthritis in rats. *Rheumatol Int*, Vol. 14, No. 2, pp. 77-81.

Shejkh, Z.; Ahmad, R.; Shejkh, N.; Ali, R. (2007). Enhanced recognition of reactive oxygen species damaged human serum albumin by circulating systemic lupus erythematosus autoantibodies. *Autoimmunity*, Vol. 40, No. 7, pp. 512- 520.

Schett, G.; Tohidast- Akrad, M.; Steiner, G. & Smolen, J. (2001). Commentary. The stressed synovium. *Arthritis Res*, Vol. 3, No. 2, pp. 80-86.

Schneider, N.; Mouithys- Mickalad, A.L.; Lejeune, J.- P.; Deby-Dupont, G.P.; Hoebeke, M. & Serteyn, D.A. (2005). Synoviocytes, not chondrocytes, release free radical after cycles of anoxia/ re-oxygenation. Biochimical *and Biophysical Research Communications*, Vol. 334, No. , pp. 669- 673.

Stefanovic- Racic, M.; Mollers, M.O.; Miller, M.A.& Evans, C.H. (1997). Nitric oxide and proteoglycan turnover in rabbit articular cartilage. *J Orthop Res*, Vol. 15, No. 3, pp. 442- 449.

Šimelyte, E.; Boyle, D. L. & Firestein, G.S. (2004). DNA mismatch repair enzyme expression in synovial tissue. *Ann Rheum Dis*, Vol. 63, No. 12, pp. 1695-1699.

Vasanthi, P.; Nalini, G.; Rajasekhar, R. (2009). Status of oxidative stress in rheumatoid arthritis. *Int J Rheum Dis*, Vol. 12, No., pp. 29- 33.

Taysi, T.; Gul, M.; Saris, R.A.; Akcay, F. & Bakan, N. (2002 a). Serum oxidant/ antioxidant status of patients with systemic lupus erythematosus. *Clin Chem Lab Med,* Vol. 40, No. 7, pp. 684- 688.

Taysi, T.; Polat, F.; Gul, M.; Saris, R.A. Bakan, N. (2002 b). Lipid peroxidation, some extracellular antioxidants, and antioxidant enzymes in serum of patients with rheumatoid arthritis. *Rheumatol Int,* Vol. 21, No. 5, pp. 200- 204.

Tewthanom, K.; Janwityanuchit, S.; Totemchockchyakarn, K. & Panomvana, D. (2008). Correlation of lipid peroxidation and glutatione levels with severity of systemic lupus erythematosus: A pilot study from single center. *J Pharm Pharmaceut Sci,* Vol. 11, No. 3, pp. 30-34.

Van de Loo, F.A.J.; Bennink, M.B.; Arntz, O.J.; Smeets, R.L.; Lubberts, E.; Joosten, L.A.D., van Lent, P.L.E.M.; Coenen- de Roo, Ch. J.J.; Cuzzocrea, S,; Segal, B.H.; Holland, S.M. & van den Berg, W. B. (2003). Deficiency of NADPH oxidase components p47phox and gp91phox caused granulomatosus synovitis and increased tissue destruction in experimental arthritis model. *Am J Pathol,* Vol. 163, No. 4, pp. 1525- 1536.

Vasanthi, P.; Nalini, G. & Rajasekhar, G. (2009). Status of oxidative stress in rheumatoid arthritis. *Int J Rheum Dis,* Vol. 12, No. 1 , pp. 29- 33.

Yoshida, S.; Katoh, T.; Tetsuka, T.; Uno, K.; Matsui, N. & Okamoto, T. (1999). Involvement of thioredoxin in rheumatoid arthritis: its costimulatory roles in the TNF- α- induced production of IL-6 and IL-8 from cultured synovial fibroblasts. *Immunol,* Vol. 163, No. 1, pp. 351-358.

4

The Role of Oxidative Stress in Female Reproduction and Pregnancy

Levente Lázár
1st Departmet of Obstetrics and Gynecology,
Semmelweis University Budapest,
Hungary

1. Introduction

In a healthy body, reactive oxygen species (ROS) and antioxidants remain in balance. Oxidative stress occurs when the generation of reactive oxygen species and other radical species exceeds the scavenging capacity by antioxidants of antioxidative agents in organism, due to excessive production of reactive oxidagen species and/or inadequate intake or increased utilization of antioxidants. Most ROS are formed at operation of electrone transport chains in mitochondria, endoplasmatic reticulum, plasmatic and nuclear membranes. Minor ROS amounts are generated by some enzymes through autooxidation of different molecules. Reactive oxygen specieses can also be formed by exogenous exposures such as alcohol, tobacco smoke, and environmental pollutants. Elimination of reactive oxygen species is catalysed by certain enzymes such as superoxide dismutses (SOD), catalases and peroxidases. Antioxidants (including vitamins C and E) and antioxidant cofactors (such as selenium, zinc, and copper) are capable to dispose, scavenge, or suppress ROS formation. „Oxidative stress" rises when due to some reasons the steady-state ROS concentration is increased, leading to oxidative modification of cellular constituents, resulting disturbance of cellular metabolism and regulatory pathways (Lushchak. 2011). Cellular ROS and their control by antioxidants are involved in the physiology of the female reproductive system. Physiological ROS levels play an important regulatory role through various signalling and transduction pathways in folliculogenesis, oocyte maturation, corpus luteum, uterine function, embryogenesis, embryonic implantation and fetoplacental development. Imbalances between antioxidants and ROS production are considered to be responsible for the initiation or development of pathological processes affecting female reproductive processes.

The establishment of pregnancy requires a receptive uterus able to respond to a variety of biochemical and molecular signals produced by the developing conceptus, as well as specific interactions between the uterine endometrium and the extra-embryonic membranes. Therefore, placental development and function are prerequisites for an adequate supply of nutrients and oxygen to the fetus and successful establishment of pregnancy. Oxidative stress has been proposed as the causative agent of female sterility, recurrent pregnancy loss and several pregnancy-related disorders as preeclampsia, intra-uterine growth restriction (IUGR) and gestational diabetes.

2. Effect of oxidative stress on female reproductive system

The female reproductive system is a complex multiorgan system which require an optimal biological environment. Aerobic metabolism utilizing oxygen is essential for reproductive homeostazis. Aerobic metabolism is associated with the generation of prooxidant molecules called ROS including hydroxyl radical, superoxide anion, hydrogen peroxide, and nitric oxide. The balance between the prooxidants and antioxidants maintain the cellular homeostasis, whenever there is an imbalance in this equilibrum leading to enhanced steady-state level a state of oxidative stress is initiated. Free radicals are key signal molecules modulating reproductive functions by the influence of the endometrial and fallopian tube function, maturation of oocytes, sperm, implantation of the preembryo and early embryo development (Figure 1.)

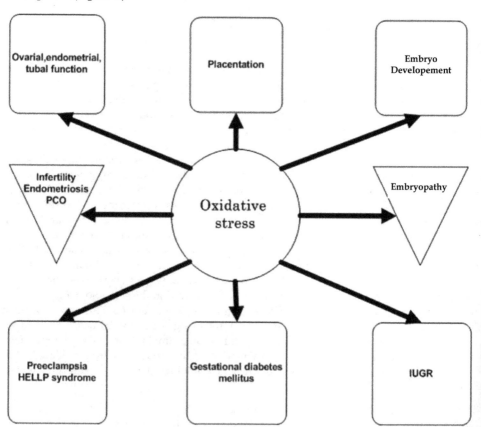

Fig. 1. Oxidative stress in female reproduction

2.1 Ovarian function

Aerobic metabolism utilizing oxygen is essential for developement of the gametes, also free radicals play a significant role in physiological processes within the ovary. The expression of various biomarkers of oxidative stress has been investigated in normal cycling human ovaries (Maruyama et al. 1997, Matsui et al., 1996), justifiing the regulatory role of ROS and antioxidants in oocyte maturation, folliculogenesis, ovarian steroidogenesis and luteolysis (Shiotani et al., 1991; Behrman et al., 2001; Sugino et al., 2004). Studies demonstrate intensified lipid peroxidation in the preovulatory Graafian follicle (Paszkovski et al., 1995). The significance of reactive oxygen specieses and antioxidant enzymes as copper zinc superoxide dismutase (Cu, Zn-SOD), manganese superoxide dismutase (Mn-SOD), and glutathione peroxidase, in oocyte maturation was provided by Riley et al. (1991) using immmunohistochemical localization and mRNA expression (Tamate et al., 1995). The antioxidant enzymes neutralize reactive oxygen specieses and protect the oocyte. In corpora lutea collected from pregnant and nonpregnant patients, it was observed that during normal situations Cu–Zn SOD expression rise from early luteal to midluteal phase and decrease during regression of the corpus luteum. Studies investigating the correlation between adrenal-4 binding protein (Ad4BP) and superoxide dismutase expression also suggest an association between oxidative stress and ovarian steroidogeneis (Matsui et al.,1996). Antibody to adrenal 4-binding protein (Ad4BP) was utilized to localize Ad4BP in the nuclei of theca and granulosa cells. Ad4BP is a steroidogenic transcription factor that induces transcription of the steroidogenic P450 enzyme. Both human granulosa and luteal cells respond to hydrogen peroxide with an extirpation of gonadotropin action and inhibition of progesterone secretion (Sabuncu et al., 2001). The production of both progesterone and estradiol hormones is reduced when hydrogen peroxide is added to a culture of human chorionic gonadotropin-stimulated luteal cells (Agrawal et al., 2006).

Levels of three oxidative stress biomarkers, conjugated dienes, lipid hydroperoxide and thiobarabituric acid reactive substances were found significantly lower in the follicular fluid compared with serum levels (Jozwik et al., 1999). The preovulatory follicle has a potent antioxidant defense, which is depleted by the intense peroxidation (Jozwik et al., 1999). Also the antioxidant enzymes glutathione peroxidase and Mn-SOD are considered to be markers for cytoplasmic maturation in metaphase II oocytes (El Mouattasim et al., 1999).

2.2 Changes in endometrium

Cyclical changes in the endometrium are accompanied by changes in the expression of antioxidants. Enzymes, such as thioredoxin, have a higher expression in the early secretory phase (Murayama et al, 1997). There is also a cyclical variation in the expression of superoxide dismutase in the endometrium. Superoxide dismutase activity decreased in the late secretory phase while ROS levels increased and ROS triggered the release of prostaglandin F2 α (Sugino et al., 1996). Estrogen or progesterone withdrawal led to increased expression of cyclooxygenase-2 (COX-2). Stimulation of the cyclooxygenase enzyme is brought about by ROS via activation of the transcription factor NF-κB, suggesting a mechanism for menstruation (Sugino et al., 2004).

Nitrogen monoxide (.NO) has also important role in decidualisation and preparation of the endometrium for implantation by regulatation of the endometrial, myometrial and microvascular functions. Expression of endothelial and inducible NO synthase (NOS) have been demonstrated in the human endometrium (Tseng et al., 1996), and the endometrial vessels (Taguchi et al,. 2000). Highest levels of transcripts of endothelial NOS mRNA have been reported in the late secretory phase of the endometrium (Tseng et al., 1996). These changes have been hypothesized to be important in the genesis of menstruation and endometrial shedding.

2.3 Fallopian tube function

Several studies demonstrated the presence of cytokines, prostaglandins, metabolites of lipid peroxidation and ROS in fluid samples of fallopain tube (Tamate et al., 1985). The eqilibrum of these components serves as an optimal milieu for fertilization and the transport of the preembryo. An endogenous nitrogen monoxide system exists in the fallopian tubes. Nitric oxide has a relaxing effect on smooth muscle and it has similar effects on tubular contractility. Deficiency of NO may lead to tubal motility dysfunction, resulting in retention of the ovum, delayed sperm transport and infertility. Increased NO levels in the fallopian tubes are cytotoxic to the invading microbes and also may be toxic to spermatozoa (Rosselli et al., 1995), leading to infetility.

2.4 Embryo implantation and placenta

The human embryo undergoes interstitial implantation by invading the maternal decidua at the blastocyst stage (Riley& Behrman, 1991). Although placental villi are bathed in maternal blood in the hemochorial placenta (Tamate et al., 1995), prior to 10 weeks of gestation maternal blood flow to the placenta is blocked by extravillous trophoblasts. Placentation is initiated when the blastocyst makes contact with the epithelial lining of the uterus shortly after implantation. Placental villi which consist of a mesenchymal core surrounded by a monolayer of mononuclear villous cytotrophoblast stem cells which either fuse to form the overlying multinucleated syncytiotrophoblast or, in anchoring villi, differentiate into extravillous trophoblasts which grow out from the villous and spread laterally around the placenta (Irving et al., 1995).

Invasive extravillous trophoblasts play an important role in adapting the decidua to sustain pregnancy. Extravillous trophoblasts invade the walls of the uterine spiral arteries and adapt these vessels into large bore conduits capable of delivering the increased blood supply required in the second and third trimesters (Robertson et al., 1967; Zhou et al., 1997). As the extravillous trophoblasts invade the spiral arteries early in pregnancy they form plugs which occlude the spiral arteries and prevent maternal blood from entering the intervillous space, creating a physiologic hypoxic environment (Hustin &nd Schaaps,1987; Jaffe et al., 1997; Burton et al., 1999).

Early placental and embryonic development occurs in a state of low oxygen in histiotroph manner (Evans et al., 2004). The early gestation placenta is poorly protected against oxidative damage, as the antioxidant enzymes Cu,Zn-SOD and Mn-SOD are not expressed by the syncytiotrophoblast until approximately 8–9 weeks of gestation (Watson et al., 1997). Premature perfusion of this space during this first 10 weeks of development increases the risk of pregnancy loss (Jauniaux et al., 2000). The low oxygen environment during early placental development is essential for normal placental angiogenesis, and this angiogenesis is promoted by hypoxia-induced transcriptional and post-transcriptional regulation of angiogenic factors, as vascular endothelial growth factor and placental growth factor (Charnock-Jones&Burton, 2000).

The partial oxygen tension in the intervillous space declines from the second to the third trimester, reaching about 40 mm Hg in the third trimester (Soothill et al., 1986). The exact mechanism by which trophoblasts sense oxygen tension is currently unclear; however, several potential pathways have been identified. Many of these pathways utilize the ROS formation, but it is currently unclear whether hypoxia results an increase or decrease in their cellular levels (DeMarco & Caniggia, 2002). In hypoxic conditions, trophoblast oxygen sensing mechanisms utilize several different pathways to control gene expression. These pathways often utilize redox-sensitive transcription factors, of which the hypoxia inducible factor (HIF) family are the best characterized. HIF-1α is a transcription factor and master regulator of the cellular response to low oxygen levels (Majmundar et al., 2010), showing prominent expression in first trimester villi (Wang&Semenza, 1993). HIF-1α regulates the expression of genes such as p53, p21, and Bcl-2 required for cells to adapt to a low oxygen environment and apoptosis. HIF-1α is able to be stabilized under normoxic conditions by a variety of growth factors and cytokines including epidermal growth factor (EGF), insulin, heregulin, insulin-like growth factors 1 and 2, transforming growth factor, and interleukin-1 (Zelzer et al., 1998; Feldser et al., 1999; Hellwig-Burgel et al., 1999; Laughner et al., 2001; Fukuda et al., 2002; Stiehl et al., 2002).

Several other transcription factors involved in trophoblast differentiation are responsive to hypoxia. The transcription factors Id1 stream stimulatory factor-1 and -2 (USF1 and USF2) mediate the effects of Mash2 are all up-regulated in 2% oxygen in comparison to 20% oxygen (Jiang et al., 2000; Jiang & Mendelson, 2003). The up-regulation of Mash2, USF1 and USF2 may inhibit cytotrophoblast fusion into syncytiotrophoblast (Jiang et al., 2000; Jiang& Mendelson, 2003). The elevation of intracellular Ca^{2+} is believed to activate an HIF-1-independent signalling pathway that involves the transcription factor activator protein-1 (AP-1), with cooperation between the HIF-1 and AP-1 pathways allowing fine regulation of gene expression under hypoxia (Laderoute et al., 2002; Salnikow et al., 2002). AP-1 is a dimeric transcription factor composed from the products of the Jun and Fos proto-oncogenes (c-Jun, JunB, JunD, c-Fos, FosB, Fra-1 and Fra-2) (Dakour et al., 1999). AP-1 transcription factors are believed to play an important role in trophoblast differentiation. In the villus, AP-1 transcription factor expression is limited; however, extravillous trophoblasts express c-Jun, JunB, c-Fos, FosB and Fra-2 both in the first trimester and later in gestation (Bamberger et al., 2004.)

The other protective system is formed by antioxidant enzymes, playing a key role in the response of trophoblast to the burst of perfusion by maternal blood. With the increase of oxygen saturation and oxidative stress the activity in intervillous space the placenta employs a number of physiologic adaptations (Burton, 2009). Levels and activity of antioxidant enzymes: catalase, glutathione peroxidase, manganese and cooper, zinc superoxide dismutase are increased within placental tissues. This response is evolved as a defense mechanism to reduce harm to placental tissues exposed to this burst of oxidative stress (Jauniaux et al., 2000).

We can summarize that the trophoblast differentiation is essential for the success of human pregnancy, and despite some conflicting experimental evidence, hypoxia appears to play a vital role in regulating trophoblast differentiation in the first trimester. The regulation of trophoblast differentiation by hypoxia is a result of complex interactions between factors associated with oxidative stress, oxygen sensing mechanisms and the release of inflammatory cytokines. Therefore, aberrations in any one of these factors, along with the temporal and spatial regulation of blood flow in the intervillous space has the potential to result in altered gene expression and trophoblast phenotype leading to fail of implantation.

3. The role of oxidative stress in embryo and fetal malformation

Basic principles of teratogenesis state that a teratogen must cause a specific malformation through a specific mechanism during a period in which the conceptus is susceptible to said mechanism (Karnofsky, 1965). Different mechanisms are responsible for malformations that are in agreement with these basic scientific principles.

A mechanism that has not been well described in teratology is oxidant induced or redox misregulation of developmental signals en route to dysmorphogenesis. The paucity of teratogenic study of redox misregulation is partially due to oxidative stress (Sies, 1985). Oxidizing and reducing equivalent imbalance in turn, leads to macromolecule damage, namely protein modification, lipid peroxidation, and DNA oxidation, and can lead to cell death. Unspecific oxidation of cellular components does not apply to basic principles of teratology or adequately explain the manifestation of teratogenic effects. If oxidative stress is the random, unspecific oxidation of cellular molecules, it does not adequately exemplify why or how specific teratogens that induce oxidative stress could cause a specific malformation. While untimely cell death during differentiation can have serious repercussions on the developing embryo, generalized cellular oxidation and subsequent apoptosis do not sufficiently describe specificity of malformations seen with most teratogens. To qualify as a plausible teratogenic mechanism, oxidative stress must be a controlled, specific event that alters cell function and/ or signal transduction pathways that would in turn cause specific dysmorphogenesis.

During particular periods in development, the embryo is more or less susceptible to oxidative stress. In early development, one-cell embryo relies on the Krebs cycle, whereas the blastocyst relies on glycolysis and anaerobic pathways as does the embryo during early organogenesis. Once the circulatory system is established, there is a higher reliance on oxidative and aerobic metabolism and more ROS are produced by mitochondria. Conversely, more antioxidants are available at this period to counteract and detoxify these reactive oxygen specieses (Hansen, 2006). Over the course of development, the delicate balance between oxidants and antioxidants can be disrupted by exogenous agents that simulate ROS production leading to oxidative stress.

Thalidomide is associated with multiple birth defects, including phocomelia (Lecutier, 1962; Taussig, 1962). The most sensitive organ to thalidomide toxicity is the limb. Although the mechanism of teratogenesis and determinants of risk remain unclear, related teratogenic xenobiotics are bioactivated by embryonic prostaglandin H synthase (PHS) producing reactive oxygen species, which cause oxidative damage to DNA and other cellular macromolecules. Similarly, thalidomide is bioactivated by horseradish peroxidase, and oxidizes DNA and glutathione, indicating free radical-mediated oxidative stress. Furthermore, thalidomide teratogenicity is reduced by the PHS inhibitor acetylsalicylic acid, indicating PHS-catalyzed bioactivation. This appears to be regulated through redox shift resulting from depletion of GSH and increased GSSG in the nucleus, and this may imply that various transcription factors are affected by thalidomide through redox regulation (Hansen et al., 2001, 2002).

Exposure to the anticonvulsant valproic acid during the first trimester of pregnancy is associated with an increased risk of congenital malformations including heart defects, craniofacial abnormalities, skeletal and limb defects, and most frequently, neural tube defects (NTDs). The mechanisms by which valproic acid induces teratogenic effects are not fully understood, although previous studies support a role for oxidative stress. Valproic acid can alter cell signaling through gene expression changes mediated through histone deacetylase inhibition (Phiel et al., 2001), and is a direct inhibitor of class I and II histone deacetylases. Several laboratories have shown that embryonic histone acetylation levels are increased following exposure to valproic acid (Menegola et al., 2005; Tung and Winn, 2010). Furthermore, studies have supported a role for histone deacetylase inhibition as a mechanism of teratogenesis as analogs of valproic acid that lack histone deacetylase inhibitory activity are less teratogenic (Gurvich et al., 2005). Gene microarray studies have also demonstrated that valproic acid targets genes regulated by histone deacetylase, including *Mt1* and *Mt2*, both of which are ROS-sensitive (Jergil et al., 2009). In addition, histone deacetylase inhibitors have also been shown to increase ROS production and induce apoptosis in several cancer cell lines (Carew et al., 2008). Therefore, alterations in gene expression and/or increases in ROS formation mediated by histone deacetylase inhibition during development may induce teratogenesis.

The widely used anticonvulsant, phenytoin, can double the incidence of structural and functional birth defects when used in pregnancy (Kaneko et al., 1991). It can induce vascular disruption, which leads to hypoxia and hypoperfusion (Danielsson et al.,1995). In addition, phenytoin results in oxidative DNA damage and dysmorphogenesis, which can be eliminated by antioxidants (Winn and Wells, 1995). Phenytoin also selectively increased NF-kB activity in targeted tissues. Blocking of these signaling events with p65 antisense oligonucleotides eliminated the associated embryopathies (Kennedy et al., 2004). Further evidence that oxidative stress is important in phenytoin mediated toxicity is exemplified by the fact that treatment with polyethylene-modified superoxide dismutase enhances embryo toxicity whereas antioxidant levels were modulated with phenytoin (Winn and Wells, 1999).

Chronic ethanol consumption can lead to the generation of ROS and, as a consequence, teratogenicity. Prolonged ethanol exposure lead to increased production of lipid peroxides and decreased expression of antioxidant enzymes. Ascorbic acid can prevent against ethanol toxicity through inhibition of ROS formation and NF-kB activation (Peng et al.,2005). Zebrafish embryos exposed to ethanol with lipoic acid and/or only partially attenuated ethanol embryo toxicity, suggesting that other mechanisms are also involved (Reimers et al., 2006).

4. Oxidative stress in pathological pregnancies

4.1 Preeclampsia

Preeclampsia is a complex multisystem disorder that occurs during the pregnancy. The disease affects 5-8% of all pregnancies and is one of the leading couses of maternal and fetal morbidity and mortality. It is characterized with hypertension and proteinuria. The systolic blood pressure ≥140 mmHg, diastolic blood pressure ≥ 90 mmHg and the proteinuria at least 300 mg in 24 h urine collection. Women with mild proteinuria generally have no symptoms. However, women with severe preeclampsia (blood presure ≥160/110 mmHg, proteinuria >2-5g/24h) may have simptoms such as renal insufficiency, liver disease, haematological and neurological disturbances. Preeclampsia is charactized by vasospasm, reduced placental perfusion and abnormal placentation. The main couse of fetal compromise is the disturbance in uteroplacental perfusion. The only cure is the delivery of the baby. With antihypertensive treatment may prolong the pregnancy, increasing the chances of the baby to survive. If the blood presure cannot be controlled, or the laboratory parameters entry in a critical value the baby must be delivered. Preeclampsia has been proposed as a two-stage disorder. In first stage the placenta produces cytotoxic factors, in the second stage the maternal response to the placental factors occurs. There are several theories regarding to the main couse of disorder: abnormal placentation, immunological background, abnormal inflamatory response, etc. Assuming preeclampsia literature we can conclude that all the theories are part of disorders etiology.

4.1.1 Oxidative agents

Some reserches suggest that the placental oxidative stress may be involved in the ethiopathogenesis of preeclampisa. As there was mentioned above, the oxidatives stress is described as an imbalance in the production of reactive oxygen specieses and the ability of atioxidant defense to scavenge them. Pregnancy is a state of oxidative stress arrising from the increased metabolic activity in the placenta and reduced scavenging power of antioxidants (Wisdom et al. 1991). During the gestation the oxygenation of the uteroplacental unit is changing. The placenta and fetus exist in a hypoxic enviromet during early pregnancy as the uterine oxygen tension is extremly low till 8-10. weeks of gestation ($pO_2 < 20$ mmHg, $5\%O_2$), prior to estabilishment of the blood flow into intervillous space. The onset of blood flow is processig from the periphery to the center of placental disc, with villous regression in the placental periphery envoluring into the chorion leave (Jauniaux et al., 2003). The developing chorioallantoic villous trees are exposed to a marked increase of pO_2 in a range of 40-80 mm Hg (Sjostedt et al., 1960; Shaaps et al., 2005; Rooth et al., 1961). This reoxygenation of the uteroplacental unit results an oxidative burst, controlled by antioxidant mechanisms.

Proposed effect of oxidative stress on placental fatty acid metabolism.

As a consequence of abnormal trophoblast invasion, and maternofetal barrier preeclampsia is characterized by induced oxidative stress and decreased antioxidants (Patil et al., 2009). In preeclamptic women, maternal circulating levels, placental tissue levels and production rate of lipid peroxides are increased and several antioxidants are markedly decreased (Serdar et al., 2003; Orhan et al., 2003). Normal pregnancy is associated with physiological hyperlipidemia (Belo et al., 2004). Physiological alterations are manifested by increased levels of triglycerides and cholesterol in pregnancy, which decreases rapidly after delivery. Preeclampsia is characterized by further elevation of serum triglycerides and serum free fatty acids (Hubel et al., 1996). (Figure 2.) Increased lipid peroxidation has been reported in preeclampsia, IUGR (Liu et al., 2005; Gupta et al., 2004, Bretelle et al., 2004).

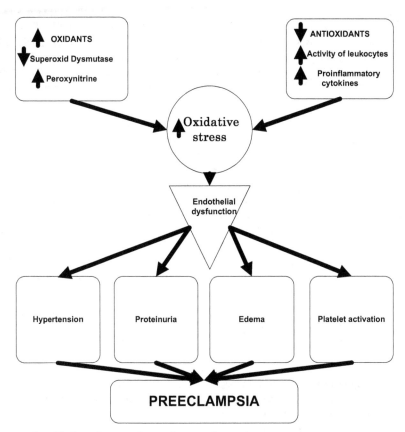

Fig. 2. Supposed oxidative stress pathway in preeclampsia

Nicotinamide adenine dinucleotide phosphate (NADPH) oxidase is an important enzyme that generates superoxide anion radical localized in the placenta syncytial microvillous membrane (Matsubara & Sato, 2001; Raijmakers et al., 2004). NADPH oxidase may play a role in placental lipid peroxidation by generating increased amounts of the superoxide anion radical. Poor antioxidant reserves can also tilt the balance in favor of prooxidation. Lipid peroxidation results in formation of primary lipid peroxidation products such as lipid hydroperoxides and secondary products such as malondialdehyde (MDA) and lipid peroxides. Lipid hydroperoxides are formed and bind to lipoproteins. They are then carried to distant sites where the hydroperoxides can cause ongoing lipid peroxidation and result in systemic oxidative stress. Increased placental production of lipid peroxides and thromboxane was demonstrated from both the trophoblast and the villous core components of placentas in patients with preeclampsia (Walsh & Wang, 1995).

Studies of women undergoing cesarean section showed significantly higher contents of lipid hydroperoxides, phospholipids, cholesterol and free 8-iso-prostaglandin F2α (8-iso-PGF2α), but not the total (free plus esterified) 8-iso-PGF2α in decidual tissues from women with preeclampsia as compared with tissues from normal pregnancies (Staff et al., 1999).

Moreover, tissue levels of free and total 8-iso-PGF2α are significantly higher in preeclamptic placenta than in normal placenta (Walsh et al., 2000). Isoprostanes like 8-iso-PGF2α are produced specifically by free radical-catalyzed peroxidation of arachidonic acid (Morrow et al., 1990). Free 8-iso-PGF2α has activities of relevance to preeclampsia, being a potent vasoconstrictor in kidney (Morrow et al., 1990) and placenta (Kwek et al., 1990), platelet activator (Minzu et al., 2002), and inducer of the release of endothelin from endothelial cells (Yura et al. 1999).

An increase in diastolic pressure correlates significantly with an increase in lipid peroxide levels, indicating that the severity of hypertension is correlated with the extent of lipid peroxidation (Aydin et al., 2004; Jain & Wise, 1995; Gupta et al., 2006). Women with preeclampsia have significantly higher mean plasma levels of malonaldehyde and significantly lower superoxide dismutase levels compared with normotensive pregnant women, (Aydin et al., 2004). The decrease in nitric oxide (NO) and superoxide dismutase (SOD) levels followed by a concomitant increase in levels of malonaledehyde, fibronectin, endothelin-1 (ET-1), and soluble-E selectin (sE-selectin) correlate with an increase in diastolic blood pressure (Aydin et al., 2004). In further studies, malonaldehyde levels in maternal erythrocytes were significantly elevated in women with developed preeclampsia. The risk of developing preeclampsia was 24-fold higher when malonaldehyde levels were above the cutoff value of 36 nmol/g (Basbug et al., 2003).

The cord plasma malonaldehyde and vitamin E levels were higher in patients with eclampsia than in patients with preeclampsia and in normotensive pregnant patients (Bowen et al. 2001).

Hyperhomocystinemia and altered eicosanoid synthesis has also been implicated in the pathophysiology of preeclampsia. Eicosanoids have vasoactive properties and enhance lipid peroxidation and decrease prostacyclin synthesis. The generation of the eicosanoid, 15-hydroxyeicosateranoic acid by the placenta was higher in women with preeclampsia than in normotensive control subjects. In preeclampsia, there is increased synthesis of thromboxane and reduced synthesis of prostacyclin. Lipid peroxides may also stimulate the cyclooxygenase enzyme to produce more thromboxane, resulting in a hypercoagulable state (Walsh, 2004).

4.1.2 Antioxidant agents

Antioxidants can be enzymatic or nonenzymatic. The enzymatic antioxidants are superoxide dismutase, thioredoxin, thioredoxin reductase, and glutathione peroxidase. The nonenzymatic antioxidants can be lipid-soluble such as vitamin E or water-soluble such as vitamin C. Serum levels of vitamin E and beta carotene (Serdar et al., 2003; Akyol et al., 2000), serum coenzyme Q10 and tocopherol levels (Palan et al., 2004), ascorbic acid were significantly reduced in pregnancies complicated by mild or severe preeclampsia, and the total antioxidant capacity was significantly reduced in pregnant women with mild and severe preeclampsia (Sagol et al., 1999). The balance between lipid peroxides and antioxidant vitamin E is tipped in favor of lipid peroxides in patients with mild and severe preeclampsia. A two-fold increase in the ratio between lipid peroxidation and antioxidant capacity was reported in the antepartum period in women with preeclampsia (Davidge et al., 1992). Significantly lower levels of vitamin C, E, and total thiol were seen in women with preeclampsia (Kharb et al., 2000). In patients with preeclampsia, antioxidants scavenge the increased free radicals, resulting in lowered antioxidant levels. Water-soluble antioxidants may function as a first line of antioxidants to scavenge excess of reactive oxygen species in plasma, whereas lipid soluble antioxidants such as tocopherol and carotene scavenge reactive oxygen species affecting the membrane lipids (Mikhail et al., 1994).

The activities of placental superoxide dismutase and glucose-6-phosphate dehydrogenase are decreased in preeclampsia compared to normal pregnancy (Poranen et al., 1996). Moreover, the activity and mRNA expression of Cu,Zn-SOD, glutathione peroxidase, and tissue levels of vitamin E are significantly lower in placental tissues from preeclampsia than from normal pregnancy (Wnag & Walsh, 1996). Glutathione and its related enzymes are antioxidants that help detoxify the increased generation of free radicals. Significantly reduced whole blood glutathione levels have been reported in women with preeclampsia and HELPP syndrome (Madazil et al., 2002; Knapen et al., 1998).

4.1.3 Leukocyte activation

The other main caracteristics of preeclampsia is the exacerbated inflamatory state (Bretelle et al., 2004; Holthe et al., 2005; Redman et al., 1999). Activated leukocytes, both monocytes and granulocytes, generate excess reactive oxygen specieses resulting in oxidative stress (Holthe et al., 2004). Compared with normotensive pregnant women, women with preeclampsia have higher levels of calprotectin, a protein involved in various physiological inflammatory processes, which is indicative of leukocyte activation (Holthe et al., 2005). The expression of surface adhesion molecules on cord blood neutrophils was significantly higher in infants born to women with preeclampsia than in infants born to the control subjects. Increased TNF secretion by leukocytes was detected in blood from patients with preeclampsia, providing further evidence of leukocyte activation (Beckman et al., 2004). TNF-α can activate the endothelial cells and upregulate the gene expression of numerous molecules such as platelet-derived growth factor, cell adhesion molecules, endothelin-1 and PAI-1. These molecules have been reported to have detrimental effects on the vasculature and also characterize preeclamptic pregnancy (Hajjar et al., 1987; van Hinsbrgh et al., 1988). Furthermore, chronic infusion of TNF-α into rats during late pregnancy results in a significant increase in renal vascular resistance and arterial pressure (Alexander et al., 2002; Giardina et al., 2002).

4.1.4 Endothelial cell dysfunction

Endothelial dysfunction is also one of the main pathogenic features of preeclampsia. The markers of endothelial dysfunction such as tissue plasminogen activator, von Willebrand factor, sE-selectin, and fibronectin are elevated in patients with preeclampsia (Aydin et al., 2004; Stubbs et al., 1984; Halligan et al., 1994). Although the exact mechanisms of vascular endothelial damage in preeclampsia are unclear, increased lipid peroxidation may lead to endothelial cell dysfunction (Davidge et al., 1996). Tumor necrosis factor (TNF), tissue factor (TF) of placental origin, endothelial nitric oxide synthase (NOS), and excessive activity of the enzyme polymerase may contribute to endothelial dysfunction. Compared with normotensive pregnant women, women with preeclampsia have reduced expression of constitutive nitrite oxidative stress -mRNA, and this lead to reduced production of NO.

NOS inhibition consequence is the increased endothelial permeability and an abnormal response of the endothelial cells to the stress (Wang et al., 2004). In preeclampsia greater nitrotyrosine immunostaining were found in placental villous vascular endothelium and its surrounding smooth muscle cells, and also in villous stromal cells compared to normal pregnant controls (Myatt et al., 2006). Moreover particularly intense immunoreactivity of nitrotyrosine was measured within the invasive cytotrophoblasts in placental biopsies and vascular endothelium in the floating villi obtained from women with preeclampsia (Many et al., 2000).

4.1.5 Vascular developement

Aberrant placental vasculature development and abnormal placental blood flow are characterized by increased impedance in Doppler velocimetry (Farag et al., 2004). These abnormalities significantly correlated with expression of tissue factor in the placenta of women with severe preeclampsia (Di Paolo et al., 2003). The expression of tissue factor was found to be markedly increased in the endothelial cells within the basal deciduus. Doppler impedance modifications were significantly correlated to the endothelial cell activation. Tumor necrosis factor, a circulating cytokine, has also been implicated as causing endothelial dysfunction in preeclampsia (Hung et al., 2004). Significantly higher tissue levels of tumor necrosis factor were demonstrated in the placenta from women with preeclampsia (Wang et al., 1996). Higher levels of tumor necrosis factor lead to increased generation of E-selectin, a marker of endothelial activation of umbilical endothelial cells.

4.2 Oxidative stress in gestational diabetes (GDM)

Gestational diabetes is defined as a carbohydrate intolerance of variable severity, which begins, or is identified during the pregnancy (Lopez et al. 2011). The prevalence of gestational diabetes mellitus is around 5% of all pregnancies (Ben Haroush et al., 2004). The presence of this disease increases the risk of macrosomia, perinatal morbido-mortality (Ostlund et al., 2003) and subsequent developement of type 2 diabetes mellitus. The pathophysiology of gestational diabetes remain unclear. Pregnant women with gestational diabetes have a reduction in insulin sensitivity (Catalano et al., 1999), hypergicemia, and hyperlipidemia. Oxidative stress implication in developement of the disease is a result of imbalance between the increase in the formation of reactive oxidative substances (Brownlee, 2001; Maddux et al., 2001) and the insufficience of antioxidative defence mechanisms (Chen et al., 2003).

4.2.1 Induction of oxidative stress in gestational diaetes pathways

Hyperglycemia induces oxidative stress and cell and tissue damage through several metabolic mechanisms.

These include the formation of advanced glycation endproducts (AGE), activation of protein kinase C (PKC), the hexosamine pathway, and increased reactive oxygen speciesies production in the mitochondria. An important source of free radicals in diabetes is the interaction of glucose with proteins. Maillard reaction that form by nonenzymatic glycation through covalent attachment of highly reactive aldehyde or ketone groups of reducing

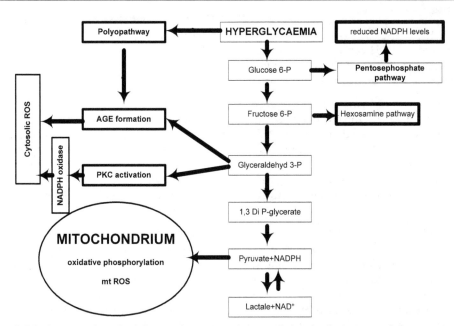

Fig. 3. Mechanisms by which hyperglycemia induces cellular dysfunction and damage.

carbohydtrates and the free amino groups on proteins, lipids, and nucleic acids. Elevated concentration of glucose leads to enhanced formation glycolytic products which together with the tricarboxylic acid (TCA) cycle intermediates provide glycation of intracellular proteins. The interaction of aldehyde groups of glucose with free amino groups on proteins generates a Schiff's base. Extracellular AGE can bind to the AGE receptor (RAGE), a multiligand member of the immunoglobulin superfamily. Engagement of RAGE by AGE results in activation of intracellular signaling molecules resulting in oxidative stress and inflammation. Since oxidative stress induction and inflammation are closely associated with gestational diabetes (Coughlan et al., 2001; Lappas et al., 2004), it is plausible that the AGE-RAGE system could play a role in the pathogenesis of this metabolic disease. During pregnancy, the AGE-RAGE axis may be involved in oxidative and inflammatory responses. Specifically, AGE-BSA stimulated the release of the pro-inflammatory cytokines IL-1β, IL-6, IL-8, and tumor necrosis factor (TNF)-α and prostaglandins PGE_2 and $PGF_{2\alpha}$. NF-κB and MAPK activate expression of several pro-inflammatory genes, including pro-inflammatory cytokines, the adhesion molecules vascular cell adhesion molecule (VCAM)-1, and intercellular cell adhesion molecule (ICAM)-1, and RAGE causing cellular inflammation. This is consistent with gestational diabetes being closely associated with low-grade inflammation (Kirwan et al., 2002) and atherosclerosis (Anastasiou et al., 1998; Hannemann et al., 2002). Additionally, the activation of cytokines by AGE in human placenta may also be involved in insulin resistance associated with gestational diabetes (Colomiere et al., 2009).

Hyperglycemia also activats hexosamine biosynthetic pathway (Rajapakse et al., 2009). This pathway of glucose metabolism uses fructose-6-phosphate derived from glycolysis to metabolize glucosamine-6-phosphate by glucosamine-6-phosphate amidotransferase.

Glucosamine-6-phosphate is a competitive inhibitor of glucose-6-phosphate dehydrogenase (G6PDH). Glucosamine-6 phosphate produced in the hexosamine pathway, leads to decreased NAPDH concentrations, diminished cellular GSH levels, and elevated oxidative stress. The activity of G6PDH also rapidly increases in response to intracellular reactive oxygen specieses production (Jian et al., 2003).

NADPH oxidase is a membrane enzyme complex accounting for ROS generation by electron transport chain and the enzyme is especially important in redox signaling. Under diabetic conditions it can be stimulated by AGE, insulin, and angiotensin II. Hypoxia possibly induces all these stimuli, which can activate NADPH oxidase. Once activated in response to high glucose levels, NADPH oxidase catalyzes the transfer of electrons from NADPH to molecular oxygen to produce O_2^- convetred to H_2O_2. High glucose levels lead to generation of reactive oxygen specieses by stimulation of NADPH oxidase (Gao et al., 2009; Gupte et al., 2010).

Under physiological conditions reactive oxygen specieses are eliminated by cellular defense mechanisms, including diverse enzymes and vitamins. Imbalance of reactive oxygen specieses production and antioxidant systems of a cell can lead to an upregulation of expression of antioxidant enzyme encoding genes. Hyperglycemia causes excessive reactive oxygen specieses formation, thus activating the Nrf2/ARE pathway (Xue et al., 2008). NADPH oxidase was shown to be higher expressed and activated in endothelial cells of pregnant women with GDM (Sankaralingam et al., 2009). Protein kinase C (PKC) promotes the activation of mitochondrial NADPH oxidase, thereby leading to oxidative stress events. Once stimulated, NADPH oxidase reduces glutathione levels and impairs the cellular antioxidant defense systems (King et al., 2004).

4.2.2 Oxidant species

Maternal MDA levels in serum and plasma are increased in gestational diabete mellitus women compared to normal glucose tolerant (NGT) pregnant women (Chaudhari et al., 2003; Surapanieni et al., 2008). Higher levels of lipid peroxidation are evident in patients with poor glycemic control. Proteins undergo oxidative damage, they become increasingly susceptible to proteolytic degradation. Erythrocytes contain proteolytic enzymes that can degrade oxidatively damaged proteins such as hemoglobin, thus preventing the accumulation of nonfunctional proteins and protein fragments. GDM is associated with higher levels of maternal erythrocyte proteolytic activity than NGT controls (Kamath et al., 1998).

4.2.3 Antioxidants

The level of total superoxide dismutase in placental tissues of gestational diabetes mellitus (both diet- and insulin-controlled) patients is lower (Kinalski et al., 2001) or did not significantly change (Biri et al., 2006; Lappas et al., 2010). Relative ratio of Cu,Zn-SOD to 8-isoprostane or protein carbonyl was lower in gestational diabetes mellitus placentas, suggesting that the increase in superoxide deismutase is not sufficient to compensate for the developed oxidative stress (Coughlan et al., 2004).

Oxidative stress plays a significant role in both NO overproduction and loss of NO bioavailability (Gloire et al., 2006; Xia et al., 2007). Oxidative stress leads to NOS-dependent increases in NO production in different tissues. Reduction in NO-induced stress is related to diabetes-induced endothelial dysfunction, NOS activation can also be induced by diabetes. NO induces the expression of antioxidant enzymes Mn- and Cu,Zn-SODs, and heme oxygenase-1 and increases intracellular glutathione concentration (Moellering et al., 1999). Although NO stimulates O_2^--induced lipoperoxidation in membranes, it can also mediate protective reactions to inhibit O_2^- and $ONOO^-$ induced lipoperoxidation (Rubbo et al., 1994). NO production has been found increased in the placenta, placental veins and arteries, and in umbilical vein endothelial cells from estational diabetes mellitus patients (Figueroa et al., 2000; vonMadach et al., 2003). NOS expression is also altered, as NOS has been found overexpressed in the placenta and eNoxidative stress increased in umbilical vein endothelial cells from estational diabetes mellitus patients (SanMartin et al., 2006). In estational diabetes mellitus, increases in reactive oxygen specieses and NO production, evident in the placenta and umbilical vessels, lead to peroxynitrite formation. In platelets from estational diabetes mellitus patients, elevated NOS activity and peroxynitrite production have been reported, possibly associated with platelet dysfunction and membrane damage due to increased lipid peroxidation (Mazzanti et al. , 2004). Strong protein nitration is found in term placentas from diabetic rats. Collectively, these data provide evidence of reactive nitrogen species (RNS)-induced damage in estational diabetes mellitus in the placenta and the vasculature of the mother, the placenta, and the umbilical cord, produced as a resulting consequence of exacerbated NO and ROS production.

4.2.4 Role of oxidative stress in gestational diabetes-induced teratogenesis

Diabetes in pregnancy is associated with suboptimal decidualization (Garris et al., 1988). NO plays a key role in decidualization and embryo implantation (Norwitz et al., 2001). It increases vascular permeability, vasodilation, and blood flow in the uterus, and is a component of the decidual cell reaction (Valdes et al., 2009). Diabetes during pregnancy is associated with embryonic dysmorphogenesis. Due to its capacity to regulate cell survival, apoptosis, differentiation, oxidative and nitrosative stresses play a significant role in embryo organogenesis. Low and high levels of NO can lead to embryonic maldevelopment, possibly due to an improper regulation of apoptotic events. During embryo and fetal development, NO has been found to be relevant in regulating differentiation of lung branching morphogenesis, cephalic morphogenesis, heart development, and nephrogenesis (Bloch et al., 1999; Tain et al., 2010).

Transcription factors, as paired box (PAX)-3 and peroxisome proliferator-activated receptor (PPAR) δ, has been found to be involved in the induction of both neural tube and heart malformations, the most common malformations in gestational diabetes pregnants (Higa et al., 2007; Loeken et al., 2006). Different antioxidants such as α-tocopherol and gluthatione ethyl ester increase expression of PAX-3 and prevent apoptosis and the induction of hyperglycemia-induced neural tube and heart defects (Chang et al., 2003; Morgan et al. 2008). The higher 8-isoprostane levels observed in the offspring of diabetic animals (Wentzel & Erikkson, 2002; Wentzel et al., 1999) have its own teratogenic potency. Diabetic embryopathy is also associated with inhibition of GAPDH activity resulting from an excess of reactive oxygen speciesesin the embryo (Wentzel et al., 2002). Oxidative glucose metabolism is low and about 80% of the glucose used by the placenta. The effect of oxidative stress on placental glucose metabolism is not known. However, in nongestational tissues, there is certainly evidence demonstrating oxidative stress regulates GLUT-1 and/or GLUT-3 dependent glucose uptake and transport. On the other hand estational diabetes mellitus placenta is less sensitive to oxidative stress due to the heightened level of antioxidants. Under normal conditions, physiological levels of reactive oxygen speciesespromote and stimulate adequate insulin signaling. The insulin signaling pathway leads to low levels of reactive oxygen species production itself and ROS act as second messengers of which disposal impairs insulin signaling. Insulin-induced reactive oxygen specieses production is accounted for by activation of the NADPH oxidase NOX4 through PI3K. The reactive oxygen speciesespathway subsequently activates kinases or induces gene expression by redox-sensitive transcription (Omroy, 2007).

4.3 Intrauterine Growth Restriction (IUGR)

Fetal growth depends on the interactions of genetic and epigenetic determinants functioning against an environment of maternal, fetal, and placental influences (Gardosi et al., 1992). Intrauterine growth restriction (IUGR) manifests as a variable syndrome of suboptimal growth and body disproportions rather than a well-defined etiologic entity. Causes for IUGR are diverse and include aneuploidies, non-aneuploid syndromes, infections, metabolic factors and placental disorders. IUGR places the fetus and neonate at risk of death or disability in the perinatal period (Baschat et al. 2000; Bernstein et al. 2000) and predisposes the child to a lifelong increased risk for hypertension, cardiovascular disorders and renal disease, among others (Murphy et al., 2006). A common definition is an estimated fetal weight less than the 10th percent for gestational age. Diminished fetal arterial and venous Doppler flows in key vascular beds predict worsening fetal acid base status (Rizzo et al., 2001; Baschat et al. 2004) and such findings frequently lead to delivery of a markedly premature baby to avoid in utero demise.

A diverse number of stimuli and mediators contribute to the observed injury to the chorioallantoic villi but oxidative stress is high on the list as an injurious agent (Hung et al., 2002). The production of reactive oxygen species during oxidative stress is linked with tissue injury in many diseases (Ryter et al., 2007). Paper shows that the placentas of pregnancies with IUGR exhibit overt signs of oxidative stress, with reduced protein translation and particular reductions in key signalling proteins pathways (Yung et al., 2008). Moreover, the syncytiotrophoblast shows signs of endoplasmatic reticulum (ER) stress by activating the unfolded protein response, which leads to an ER signal for enhanced apoptosis. The identified dysregulation of protein translation, signalling pathways and trophoblast turnover in placentas of pregnancies with IUGR (Burton et al., 2009). Hypoxia, ischaemia/reperfusion, or both may contribute to placental injury through mechanisms other than reactive oxygen species generation, as variable blood flow to organs also activates the complement cascade (Levy et al., 2000; Hung et al., 2002; Heazell et al,. 2008). Activation of the complement cascade injure the feto-placental unit (Girardi et al,. 2003).

Clarifying the role of complement activation in pregnancies complicated by IUGR, and in placental dysfunction generally, may lead to new approaches to treatment for IUGR, as therapeutic options to modulate complement receptors and complement activity are on the horizon.

5. Role of antioxidant supplementation in pregnancy

Clinical and research centers are investigating the usefulness of antioxidant supplementation and their role in prevention of pathological pregnancies. Antioxidant supplementation, for example vitamin C and vitamin E, has been shown to have beneficial effects in preventing luteal phase deficiency and resultant increased pregnancy rate (Hemi et al., 2003; Crha et al., 2003). Meta-analysis investigating the intervention of vitamin-C supplementation in pregnancy was inconclusive (Rumbold et al, 2005). Another meta-analysis of women taking any of the vitamin supplements started prior to 20 weeks' gestation revealed no reduction in total fetal losses, or in early and late miscarriage, having used the fixed-effects model. Improved pregnancy rates were also reported with combination therapy with the antioxidants pentoxifylline and vitamin-E supplementation for 6 months in patients with thin endometria who were undergoing in vitro fertilization with oocyte donation (Ledee-Bataille et al., 2002). Supplementation with vitamin E has also been reported to prevent the deleterious effects of ethanol toxicity on cerebral development in the animal model (Peng et al., 2005). There are essential differences among the population groups and the dosage and duration of supplementation for prevention of preeclampsia. Although many advances are being made in the field of antioxidants therapy, the data are still debatable and need further controlled evaluations in larger populations (Ashok et al., 2006).

6. Conclusions

The establishment of pregnancy requires a harmonic hormonal, ovarial and fallopian tube function, a receptive uterus able to respond to a variety of biochemical and molecular signals produced by the developing conceptus, as well as specific interactions between the uterine endometrium and the extra-embryonic membranes. Therefore, the fetal, placental development and function are prerequisites for an adequate supply of nutrients and oxygen to the fetus and successful establishment of pregnancy. Oxidative stress is a complex system, affecting in a complex way the female fertility, and pregnancy outcome. The inbalance of the oxidative agents and antioxidants has been proposed as the causative agents of female sterility, recurrent pregnancy loss and several pregnancy-related disorders, most notably preeclampsia, intra-uterine growth restriction (IUGR) and gestational diabetes.

Preeclampsia is characterized by increased oxidative stress due to the imbalance between lipid peroxidation and antioxidant defense mechanisms, leading to endothelial dysfunction and free radical mediated cell injury. Other maternal factors including activated neutrophils and imbalance between anticoagulants and procoagulants aggravate the oxidative stress and endothelial dysfunction this plays crucial role in developement of the disease. There is no doubt that both hypoxia and hypoxia-reperfusion lead to reactive oxygen species production, but both may also arise from the same underlying problem of impaired conversion of the spiral arteries. The effects of the reduced trophoblast invasion associated with complicated pregnancies can easily be superimposed on this basic model. Reduced invasion will leave the spiral arteries vasoreactive, and thus more likely to undergo spontaneous transient vasoconstriction. They will be more responsive to endogenous and exogenous vasoactive stimuli. Partial obliteration of their lumens by atherotic changes will also impair flow. Excessive production of inflammatory cytokines, deportation of apoptotic microvillous placental fragments, activation of maternal leukocytes and platelets, or depletion of NO production may then cause or contribute to the maternal endothelial response. The degree of the oxidative stress will likely reflect the extent of the maternal vascular pathology.

There are a number of pathways that may contribute to oxidative stress observed in the gestational diabetes mellitus placenta. In the placenta, reactive oxygen species and reactive nitrite spesies are an important source of growth and signaling factors, and are susceptible to ROS-mediated apoptosis. The placenta is endowed with many antioxidants, some of which are increased in gestational diabetes mellitus. However, there is much data to indicate that maternal diabetes during pregnancy may induce oxidative stress in the newborn that may entail biochemical disturbances of the fetus (Hung et al., 2006). Given that The placenta provides the interface of the maternal and fetal circulations, it may play a crucial role in protecting the fetus from adverse effects of the maternal diabetic milieu (Lappas et al., 2011).

The investigation of oxidtive stress is inevitabile for better understanding of aerobe organism function. Evaluation of environmental factors effect on oxidative stress molecular pathways can sereve possible solutions for female reproductive malfunctions. The are several fertility and pregnancy related disease, as unexplained infertility, preeclampsia, HELLP syndrome where the reactive oxidative species and antioxidant mechanism play key role in pathogenesis of the disease. The antioxidant supplementation, avoidance of different enviromental factors, as polluted comestibles may lead to decrease of infertility rate and incidence of pregnancy related disorders.

7. References

Agarwal A, Allamaneni SS. (2004). Role of free radicals in female reproductive diseases and assisted reproduction. *Reprod Biomed Online*, 9, 338-47.

Allen RG. (1991). Oxygen-reactive species and antioxidant response during development. The metabolic paradox of cellular differentiation. *Soc Exp Biol Med*, 196, 117–129.

Anastasiou E, Lekakis JP, Alevizaki M, Papamichael CM, Megas J, Souvatzoglou A, Stamatelopoulos SF. (1998). Impaired endothelium-dependent vasodilatation in women with previous gestational diabetes. *Diabetes Care*, 21, 2111–2115.

Behrman HR, Kodaman PH, Preston SL, Gao S. (2001). Oxidative stress and the ovary. *J Soc Gynecol Investig*, 8, S40–S42.

Biri A, Onan A, Devrim E, Babacan F, Kavutcu M, Durak I. (2006). Oxidant status in maternal and cord plasma and placental tissue in gestational diabetes. *Placenta* 27, 327–332.

Bloch W, Fleischmann BK, Lorke DE, Andressen C, Hops B, Hescheler J, Addicks K. (1999). Nitric oxide synthase expression and role during cardiomyogenesis. *Cardiovasc Res*, 43: 675–684.

Carew JS, Giles FJ and Nawrocki ST (2008) Histone deacetylase inhibitors: mechanisms of cell death and promise in combination cancer therapy. *Cancer Lett* 269:7-17.

Chang TI, Horal M, Jain SK, Wang F, Patel R, Loeken MR. (2003). Oxidant regulation of gene expression and neural tube development: insights gained from diabetic pregnancy on molecular causes of neural tube defects. *Diabetologia*, 46, 538–545.

Chaudhari L, Tandon OP, Vaney N, Agarwal N. (2003). Lipid peroxidation and antioxidant enzymes in gestational diabetics. *Indian J Physiol Pharmacol* 47, 441–446.

Colomiere M, Permezel M, Riley C, Desoye G, Lappas M. (2009). Defective insulin signaling in placenta from pregnancies complicated by gestational diabetes mellitus. *Eur J Endocrinol* 160, 567–578.

Coughlan MT, Oliva K, Georgiou HM, Permezel JM, Rice GE. (2001). Glucose-induced release of tumour necrosis factor-alpha from human placental and adipose tissues in gestational diabetes mellitus. *Diabet Med*, 18, 921–927.

Coughlan MT, Vervaart PP, Permezel M, Georgiou HM, Rice GE. (2004). Altered placental oxidative stress status in gestational diabetes mellitus. *Placenta*, 25, 78–84.

Crha I, Hrubá D, Ventruba P, Fiala J, Totusek J, Visnová H. (2003). Ascorbic acid and infertility treatment. *Cent Eur J Public Health*, 11,63–67.

El Hage S, Singh SM. (1990). Temporal expression of genes encoding free radical-metabolizing enzymes is associated with higher mRNA levels during in vitro development in mice. *Dev Genet*, 11, 149–159.

El Mouatassim S, Guerin P, Menezo Y. (1999). Expression of genes encoding antioxidant enzymes in human and mouse oocytes during the final stages of maturation. *Mol Hum Reprod*, 5, 720–725.

Evans MD, Dizdaroglu M, Cooke MS. (2004). Oxidative DNA damage and disease: induction, repair and significance. *Mutat Res*, 567, 1-61.

Figueroa R, Martinez E, Fayngersh RP, Tejani N, Mohazzab HK, Wolin MS. (2000). Alterations in relaxation to lactate and H(2)O(2) in human placental vessels from gestational diabetic pregnancies. *Am J Physiol Heart Circ Physiol*, 278, 706–713.

Gao L, Mann GE. (2009). Vascular NAD(P)H oxidase activation in diabetes: a double-edged sword in redox signalling. *Cardiovasc Res*, 82, 9–20.

Garris DR. (1988). Effects of diabetes on uterine condition, decidualization, vascularization, and corpus luteum function in the pseudopregnant rat. *Endocrinology*, 122, 665–672.

Gloire G, Legrand-Poels S, Piette J. (2006). NF-kappaB activation by reactive oxygen species: fifteen years later. *Biochem Pharmacol*, 72, 1493–1505.

Gupta S, Agarwal A, Sharma RK. (2005). The role of placental oxidative stress and lipid peroxidation in preeclampsia. *Obstet Gynecol Surv*, 12, 807-816.

Gupta S, Labinskyy N, Gupte R, Csiszar A, Ungvari Z, Edwards JG. (2010). Role of NAD(P)H oxidase in superoxide generation and endothelial dysfunction in Goto-Kakizaki (GK) rats as a model of nonobese NIDDM. *PLoS One*, 5, 11800.

Gurvich N, Berman MG, Wittner BS, Gentleman RC, Klein PS and Green JBA (2005) Association of valproate-induced teratogenesis with histone deacetylase inhibition in vivo. *FASEB J* 19,1166-1180.

Halliwell B, Gutteridge JMC, Cross C. (1992). Free radicals, antioxidants and human diseases: where are we now? *Lab Clin Med*, 119, 598–613.

Halliwell B, Gutteridge JMC. (1995). Free radicals in biology and medicine, *Clarendon Press*, Oxford.

Halliwell B. (1990). How to characterize a biological antioxidant. *Free Radic Res Commun*, 9, 1–32.

Hannemann MM, Liddell WG, Shore AC, Clark PM, Tooke JE. (2002). Vascular function in women with previous gestational diabetes mellitus. *J Vasc Res*, 39, 311–319.

Hass M, Iqbal J, Clearch LB, Brank L, Massaro D.(1989). Cu, Zn superoxide dismutase, isolation and sequence of full-length cDNA and studies of enzyme induction. *J Clin Invest*, 83, 1241–1245.

Henmi H, Endo T, Kitajima Y, Manase K, Hata H, Kudo R. (2003). Effects of ascorbic acid supplementation on serum progesterone levels in patients with a luteal phase defect. *Fertil Steril*, 80, 459–461.

Higa R, Gonzalez E, Pustovrh MC, White V, Capobianco E, Martinez N, Jawerbaum A. (2007). PPARdelta and its activator PGI2 are reduced in diabetic embryopathy: involvement of PPARdelta activation in lipid metabolic and signalling pathways in rat embryo early organogenesis. *Mol Hum Reprod*, 13, 103–110.

Hung TH, Burton GJ. (2006). Hypoxia and reoxygenation: a possible mechanism for placental oxidative stress in preeclampsia. *Taiwan J Obstet Gynecol*, 3,189-200.

Jain M, Brenner DA, Cui L, Lim CC, Wang B, Pimentel DR, Koh S, Sawyer DB, Leopold JA, Handy DE, Loscalzo J, Apstein CS, Liao R. (2003). Glucose-6-phosphate dehydrogenase modulates cytosolic redox status and contractile phenotype in adult cardiomyocytes. *Circ Res*, 93, 9–16.

Jauniaux E, Watson AL, Hempstock J, Bao YP, Skepper JN, Burton GJ. (2000). Onset of maternal arterial blood flow and placental oxidative stress. A possible factor in human early pregnancy failure. *Am J Pathol*, 157, 2111–2122.

Jergil M, Kultima K, Gustafson A, Dencker L and Stigson M (2009) Valproic acid–induced deregulation in vitro of genes associated in vivo with neural tube defects. *Toxicol Sci* 108,132-148.

Jiang B, Mendelson R. (2003). USF1 and USF2 mediate inhibition of human trophoblast differentiation and CYP19 gene expression by Mash-2 and hypoxia. *Mol Cell Biol*, 23, 6117-6128.

Jozwik M, Wolczynski S, Szamatowicz M. (1999). Oxidative stress markers in preovulatory follicular fluid in humans. *Mol Hum Reprod*, 5, 409-413.

Kamath U, Rao G, Raghothama C, Rai L, Rao P. (1998). Erythrocyte indicators of oxidative stress in gestational diabetes. *Acta Paediatr*, 87, 676–679.

Karowicz-Bilinska A, Suzin J, Sieroszewsk P. (2002). Evaluation of oxidative stress indices during treatment in pregnant women with intrauterine growth retardation. *Med Sci Monit*, 3, 211-6.

Kinalski M, Sledziewski A, Telejko B, Kowalska I, Kretowski A, Zarzycki W, Kinalska I. (2001). Lipid peroxidation, antioxidant defence and acid-base status in cord blood at birth: the influence of diabetes. *Horm Metab Res*, 33, 227–231.

King GL, Loeken MR. (2004). Hyperglycemia-induced oxidative stress in diabetic complications. *Histochem Cell Biol*, 122, 333–338.

Kirwan JP, Hauguel-De Mouzon S, Lepercq J, Challier JC, Huston-Presley L, Friedman JE, Kalhan SC, Catalano PM. TNF-alpha is a predictor of insulin resistance in human pregnancy. *Diabetes*, 51, 2207–2213.

Kohen R, Nyska A. (2002). Oxidation of biological systems: oxidative stress phenomena, antioxidants, redox reaction and methods for their quantitation. *Toxicol Pathol*, 30, 620–650.

Lappas M, Hiden U, Desoye G, Froehlich J, Mouzon SH, Jawerbaum A. (2011). The Role of Oxidative Stress in the Pathophysiology of Gestational Diabetes Mellitus. *Antioxid Redox Signal*. (epubl).

Lappas M, Mitton A, Permezel M. (2010). In response to oxidative stress, the expression of inflammatory cytokines and antioxidant enzymes are impaired in placenta, but not adipose tissue, of women with gestational diabetes. *J Endocrinol*, 204, 75–84.

Lappas M, Permezel M, Ho PW, Moseley JM, Wlodek ME, Rice GE. (2004). Effect of nuclear factor-kappa B inhibitors and peroxisome proliferator-activated receptor-gamma ligands on PTHrP release from human fetal membranes. *Placenta*, 25, 699–704.

Lédée-Bataille N, Olivennes F, Lefaix JL, Chaouat G, Frydman R, Delanian S. (2002). Combined treatment by pentoxifylline and tocopherol for recipient women with a thin endometrium enrolled in an oocyte donation programme. *Hum Reprod*, 17, 1249–1253.

Loeken MR. (2006). Advances in understanding the molecular causes of diabetes-induced birth defects. *J Soc Gynecol Investig* 13, 2–10.

López-Tinoco C, Roca M, García-Valero A, Murri M, Tinahones FJ, Segundo C, Bartha JL, Aguilar-Diosdado M. (2011). Oxidative stress and antioxidant status in patients with late-onset gestational diabetes mellitus. *Acta Diabetol*, (epubl)

Lushchak VI. (2011). Adaptive response to oxidative stress: Bacteria, fungi, plants and animals. Comp Biochem Physiol C *Toxicol Pharmacol* 153,175-90.

Lushchak VI. (2011). Environmentally induced oxidative stress in aquatic animals. *Aquat Toxicol* 101,13-30.

Maruyama T, Kitaoka Y, Sachi Y, Nakanoin K, Hirota K, Shiozawa T, Yoshimura Y, Fujii S, Yodoi J. (1997). Thioredoxin expression in the human endometrium during the menstrual cycle. *Mol Hum Reprod*, 3, 989–993.

Matsui M, Oshima M, Oshima H, Takaku K, Maruyama T, Yodoi J, Taketo MM. (1996). Early embryonic lethality caused by targeted disruption of the mouse thioredoxin gene. *Dev Biol*, 178, 179–185.

Mazzanti L, Nanetti L, Vignini A, Rabini RA, Grechi G, Cester N, Curzi CM, and Tranquilli AL. (2004). Gestational diabetes affects platelet behaviour through modified oxidative radical metabolism. *Diabet Med*, 21, 68–72.

Menegola E, Di Renzo F, Broccia ML, Prudenziati M, Minucci S, Massa V and Giavini E (2005) Inhibition of histone deacetylase activity on specific embryonic tissues as a new mechanism for teratogenicity. *Birth Defects Res B Dev Reprod Toxicol* 74, 392-398.

Moellering D, Mc Andrew J, Patel RP, Forman HJ, Mulcahy RT, Jo H, and Darley-Usmar VM. (1999). The induction of GSH synthesis by nanomolar concentrations of NO in endothelial cells: a role for gamma-glutamylcysteine synthetase and gamma-glutamyl transpeptidase. *FEBS Lett*, 448, 292–296.

Morgan SC, Relaix F, Sandell LL, Loeken MR. (2008). Oxidative stress during diabetic pregnancy disrupts cardiac neural crest migration and causes outflow tract defects. *Birth Defects Res A Clin Mol Teratol*, 82, 453–463.

Myatt L, Cui X. (2004). Oxidative stress in the placenta. *Histochem Cell Biol*, 122, 369-382.

Norwitz ER, Schust DJ, Fisher SJ. (2001). Implantation and the survival of early pregnancy. *N Engl J Med*, 345, 1400–1408.

Ornoy A, Zaken V, Kohen R. (1999). Role of reactive oxygen species (ROS) in the diabetes-induced anomalies in rat embryos in vitro: reduction in antioxidant enzymes and LMWA may be the causative factor for increased anomalies. *Teratology*, 60, 376–386.

Ornoy A. (2007). Embryonic oxidative stress as a mechanism of teratogenesis with special emphasis on diabetic embryopathy. *Reprod Toxicol*. 24, 31-41.

Papaccio G, Pisanti FA, Frascatore S. (1986). Acetyl homocysteine thiolactone induced increase: superoxide dismutase counteracts the effect of sub diabetogenic doses of streptozocin. *Diabetes*, 35, 470–474

Paszkowski T, Traub AI, Robinson SY, McMaster D.(1995). Selenium dependent glutathione peroxidase activity in human follicular fluid. *Clin Chim Acta*, 236, 173–180.

Peng Y, Kwok KH, Yang PH, Ng SS, Liu J, Wong OG, He ML, Kung IT, Lin MC. (2005). Ascorbic acid inhibits reactive oxygen speciesproduction, NF-kappa B activation and prevents ethanol-induced growth retardation and microencephaly. *Neuropharmacology*, 48, 426–434.

Phiel CJ, Zhang F, Huang EY, Guenther MG, Lazar MA and Klein PS (2001) Histone Deacetylase Is a Direct Target of Valproic Acid, a Potent Anticonvulsant, Mood Stabilizer, and Teratogen. *J Biol Chem* 276, 36734-36741.

Rajapakse AG, Ming XF, Carvas JM, Yang Z. (2009). The hexosamine biosynthesis inhibitor azaserine prevents endothelial inflammation and dysfunction under hyperglycemic condition through antioxidant effects. *Am J Physiol Heart Circ Physiol*, 296, 815–822.

Riley JC, Behrman HR. (1991). Oxygen radicals and reactive oxygen species in reproduction. *Proc Soc Exp Biol Med*, 198, 781–791.

Rosselli M, Dubey RK, Imthurn B, Macas E, Keller PJ. (1995). Effects of nitric oxide on human spermatozoa: evidence that nitric oxide decreases sperm motility and induces sperm toxicity. *Hum Reprod*, 10, 1786-1790.

Rubbo H, Radi R, Trujillo M, Telleri R, Kalyanaraman B, Barnes S, Kirk M, Freeman BA. (1994). Nitric oxide regulation of superoxide and peroxynitrite-dependent lipid peroxidation. Formation of novel nitrogen-containing oxidized lipid derivatives. *J Biol Chem*, 269, 26066-26075.

Rumbold A, Crowther CA. (2005). Vitamin C supplementation in pregnancy. *Cochrane Database Syst Rev*, (CD004072).

Sabuncu T, Vural H, Harma M. (2001). Oxidative stress in polycystic ovary syndrome and its contribution to the risk of cardiovascular disease. *Clin Biochem*, 34, 407–413.

San Martin R, Sobrevia L. (2006). Gestational diabetes and the adenosine/L-arginine/nitric oxide (ALANO) pathway in human umbilical vein endothelium. *Placenta*, 27, 1–10.

Sankaralingam S, Xu Y, Sawamura T, Davidge ST. (2009). Increased lectin-like oxidized low-density lipoprotein receptor-1 expression in the maternal vasculature of women with preeclampsia: role for peroxynitrite. *Hypertension*, 53, 270–277.

Shiotani M, Noda Y, Narimoto K, Imai K, Mori T, Fujimoto K, Ogawa K. (1991). Immunohistochemical localization of superoxide dismutase in the human ovary. *Hum Reprod*, 6, 1349–1353.

Siddiqui IA, Jaleel A, Tamimi W, Al Kadri HM. (2010). Role of oxidative stress in the pathogenesis of preeclampsia. *Arch Gynecol Obstet*, 282, 469-474.

Sugino N, Karube-Harada A, Sakata A, Takiguchi S, Kato H. (2002). Nuclear factor-kappa B is required for tumor necrosis factor-alpha-induced manganese superoxide dismutase expression in human endometrial stromal cells. *J Clin Endocrinol Metab*, 87, 3845–3850.

Sugino N, Karube-Harada A, Taketani T, Sakata A, Nakamura Y. (2004). Withdrawal of ovarian steroids stimulates prostaglandin F2alpha production through nuclear factor-kappaB activation via oxygen radicals in human endometrial stromal cells: potential relevance to menstruation. *J Reprod Dev*, 50, 215–225.

Sugino N, Shimamura K, Takiguchi S, Tamura H, Ono M, Nakata M, Nakamura Y, Ogino K, Uda T, Kato H. (1996). Changes in activity of superoxide dismutase in the human endometrium throughout the menstrual cycle and in early pregnancy. *Hum Reprod*, 11, 1073-1078.

Surapaneni KM, Vishnu Priya V. (2008). Antioxidant enzymes and vitamins in gestational diabetes. *J Clin Diagn Res*, 2, 1081–1085.

Taguchi M, Alfer J, Chwalisz K, Beier HM, Classen-Linke I. (2000). Endothelial nitric oxide synthase is differently expressed in human endometrial vessels during the menstrual cycle. *Mol Hum Reprod*, 6, 185-190.

Tain YL, Hsieh CS, Lin IC, Chen CC, Sheen JM, and Huang LT. (2010). Effects of maternal L-citrulline supplementation on renal function and blood pressure in offspring exposed to maternal caloric restriction: the impact of nitric oxide pathway. *Nitric Oxid*, 23, 34–41.

Tamate K, Sengoku K, Ishikawa M. (1995). The role of superoxide dismutase in the human ovary and fallopian tube. *J Obstet Gynaecol*, 21, 401–409.

Telfer JF, Lyall F, Norman JE, Cameron IT. (1995). Identification of nitric oxide synthase in human uterus. *Hum Reprod*, 10, 19-23.

Tseng L, Zhang J, Peresleni T, Goligorsky MS. (1996). Cyclic expression of endothelial nitric oxide synthase mRNA in the epithelial glands of human endometrium. *J Soc Gynecol Investig*, 3, 33-38.

Tung EWY and Winn LM (2010) Epigenetic modifications in valproic acid-induced teratogenesis.*Toxicol Appl Pharmacol* 248, 201-209.

Valdes G, Kaufmann P, Corthorn J, Erices R, Brosnihan KB, Joyner-Grantham J. (2009). Vasodilator factors in the systemic and local adaptations to pregnancy. *Reprod Biol Endocrinol*, 7, 79.

von Mandach U, Lauth D, and Huch R. (2003). Maternal and fetal nitric oxide production in normal and abnormal pregnancy. *J Matern Fetal Neonatal Med*, 13, 22–27.

Wentzel P and Eriksson UJ. (2002). 8-Iso-PGF(2alpha) administration generates dysmorphogenesis and increased lipid peroxidation in rat embryos in vitro. *Teratology*, 66, 164–168.

Wentzel P, Ejdesjö A, Eriksson UJ. (2003). Maternal diabetes in vivo and high glucose in vitro diminish GAPDH activity in rat embryos. *Diabetes*, 52, 1222–1226.

Wentzel P, Welsh N, Eriksson UJ. (1999). Developmental damage, increased lipid peroxidation, diminished cyclooxygenase-2 gene expression, and lowered prostaglandin E2 levels in rat embryos exposed to a diabetic environment. *Diabetes* 48, 813–820.

Xia Y. (2007). Superoxide generation from nitric oxide synthases. *Antioxid Redox Signal* 9, 1773–1778.

Xue M, Qian Q, Adaikalakoteswari A, Rabbani N, Babaei-Jadidi R, Thornalley PJ. (2008). Activation of NF-E2-related factor-2 reverses biochemical dysfunction of endothelial cells induced by hyperglycemia linked to vascular disease. *Diabetes* 57, 2809–2817.

Zimmerman EF, Potturi RB, Resnick E, Fisher JE. (1994). Role of oxygen free radicals in cocaine-induced vascular disruption in mice. *Teratology*, 49, 192–201.

5

Circulating Advanced Oxidation Protein Products, Nε-(Carboxymethyl) Lysine and Pro-Inflammatory Cytokines in Patients with Liver Cirrhosis: Correlations with Clinical Parameters

Jolanta Zuwala-Jagiello[1],
Eugenia Murawska-Cialowicz[2] and Monika Pazgan-Simon[3]
[1]*Department of Pharmaceutical Biochemistry, Wroclaw Medical University,*
[2]*Department of Physiology and Biochemistry, University of Physical Education, Wroclaw,*
[3]*Clinic of Infectious Diseases, Liver Diseases and Acquired Immune Deficiency,*
Wroclaw Medical University,
Poland

1. Introduction

Patients with chronic liver disease are characterized by hepatic inflammation and the destruction of hepatocytes. Viral antigen-specific cytotoxic T lymphocytes, polyclonal cytokines, immune modulators, and oxidized biomolecules have been shown to induce damage and destruction of hepatocytes in these patients (Tsutsui *et al.*, 2003). The contribution of oxidative stress *per se* to the chronic inflammatory state has been suggested, and consistent evidence has been afforded that both monocyte/macrophage activation and a defect in antioxidant systems occur early in the course of chronic liver failure and gradually increase with its progression to end-stage liver disease (Kirkham, 2007; Videla, 2009). Oxidative stress lead to formation of glycoxidation products, including advanced glycation endproducts (AGEs - among them Nε-(carboxymethyl)lysine (CML) is best known), and advanced oxidation protein products (AOPPs). AOPPs can be formed *in vitro* by exposure of serum albumin to hypochlorous acid. *In vivo*, plasma AOPPs are mainly carried by albumin and their concentrations are closely correlated with the levels of dityrosine. Within the heterogeneous group of AGEs, Nε-(carboxymethyl)lysine has been identified as a major AGEs *in vivo* (Reddy *et al.*, 1995). Plasma concentrations of AGEs (closely correlating with AOPPs levels) increase with progression of chronic diseases (Witko-Sarsat *et al.*, 1996; 1998), therefore CML has been considered as liver disease-related biomarker for oxidative stress (Sebeková *et al.*, 2002; Yagmur *et al.*, 2006).

The receptor for advanced glycation endproducts (RAGE) is a signal transduction receptor that binds both AGEs and AOPPs. RAGE is expressed by various cell types, including monocytes/macrophages, endothelial cells, smooth muscle cells and renal cells (Miyata *et al.*, 1994). Advanced glycation endproducts have been found to act as pro-inflammatory factors (Sparvero *et al.*, 2009). Nevertheless, AOPPs are believed to be more closely related to inflammation (Alderman *et al.*, 2002; Baskol *et al.*, 2006; Fialova *et al.*, 2006; Witko-Sarsat *et al.*, 2003; Yazici *et al.*, 2004) than AGEs, whose receptor for advanced glycation endproducts participates in AOPPs-mediated signal transduction (Kalousová *et al.*, 2003; 2005). These interactions enhance reactive oxygen species formation, with activation of nuclear factor NF-κB and release of pro-inflammatory cytokines (Bierhaus *et al.*, 2006; Hyogo & Yamagishi, 2008; Saito & Ishii, 2004). Moreover, the monocyte/macrophage RAGE can be up-regulated by tumor necrosis factor-α (TNF-α) (Miyata *et al.*, 1994). Peripheral blood monocytes showed activity and elevated expression of TNF-α which correlated with liver disease severity (Hanck *et al.*, 2000). The concentrations of advanced oxidation protein products are high in liver cirrhosis of various etiologies (Zuwala-Jagiello *et al.*, 2009) and can reflect hemodynamic alterations in the liver (Guo *et al.*, 2008). This is accompanied by the activation of monocytes and increased expression of TNF-α (Giron-González *et al.*, 2004). High serum levels of TNF-α and interleukin-6 (IL-6) have been found in cirrhotic patients with ascites in the absence of demonstrable infection (Tilg *et al.*, 1992; Zeni *et al.*, 1993).

The accumulation of AGEs has been linked to vascular lesions in diabetes, chronic renal insufficiency, and atherosclerosis. Activation of NF-κB, mediated by RAGE, promotes expression of the cytokines, as well as pro-inflammatory adhesion molecules (Basta *et al.*, 2002; Bierhaus *et al.*, 1998; Esposito *et al.*, 1989), what may enhance interaction of cirrhotic vasculature with circulating monocytes (Cybulsky *et al.*, 1991; Li *et al.*, 1993). Recently, it has also been shown that AOPPs activates vascular endothelial cells *via* RAGE-mediated signals (Guo *et al.*, 2008).

Endothelial activation plays an active role in the modifications of circulatory status of cirrhotic patients (Genesca *et al.*, 1999). The circulatory changes are more evident in advanced stages of liver cirrhosis, such as those represented by the presence of ascites or hepatorenal syndrome (Porcel *et al.*, 2002). We have demonstrated that elevated levels of AOPPs modified–albumin (AOPPs-albumin) are related to the severity of liver cirrhosis (Zuwala-Jagiello *et al.*, 2009; 2011). The role of AOPPs-albumin in liver cirrhosis and portal hypertension has not yet been studied. The effects of pro-inflammatory cytokines on the vessels and on liver function would influence the liver cirrhosis, with higher plasma levels of AOPPs indicating a poor prognosis. In the present study, plasma levels of AOPPs-albumin, as well as of Nε (carboxymethyl)lysine modified-albumin (CML-albumin) and pro-inflammatory cytokines, such as TNF-α and IL-6, have been analyzed in cirrhotic patients and were found to be correlated with clinical parameters of liver dysfunction.

2. Patients and methods

2.1 Patients

This study was performed on 129 patients with chronic liver disease admitted to the Clinic of Infectious Diseases, Liver Diseases and Acquired Immune Deficiency for evaluation. The experimental group consisted of 68 men and 61 women with age of 18–74 years (median age was 66). The control group contained 40 healthy subjects (23 men and 17 women) with age of 19–56 (median age was 55). Blood samples were collected in the Department of

Physiology and Biochemistry, University of Physical Education in Wroclaw. Clinical and biochemical characteristics of the study group are reported in detail in Table 1.

	Healthy controls	All patients	Non-cirrhotic patients	Cirrhotic patients
(n)	40	129	41	88
Male:Female ratio	23:17	68:61	15:26	53:35
Age (years)	55 (19–56)	66 (18–74)	56 (18–69)	55 (21-74)
Etiology (n)				
Virus hepatitis		62	18	44
Alcohol		54	21	33
Biliary		13	2	11
Albumin (g/L)	45 (36–57)	34 (16–45)	37* (29–49)	30* (16-45)
ALT (U/L)	24 (20-28)	40 (16-79)	28 (24-33)	47** (16-79)
AST (U/L)	27 (23-30)	70 (19-150)	41 (19-64)	79** (19-150)
Bilirubin (mg/dL)	0.7 (0.6-0.9)	0.98 (1.0-3.6)	0.92 (0.90-0.95)	1.6* (1.0-3.6)
γGT (U/L)	26 (25.3-27.8)	70 (41-106)	48 (41-56)	92** (78-106)
AP (U/L)	90 (50-130)	125 (100-163)	105 (100-147)	152** (141-163)
Serum creatinine (mg/dL)	0.8 (0.7-1.0)	1.4 (0.7-2.4)	0.96 (0.9-1.2)	1.39* (0.7-2.4)
Serum sodium (mEq/L)	140 (138-141)	137 (129-142)	136 (129-138)	130** (129-142)

Table 1. Clinical and biochemical characteristics of the study subjects. Statistical significance: $*P < 0.05$; $** P < 0.01$ vs. healthy controls. AST, aspartate aminotransferase; ALT, alanine aminotransferase; AP, alkaline phosphatase; γGT, γ-glutamyltransferase; INR, normalised international ratio; MELD, model of end-stage liver disease.

The diagnosis of liver cirrhosis was based on clinical, laboratory and ultrasonographic findings or histological criteria. Alcohol-related liver cirrhosis was diagnosed in 33 of patients, primary biliary cirrhosis in 11, cirrhosis caused by hepatitis C virus in 30, whereas cirrhosis caused by hepatitis B virus in 14 patients. The Child-Pugh score was used to assess the severity of liver disease. Three biochemical variables [serum albumin, bilirubin, and prothrombin time (international normalized ratio, INR)] in addition to the two clinical characteristics (presence or absence of ascites and clinical signs of encephalopathy) were determined. Patients were scored as follows: 5-6 as class (group) A, 7-9 as class (group) B and 10-15 as class (group) C. The patients with cirrhosis were divided into compensated (Child-Pugh class A) and decompensated (Child-Pugh classes B and C) groups. At the time of the study no Child-Pugh A patients showed clinical features of decompensated liver cirrhosis (ascites or hepatic encephalopathy). At enrollment, esophageal varices were detected by endoscopy in 88% of patients, ascites and hepatic encephalopathy grade were present by physical examination in 53 (60%) and 23 (26%) patients, respectively.

Exclusion criteria were concurrent use of antioxidant drugs; co-existing diseases like diabetes mellitus, chronic kidney disease, cardiovascular disease, hepatocellular carcinoma; gastrointestinal bleeding, bacterial infection, and blood transfusion within previous two weeks.

Patients had not been receiving diuretic, antibiotic, vasoactive drug (nitrates, β-blockers), and lactulose or lactitiol therapy during the eight days before inclusion in the study. After 2 h of bed rest, blood pressure was determined with an automatic digital sphygmomanometer and blood samples were collected in ice-cooled, ethylenediaminetetraacetic acid (EDTA)-containing tubes for the determination of plasma renin activity, antidiuretic hormone, and plasma AOPPs or Nε-(carboxymethyl)lysine, in tubes with no additive for routine biochemical study and aldosterone and cytokine concentrations. All samples were separated immediately by centrifugation at 4°C and stored at -80°C until further analysis.

The consent of the Bioethics Committee of the Wroclaw Medical University was obtained and all patients were informed about the character of analyses made. Studies were conducted in compliance with the ethical standards formulated in the Helsinki Declaration of 1975 (revised in 1983).

2.2 Determination of circulating AOPPs

In vivo plasma levels of AOPPs closely correlate with levels of dityrosine, a hallmark of oxidized proteins and with pentosidine, a marker of protein glycation closely related to oxidative stress. A new chromogen is found which caused increased absorbance at 340 nm and its spectrophotometric determination is proposed as a novel index of oxidative stress measuring the level of AOPPs (Witko-Sarsat et al., 1996). Two- hundred microliters of plasma diluted 1:5 in 20 mM phosphate buffer pH 7.4 containing 0.9% sodium chloride (PBS), or chloramine-T standard solutions (0 to 100 μmol/L), were placed in each well of a 96-well microtiter plate (Becton Dickinson Labware, Lincoln Park, NJ, USA), followed by 20 μL of 10% acetic acid. Ten microliters of 1.16 M potassium iodide (Sigma-Aldrich Co. LLC, Canada) were then added, followed by 20 μL of 10% acetic acid. The absorbance of the reaction mixture was immediately read at 340 nm in a microplate reader against a blank containing 200 μL of PBS, 10 μL of KI and 20 μL of 10% acetic acid. The chloramine-T absorbance at 340 nm was linear within the range of 0 to 100 μmol/L. The ratio of AOPPs concentration to albumin level (AOPPs-albumin) was expressed in micromoles of AOPPs per gram of albumin (μmol/g). The ratio of AOPPs to albumin content allows the evaluation of whether the proportion of oxidatively modified albumin is altered. Coefficient of variation (CV) served as an indicator of precision. Intra-day and inter-day CV values were <10%.

2.3 Determination of circulating Nε-(carboxymethyl) lysine

Plasma Nε-(carboxymethyl)lysine (CML) levels were determined using a specific competitive ELISA kit [CircuLex CML/Nε-(carboxymethyl)lysine ELISA Kit (CycLex Co., Ltd, Nagano, Japan)]. Measurements were performed in duplicate and the results were averaged. The ratio of CML concentration to albumin level (CML-albumin) was expressed in micrograms of CML per gram of albumin (μg/g).

2.4 Laboratory determinations

Biochemical parameters were measured using routine laboratory methods. Serum high-sensitivity C-reactive protein (hs-CRP) level was determined with a high-sensitivity nephelometric method using the Beckman Image Immunochemistry system (Beckman Instruments, Fullerton, CA), which has a minimum level of detection of 0.2 mg/L. Serum levels of TNF-α and IL-6 were assayed with enzyme-linked immunosorbent assay (ELISA) kits (R&D Systems Inc., Minneapolis, MN, USA) according to the manufacturer's instructions. The minimum levels of detection were 1.6 pg/mL and < 0.70 pg/mL for TNF-α and IL-6, respectively. The intra- and interassay coefficients of variation for measurements of CRP, IL-6, and TNF-α were 2.7%, 4.3%, and 5.0%, respectively, and 3.0%, 5.5, and 6.9%, respectively.

Aldosterone (Aldoctk-2-P2714; Sorin Biomedica Diagnostics, Barcelona, Spain. Normal values, 35-150 pg/mL) and plasma renin activity (Clinical Assays, Baxter, Cambridge, Mass., USA. Normal values, 0.4-2.3 ng mL^{-1} h^{-1}) were measured by specific radioimmunoassays. Antidiuretic hormone was also tested by a commercial radioimmunoassay (Buhlman Laboratories, Basel, Switzerland. Normal values, less than 1 pg/mL).

2.5 Measurement of the total antioxidant status of plasma

The plasma antioxidant capacity was measured using a commercially available total antioxidant status TAS kit (Randox Laboratories, Crumlin, UK). The TAS assay is based on the inhibition by antioxidants of the absorbance of the radical cation of 2,2'-azinobis-(3-ethylbenzothiazoline-6-sulfonate) (ABTS) formed by the interaction of ABTS with ferrylmyoglobin radical species. Upon the addition of a plasma sample, the oxidative reactions initiated by the hydroxyl radicals present in the reaction mix are suppressed by the antioxidant components of the plasma, preventing the color change and thereby providing an effective measure of the total antioxidant capacity of the plasma. The assay has excellent precision values, lower than 3%, and the results are expressed as mmol/L.

2.6 Model for End-stage Liver Disease (MELD) score

The model for end-stage liver disease (MELD) score was calculated from the following equation:

9.57 × log$_e$ (creatinine mg/dL) + 3.78 × log$_e$ (total bilirubin mg/dL) + 11.2 × log$_e$ (international normalized ratio-INR) + 6.43 (constant for liver disease etiology).

The maximal creatinine concentration considered in the MELD score is 4.0 mg/dL (Huo *et al.*, 2006).

2.7 Statistical analysis

Results are expressed as median (25th percentile–75th percentile). Frequency data were compared using the χ2 test or the Fischer's exact test when necessary. Because many of the variables analyzed did not have a normal distribution as determined by the Kolmogorov-Smirnov test, nonparametric tests were used for comparison of data. The Mann- Whitney U test and the Kruskal-Wallis test were used to analyze differences among two or more groups, respectively. Multivariate analysis by conditional logistical regression with a forward stepwise method was performed to find independent variables associated with the presence of ascites and low mean arterial pressure. Regression analysis to determine significant correlations among different parameters was performed using the Spearman correlation coefficient. Statistical significance was established at $P < 0.05$

3. Results

3.1 AOPPs-albumin plasma concentrations in patients with chronic liver disease and healthy controls

We analyzed 129 patients (68 males/61 females, median age 66 years, range 18–74 years) with chronic liver disease. The distribution of the stages of liver cirrhosis as defined according to the Child–Pugh score, and measurements of AOPPs-albumin and CML-albumin plasma concentrations are presented in Fig.1. The concentration of AOPPs-albumin

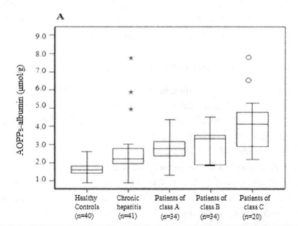

	Healthy Controls	Chronic hepatitis	Patients of class A	Patients of class B	Patients of class C
AOPPs-albumin (μmol/g)	1.7 (0.8-2.7)	2.1* (0.9-3.0)	2.8* (1.3-4.4)	3.2 (1.9-4.5)	4.1**⁺ (2.3-5.2)

Significance between groups: *P < 0.05; ** P < 0.01 vs. healthy controls; ⁺ P < 0.05 vs. patients of class A

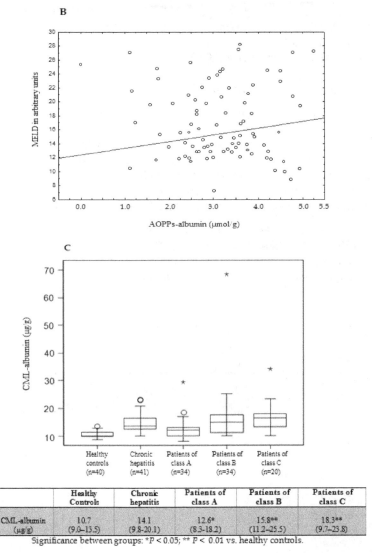

	Healthy Controls	Chronic hepatitis	Patients of class A	Patients of class B	Patients of class C
CML-albumin (µg/g)	10.7 (9.0–13.5)	14.1 (9.8–20.1)	12.6* (8.3–18.2)	15.8** (11.2–25.5)	18.3** (9.7–23.8)

Significance between groups: *P < 0.05; ** P < 0.01 vs. healthy controls.

Fig. 1. (A) AOPPs-albumin serum concentrations in 129 patients with chronic liver disease, according to Child's stage of cirrhosis, and in an control group of 40 healthy blood donors. P values are given in the table. Comparisons between subgroups are illustrated with box plot graphics, where the dotted line indicates the median per group, the box represents 50% of the values, and horizontal lines show minimum and maximum values of the calculated non-outlier values; asterisks and open circles indicate outlier values. (B) AOPPs-albumin serum concentrations in patients with cirrhosis are correlated with the MELD (model of end-stage liver disease) score (r = 0.43, P < 0.01, Spearman rank correlation test). (C) CML-albumin serum concentrations increase with the stage of liver cirrhosis in patients with chronic liver disease. P values are given in the table.

in healthy subjects was 1.7 µmol/g (range 0.8-2.7 µmol/g, P < 0.05). In patients with chronic liver disease, AOPPs-albumin plasma concentrations were 1.3-fold higher. In healthy controls, the plasma AOPPs or CML were similar to those in control groups in other studies (Sebeková *et al.*, 2002; Kalousová *et al.*, 2003).

3.2 AOPPs-albumin and liver cirrhosis

AOPPs-albumin plasma concentration was significantly higher in patients with liver cirrhosis (n = 88, median 2.4 µmol/g, range 1.3-5.6 µmol/g) compared to patients with chronic liver disease without cirrhosis (n = 41, median 2.1 µmol/g, range 0.9-3.0 µmol/g) (*P* < 0.05, Fig.1A). Patients with Child-Pugh class C exhibited significantly higher plasma concentrations of AOPPs-albumin than patients with Child-Pugh class A and controls (*P* < 0.05, *P* < 0.01, respectively) (Fig.1A). There was no significant difference in AOPPs concentrations between control subjects and Child-Pugh B cirrhotic patients.

Differences in plasma AOPPs-albumin or CML-albumin were not significant in patients with liver cirrhosis of various etiologies (Table 2). Only in the group with primary biliary cirrhosis AOPPs-albumin were decreased (n = 11, median 1.3 µmol/g, range 0.80-2.2 µmol/g), though it should be consider with caution since small number of subjects included in this group.

	AOPPs-albumin (µmol/g)	CML-albumin (µg/g)	TAS (mmol/L)
Healthy controls (n=40)	1.7 (0.80-2.7)	10.7 (9.0–13.5)	1.31 (1.12-1.5)
Viral hepatitis-related cirrhosis (n=44)	3.09* (1.5-5.2)	13.3* (11.6-18.1)	0.65* (0.48-0.75)
Alcohol-related cirrhosis (n=33)	2.9* (1.6-4.3)	16.3* (13.3-25.5)	0.71* (0.60-0.73)
Primary biliary cirrhosis (n=11)	1.3 (0.80-2.2)	11.8 (8.3-12.5)	0.98 (0.76-0.83)

Biliary etiology shows lower AOPPs-albumin levels compared with other etiologies of liver disease. Significance levels between groups: *P < 0.05 vs. healthy controls.
AOPPs, advanced oxidation protein products; CML, Nε-(carboxymethyl)lysine; TAS, total antioxidant status.

Table 2. Plasma AOPPs-albumin, CML-albumin and TAS in liver cirrhosis patients of various etiologies.

The MELD scores were determined in the 88 patients with liver cirrhosis (Table 3). These were higher in the Child-Pugh C cirrhotic patients than in the Child-Pugh A cirrhotic patients (p<0.01). Significant correlations between AOPPs levels and MELD scores (r = 0.43, P < 0.01; Fig. 1B) were observed among the cirrhotic patients belonging to all three Child-Pugh classes.

	Healthy controls	Non-cirrhotic patients	Patients of class A	Patients of class B	Patients of class C
(*n*)	40	41	34	34	20
Age (years)	55 (19–56)	56 (18–69)	51 (21–74)	58 (24–71)	56 (29–69)
Albumin (g/L)	45 (36–57)	40 (29–49)	34* (28–45)	30* (20–40)	25* (16–32)
Bilirubin (mg/dL)	0.7 (0.6-0.9)	0.92 (0.90-0.95)	1.01 (1.02-1.03)	1.56* (1.0-2.0)	2.15* (1.1-3.6)
AOPPs-albumin (µmol/g)	1.7 (0.8-2.7)	2.1* (0.9-3.0)	2.8* (1.3-4.4)	3.2 (1.9-4.5)	4.1***+ (2.3-5.2)
CML-albumin (µg/g)	10.7 (9.0-13.5)	14.1 (9.8-20.1)	12.6* (8.3-18.2)	15.8** (11.2-25.5)	18.3** (9.7-23.8)
Uric acid (mmol/L)	0.35 (0.25-0.39)	0.29 (0.21-0.33)	0.31 (0.26-0.34)	0.22 (0.16-0.24)	0.18* (0.19-0.31)
Vitamin C (µmol/L)	54.3 (38.3-70.3)	45.2 (31.9-58.6)	47.1 (28.3-64.0)	33.6 (20.2-45.7)	36.9 (29.2-54.2)
TAS (mmol/L)	1.3 (1.1-1.5)	1.1 (0.9-1.3)	0.9 (0.6-1.0)	0.63 (0.5-0.76)	0.53***+ (0.5-0.8)
INR	0.8-1.1	-	0.9 (0.8-1.09)	1.2 (1.1-1.3)	2.3 (1.6-2.9)
MELD score	6-8	-	7.8 (5.2-10.3)	15.9* (8.7-23.1)	24.1**+ (14.4-28.4)

Table 3. Plasma concentrations of AOPPs-albumin, CML-albumin and antioxidant parameters in patients with chronic liver disease without cirrhosis and in patients with cirrhosis. Significance between groups: *P < 0.05; ** P < 0.01 vs. healthy controls; + P < 0.05; ++P < 0.01 vs. patients of class A. AOPPs, advanced oxidation protein products; CML, Nε-(carboxymethyl)lysine; TAS, total antioxidant status; MELD, model for end-stage liver disease score.

3.3 CML-albumin plasma concentrations in patients with CLD and healthy controls

In patients with chronic liver disease, CML-albumin had a median value of 14.1 µg/g (range 9.8-20.1 µg/g). Plasma CML-albumin concentrations were higher in Child-Pugh A to C cirrhotic patients (n = 88, median 15.7 µg/g, range 8.3-25.5 µg/g) than in patients without cirrhosis, but this difference was not statistically significant (Fig. 1C). The levels of plasma CML-albumin in all liver cirrhotic patients were higher than those of the controls and this difference was statistically significant (Fig. 1C). Plasma CML-albumin in patients with Child-Pugh class C cirrhosis was only slightly elevated compared with those in Child-Pugh class A cirrhosis (P = 0.17) (Fig. 1C). There was no statistically significant correlation between CML-albumin levels and the Child-Pugh score in cirrhotic patients.

3.4 Antioxidant parameters and liver cirrhosis

As it is seen in Table 3, while all individual parameters of the antioxidant status tend to decrease, only the decrease of uric acid was statistically significant. There was a markedly decreased total antioxidant status (TAS) in patients with Child-Pugh class C cirrhosis compared to those with Child-Pugh class A cirrhosis or controls (P < 0.05, P < 0.01, respectively). Although differences between cirrhosis and chronic liver disease (n = 41, median 1.1, mmol/L, range 0.9-1.3 mmol/L) were not statistically significant, weak but significant correlation was observed between TAS and plasma AOPPs-albumin (r =-0.31, P < 0.05). We failed to find, however, any relation between circulating CML-albumin levels and TAS (r = -0.22, P = 0.059).

3.5 Markers of oxidative stress and hepatic function

The serum albumin concentration was determined in all patients (n = 129, median 34 g/L, range 16-45 g/L) and healthy control subjects (n = 40, median 45 g/L, range 36-57 g/L). Level of albumin, the main substrate in both AOPPs and CML formation (Kalousová *et al.*; 2005), was significantly depleted both in chronic liver disease (n = 41, median 37 g/L, range 29-49 g/L) and in cirrhosis (median 30 g/L, range 16-45 g/L) (Table 1). Plasma CML-albumin and foremost AOPPs-albumin showed significant associations with biochemical indices of liver function (albumin, prothrombin time, bilirubin concentration) but not with markers of liver injury – aminotransferases (Table 4). As expected, in patients with cirrhosis, AOPPs-albumin weakly but significantly correlated with the serum albumin (r = -0.38, P < 0.05).

3.6 AOPPs-albumin, CML-albumin and chronic inflammatory state in cirrhotic patients

We assessed the levels of several inflammatory markers and their association with the levels of AOPPs-albumin and CML-albumin. Serum high-sensitivity C-reactive protein (hs-CRP) levels and white blood cells (WBC) counts were significantly elevated in cirrhotic patients (Table 5). Serum TNF-α levels were higher in the Child-Pugh class C cirrhosis patients than in the Child-Pugh class A cirrhosis patients (P < 0.05) (Table 5). Moreover, TNF-α concentrations were weakly but significantly correlated with Child-Pugh score in cirrhotic group (r = 0.31, P < 0.05). The levels of serum IL-6 in cirrhotic patients were higher than those of the control group and this difference was statistically significant (P < 0.05) (Table 5). The levels of serum IL-6 in patients with Child-Pugh class C cirrhosis were higher than those in Child-Pugh class A cirrhosis, but this difference was not statistically significant.

	AOPPs-albumin (μmol/g)	CML-albumin (μg/g)
Albumin (g/L)	r = -0.38, P < 0.05	r = - 0.26, P = 0.07
ALT (U/L)	r = 0.10, P = 0.54	r = -0.14, P = 0.31
AST (U/L)	r = -0.20, P = 0.11	r = -0.15, P = 0.26
Bilirubin (mg/dL)	r = 0.23, P = 0.07	r = 0.24, P = 0.06
Prothrombin ratio (%)	r = -0.25, P < 0.05	r = - 0.43, P < 0.001
MELD score	r = 0.43, P < 0.01	r = 0.23, P = 0.07

Table 4. Correlations between plasma AOPPs-albumin and CML-albumin and selected biochemical indices of liver function and injury. ALT, alanine aminotransferase; AST, aspartate aminotransferase; MELD, model for end-stage liver disease score.

	Healthy controls (n=40)	Overall (n=88)	Patients of class A (n=34)	Patients of class B (n=34)	Patients of class C (n=20)
AOPPs-albumin (μmol/g)	1.7 (0.8-2.7)	2.4* (1.3-5.6)	2.8* (1.3-4.4)	3.2 (1.9-4.5)	4.1*+ (2.3-5.2)
CML-albumin (μg/g)	10.7 (9.0–13.5)	15.7* (8.3-25.5)	12.6* (8.3-18.2)	15.8** (11.2–25.5)	18.3** (9.7–23.8)
Albumin (g/L)	45 (36–57)	31** (16-35.4)	34* (28–45)	30* (20–40)	25* (16–32)
TNF-α (pg/mL)	32.9 (31.0-35.2)	41.5* (37.6-64.0)	36.9* (37.7-45.6)	42.0* (37.6-47.2)	51.7*+ (48.7-58.3)
IL-6 (pg/mL)	5.9 (5.4-6.8)	13.3** (6.4-39.9)	8.8* (6.4-34.6)	12.3* (6.8-33.9)	18.9* (9.0-39.9)
hsCRP (mg/L)	1.5 (0.63-2.0)	5.3** (4.9-11.0)	4.8* (5.5-7.0)	5.2** (4.9-7.7)	6.3*+ (6.8-10.1)

	Healthy controls (n=40)	Overall (n=88)	Patients of class A (n=34)	Patients of class B (n=34)	Patients of class C (n=20)
WBC (x10^9/L)	4.0 (4.2-5.0)	5.0 (1.1-8.6)	4.7 (1.1-8.6)	5.2* (3.0-7.8)	5.4**+ (2.7-8.2)
MELD score	6-8	13.5* (5.2-28.4)	7.8 (5.2-10.3)	15.9* (8.7-23.1)	24.1*+ + (14.4-28.4)

Table 5. Plasma concentrations of AOPPs-albumin, CML albumin and inflammatory markers in healthy controls and in patients with liver cirrhosis. Significance levels between groups: *P < 0.05; ** P < 0.01 vs. healthy controls; + P < 0.05, ++P < 0.01 vs. patients of class A. AOPPs, advanced oxidation protein products; CML, Nε-(carboxymethyl)lysine; TNF, tumor necrosis factor; IL, interleukin; CRP, C-reactive protein; WBC, white blood cells; MELD, model for end-stage liver disease score.

The association study revealed only a tendency toward an extremely weak but significant correlation between AOPPs-albumin and WBC in all cirrhotic patients (r = 0.23, P < 0.05). In turn, a weak but significant correlation between AOPPs-albumin levels and hs-CRP was observed among the cirrhotic patients belonging to all three Child-Pugh classes (r = 0.33, P < 0.05). There was a significant correlation between the IL-6 and the AOPPs-albumin level (r = 0.42, P < 0.05) and MELD score (r = 0.38, P < 0.05) in cirrhotic patients. As it was expected, a significant correlation between AOPPs-albumin levels and TNF-α (r = 0.48, P < 0.05) was observed in Child-Pugh class A cirrhosis patients. In the multivariate analysis the relationship between plasma AOPPs-albumin, TNF-α and Child-Pugh score was independent of age, sex and liver cirrhosis etiology (data not shown).

There was no statistically significant correlation between CML-albumin level and hs-CRP or cytokines levels in all liver cirrhotic patients (data not shown).

The ROC curve analyses are shown in Fig. 2 (sensitivity versus 1-specificity). The cut-off values of plasma AOPPs-albumin, TNF-α and IL-6 to separate cirrhotic patients from healthy controls were 3.71 μmol/g, 37.2 pg/mL, and 8.95 pg/mL, respectively.

3.7 Hemodynamic characteristics of the patients with liver cirrhosis

Hemodynamic characteristics of patients are shown in Tables 1 and 6. Cirrhotic patients had significantly lower values of mean arterial pressure (MAP) when compared with controls (Table 6). Parameters related to hemodynamic disturbances such as decreased mean arterial pressure and increased plasma renin activity and aldosterone levels deteriorated with increasing of Child-Pugh score. However, similar values of antidiuretic hormone were detected in all patients grouped according to the Child-Pugh classification.

Among clinical parameters of liver dysfunction ascites revealed significant association with plasma AOPPs-albumin, TNF-α and IL-6 (Table 6). By contrast, patients presented similar AOPPs-albumin levels when classified according to the presence or absence of encephalopathy grade I (data not shown). Ascitic patients had a more intense alteration of hemodynamic parameters (plasma renin activity, aldosterone), along with higher levels of

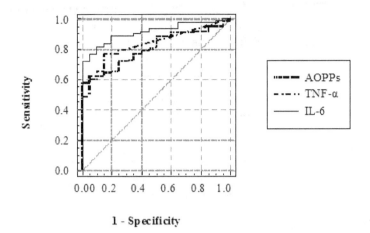

	AUC	Standard error	95% confidence interval	Sensitivity (%)	Specificity (%)	Optimal cut-off
AOPPs-albumin	0.802	0.018	0.766-0.838	57.6	100	3.71 μmol/g
TNF-α	0.902	0.015	0.872-0.932	71.7	100	37.2 pg/mL
IL-6	0.832	0.012	0.808-0.856	76.1	85	8.95 pg/mL

Fig. 2. Receiver operating characteristic (ROC) curve and optimal cut-off levels of advanced oxidation protein products (AOPPs), tumor necrosis factor-α (TNF-α), and interleukin-6 (IL-6) for distinguishing cirrhotic patients from healthy controls; AUC, area under the curve.

AOPPs-albumin, TNF-a, and IL-6, compared with patients without ascites. Cirrhotic patients who had ascites showed higher AOPPs-albumin levels (n = 53, median 3.6 μmol/g, range 1.5-5.3 μmol/g) than patients without ascites (n = 35, median 2.2 μmol/g, range 1.0-3.4 μmol/g) (P < 0.05, Table 6). AOPPs-albumin levels were not different between cirrhotic patients without ascites and controls, while patients with ascites had higher AOPPs-albumin levels than controls (median 1.7 μmol/g, range 0.8-2.7 μmol/g) (P < 0.01). High IL-6 levels showed an independent association with the presence of ascites.

To differentiate cirrhotic patients with a more intense hemodynamic alteration (vasodilatation), we divided patients according to those with low mean arterial pressure (MAP ≤83 mm Hg) and high mean arterial pressure (MAP >83 mm Hg) (Table 7). Only IL-6 levels were significantly higher in patients with more severe vasodilatation; this association was independent of other factors.

Plasma levels of AOPPs-albumin very weakly but significantly correlated with MAP (r = -0.25, P < 0.01; Fig.3). Furthermore, IL-6 levels had a significant correlation with several parameters, including plasma renin activity (r = 0.39, $P < 0.05$) and MAP (r = -0.38, $P < 0.01$), in addition to albumin (r = - 0.51, $P < 0.001$). Neither TNF-α levels nor the CML-albumin levels correlated significantly with hemodynamic parameters.

	Overall (n=88)	Patients of class A (n=34)	Patients of class B (n=34)	Patients of class C (n=20)	Cirrhosis without ascites (n=35)	Cirrhosis with ascites n=53)
Mean arterial pressure-MAP (mmHg)	83 (76-93)	89 (85-93)	83 (77-91)	76* (73-81)	88 (85-93)	77+ (73-89)
Plasma renin activity (ng mL^{-1} h^{-1})	0.6 (0.1-7.6)	0.37 (0.1-1.2)	1.9* (0.95-4.8)	6.0** 1.2-7.6)	0.48 (0.13-1.6)	1.9++ (0.51-6.2)
Aldosterone (ng/dL) 20.6 (14.0-51.0)	15.0 (5.0-71.0)	12.0 (5.0-19.0)	20.6 (14.0-/51.0)	30.0* (21.0-71.0)	13.2 (5.5-20.9)	33.0+++ (13.7-52.2)
Antidiuretic hormone (pg/mL)	4.9 (2.5-6.5)	3.8 (2.5-5.7)	4.5 (3.7-6.2)	4.8 (3.8-6.5)	4.6 (3.0-6.8)	4.5 (3.6-6.4)
AOPPs-albumin (μmol/g)	2.4 (1.3-5.2)	2.8 (1.3-4.4)	3.2 (1.9-4.5)	4.1* (2.3–5.2)	2.2 (1.0-3.4)	3.6+ (1.5-5.3)
TNF-α (pg/mL)	41.5 (37.6-64.0)	36.9 (37.7-45.6)	42.0 (37.6-47.2)	51.7* (48.7-58.3)	37.5 (35.4-46.5)	57.5+ (52.1-69.7)
IL-6 (pg/mL)	13.3 (6.4-39.9)	8.8 (6.4-34.6)	12.3 (6.8-33.9)	18.9 (9.0-39.9	6.4 (5.6-7.3)	10.8++ (8.5-12.4)

Significance between groups: *$P < 0.05$; **$P < 0.001$ vs. patients of class A. +$P < 0.05$; ++$P < 0.01$, +++$P < 0.001$ vs. liver cirrhosis without ascites. AOPPs, advanced oxidation protein products; TNF, tumor necrosis factor; IL, interleukin.

Table 6. Blood pressure, plasma renin activity and plasma concentrations of aldosterone, antidiuretic hormone and AOPPs-albumin according to the Child–Pugh class or presence of ascites.

Circulating Advanced Oxidation Protein Products, Nε-(Carboxymethyl) Lysine and Pro-Inflammatory Cytokines in
Patients with Liver Cirrhosis: Correlations with Clinical Parameters

109

	Mean arterial pressure (MAP) ≤ 83 mm Hg (n=51)	Mean arterial pressure (MAP) > 83 mm Hg (n=37)
Plasma renin activity (ng mL^{-1} h^{-1})	1.7 (0.46-5.5)	1.4 (0.37-4.4)
Aldosterone (ng/dL)	25.6 (10.6-40.5)	13.7 (5.7-21.6)
Antidiuretic hormone (pg/mL)	4.7 (3.7-6.7)	4.2 (3.3-5.9)
AOPPs-albumin (μmol/g)	3.3 (1.4-4.8)	3.01 (1.2-4.5)
IL-6 (pg/mL)	11.0+ (9.5-12.8)	6.3 (5.4-7.0)
TNF-α (pg/mL)	47.9 (43.4-58.0)	49.8 (45.1-60.3)

Table 7. Comparison between cirrhotic patients classified according to the finding of low (≤ 83 mm Hg) and high (>83 mm Hg) mean arterial pressure (MAP). Significance between groups: ^+P < 0.01 by multivariate analysis.

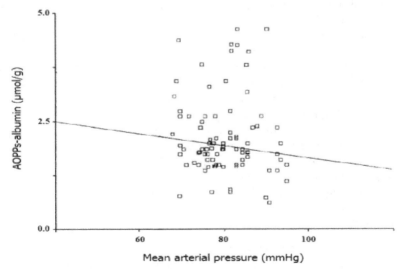

Fig. 3. AOPPs-albumin concentrations are very weakly but significantly correlated with the mean arterial pressure (MAP) in patients with chronic liver disease (r = -0.25, P < 0.01).

4. Discussion

Cirrhosis is characterized by inflammation of the liver, often caused by a rise in oxygen-derived free radicals within the liver. Under normal circumstances, the liver maintains a supply of internal antioxidants to neutralize the reactive species generated in response to viral infection and during metabolism of various endo- and exogenous compounds processed in the liver. However, when the liver antioxidants are low, or when the liver is undergone to continued oxidative insults (e.g., long-lasting alcohol abuse or infection with different hepatitis viruses), the damage from reactive species (Halliwell, 2007) may increase, resulting in inflammation and the formation of scar tissue (fibrosis) (Valko et al., 2007). The progressive decrease of antioxidant reserves, the dysfunction of liver microcirculation through nitric oxide-mediated pathways, may determine the shift to liver cirrhosis. Advanced glycation and oxidation endproducts (AGEs and AOPPs, respectively) cause oxidative stress and trigger cytokine driven inflammatory reactions in vitro. (reviewed in Yan et al., 2010). The net effects on markers of inflammation and hemodynamic changes in cirrhotic patients are unknown.

Advanced glycation endproducts are formed from the reaction of glucose and other reducing sugars with amino acid groups of proteins. This interaction generates a labile Schiff base followed by rearrangement to more stable Amadori-products, and subsequently these early glycation products may undergo further chemical rearrangements resulting in various irreversibly formed AGEs (Baynes & Thorpe, 1999). Three different mechanisms have been proposed by which AGEs lead to cirrhosis complications: 1) the binding of AGEs to the receptors for advanced glycation endproducts on different cell types including monocytes/macrophages, T lymphocytes, endothelial cells, smooth muscle cells, and activation of cell signaling pathways with subsequent modulation of gene expression, 2) intracellular AGEs formation leading to impaired cell function, and 3) the accumulation of AGEs in the extracellular matrix. CML-albumin levels (as prototype of the AGEs) were higher in cirrhosis groups than in the controls. In agreement with two previous reports (Zuwala-Jagiello et al., 2009; 2011) these results indicate that reactive oxygen species are overproduced in patients with liver cirrhosis. CML may lead to progression of cirrhosis by interaction with receptors that induce production of reactive species followed by a release of inflammatory cytokines in different cell types (Bierhaus et al., 1998; Raj et al., 2000) such as IL-6, ultimately leading to the production of CRP by the liver. A correlation to high sensitive C-reactive protein (hs-CRP) could not be demonstrated, and the suggestion that high levels of CML may activate an inflammatory response was not demonstrated by serum markers (TNF-α and IL-6). It is likely that the blood load of CML resembles only a small fraction of body's AGEs content and that the serum levels reflect particular changes in the body's AGEs pool. Thus, circulating CML-albumin may not be an adequate parameter for demonstrating effects on inflammatory response in cirrhotic patients. Most likely the focus should be on intracellular AGEs (Thornalley et al., 2003).

A role of oxidative stress in the pathogenesis of chronic liver disease has been proposed by several authors (Bandara et al., 2005; Nagata et al., 2007; Nakhjavani et al., 2011; Serejo et al., 2003; Zuwala-Jagiello et al., 2007). Studies have shown increased plasma levels of markers of lipid peroxidation and reduced plasma antioxidant content. Protein oxidation products are increasingly being used as markers instead of lipid peroxidation products in demonstrating oxidative stress (Dalle-Donne et al., 2003). AOPPs measurements reflect the reactive species generation and the degree of protein oxidation (Witko-Sarsat et al., 1996). It was reported that AOPPs generated by different oxidation patterns lead to the production of either hydrogen peroxide or nitric oxide (Servettaz et al., 2007). Nitric oxide can interact with superoxide anion-radical forming reactive nitrogen species such as peroxynitrite. These reactive nitrogen species secondarily promote important reactions such as nitrosation, oxidation or nitration, leading to impaired cellular functions and enhanced inflammatory reactions (Friedman, 2008; Iwakiri & Groszmann, 2007). AOPPs are referred to as markers of oxidative stress as well as markers of neutrophil activation in chronic disease (Witko-Sarsat et al., 2003). It has thus been shown that chlorinated oxidants of neutrophil origin may lead to oxidative stress, notably protein oxidation. Once formed, such AOPPs foci create a nidus for the amplification of oxidative stress. In addition to increased formation, decreased removal/detoxification of AOPPs may contribute to the stress. There is increasing evidence that the liver plays important roles in the elimination of AOPPs (Iwao et al., 2006). In patients with chronic liver diseases, constriction of the sinusoidal blood stream leads to the development of portal hypertension with portocaval shunts (Svistounov & Smedsrød, 2004). The hindrance of substance exchange between hepatocytes and the sinusoidal blood stream could increase plasma level of AOPPs in these patients. Therefore, the liver, especially in cirrhotic patients, cannot prevent the accumulation of AOPPs effectively. Finally, our findings extended the results of Oettl et al. (2008) which suggested that albumin is oxidatively modified in patients with advanced liver disease depending on its severity. The present finding that AOPPs-albumin accumulation coexists with decreased TAS, while the plasma concentration of CML-modified albumin remains stable, supports the contention that AOPPs-albumin is more accurate marker of oxidative stress than glycoxidation products in cirrhotic patients. An increase in reactive species formation, manifested by increased hepatic and plasma levels of AOPPs (Gorka et al., 2008; Sebeková et al., 2002; Yagmur et al., 2006; Zuwala–Jagiello et al., 2006;) and as well as decreased antioxidant levels (Jain et al., 2002; Zuwala–Jagiello et al., 2009) have been reported in patients with liver cirrhosis. Finally, our previous study found that the patients with cirrhosis were exposed to oxidative stress and the level of AOPPs was significantly related to the severity of liver cirrhosis of various etiologies (Zuwala-Jagiello et al., 2011).

The adverse effects of oxidative stress on the progression of cirrhosis may be categorized into effects on protein modifications and inflammatory response. Figure 4 presents a summary of the effects of AOPPs or AGEs and oxidative stress on markers of inflammation and hemodynamic changes in cirrhotic patients.

Very recently, advanced glycation endproducts have been found to act as pro-inflammatory factors (Sparvero et al., 2009). Nevertheless, AOPPs are believed to be more closely related to inflammation (Fialova et al., 2006) than AGEs, whose RAGE participates in AOPPs-mediated signal transduction (Kalousová et al., 2005; 2006). These interactions enhance reactive oxygen species formation, with activation of nuclear factor NF-κB and release of pro-inflammatory cytokines (Bierhaus et al., 2006; Hyogo & Yamagishi, 2008; Saito & Ishii, 2004) (Fig. 4). Moreover, the macrophage RAGE can be up-regulated by TNF-α (Miyata et al.,

1994). This is accompanied by the activation of macrophages and increased expression of TNF-α (Giron-González et al., 2004). TNF-α production is also stimulated by macrophage sensing of intestinal microflora pathogen associated molecular patterns by toll-like receptors (Riordan et al., 2003). It could therefore be consistent with observed increase levels of both AOPPs-albumin and TNF-α at an early stage of liver cirrhosis.

The hyperdynamic circulatory state associated with liver cirrhosis is characterized by vasodilatation and increased cardiac output; the arterial hypotension and relative hypovolemia caused by vasodilatation activate a number of vasoactive and neurohumoral systems (Wiest & Groszmann, 2002). TNF-α induces an endothelial activation, which can be detected by increased synthesis of nitric oxide (Spitzer, 1994). Endogenous nitric oxide, a powerful endothelium-derived vasodilator, has been implicated in hemodynamic changes present in cirrhotic patients (Garci'a-Tsao et al., 1998). Treatment with both specific anti-TNF polyclonal antibodies and thalidomide, an inhibitor of TNF-α production, significantly prevent the development of the hyperdynamic circulation and reduces portal pressure (Gatta et al., 2008). TNF-α levels in our patients were clearly different from control group and were much higher in ascitic patients. However, no differences existed between patients with high and low mean arterial pressure, and no significant correlations with hemodynamic values were found. These data suggest that, although TNF-α might be one of the inducers of nitric oxide generation in cirrhotic patients, other factors acting through different pathways probably exist.

AOPPs derived from in vivo sources stimulated endothelial cell generation of reactive oxygen species, in particular superoxide anion (Guo et al., 2008), at least in part through NADPH-oxidase (Wautier et al., 2001). However, the exact mechanisms and sources by which reactive oxygen species are generated in the vasculature are not yet known in detail. It has been observed in several experimental animal models, that the endothelium is one of the major sources for the generation of reactive oxygen species. In parallel with the vascular dysfunction the formation of superoxide anions became augmented, and removal of endothelium completely abolished the production of reactive oxygen species. (reviewed in Wright et al., 2006). In another report, Rhee et al. (2003) demonstrated that growth factor-induced H_2O_2 production (e.g. PDGF, EGF) requires the activation of phosphoinositide 3-kinase. The essential role of phosphoinositide 3-kinase is likely to provide phosphatidylinositol (3, 4, 5)-trisphosphate that recruits and activates a guanine nucleotide exchange factor of Rac, which is required for the activation of NADPH-oxidase. Thus, the generation of reactive oxygen species is largely dependent on the activation of NADPH-oxidase that is present in endothelial cell. AOPPs stimulated endothelial cell activation of the following signaling mediators: NADPH oxidase and NF-κB, the factors linked to increased expression of pro-inflammatory adhesion molecules, such as intercellular adhesion molecule-1 (ICAM-1) and vascular adhesion molecule-1 (VCAM-1) (Kim et al., 2001; Yan et al., 2010) (Fig. 4). Endothelial activation plays an active role in the modifications of circulatory status of cirrhotic patients (Grangé & Amiot, 2004). Elevated levels of AOPPs-albumin were detected in the early stages of liver dysfunction: plasma concentrations were increased in patients in Child Pugh class A, with higher values found in those in class B or C. Plasma concentrations of AOPPs-albumin were very weakly correlated with hemodynamic alterations (mean arterial pressure).

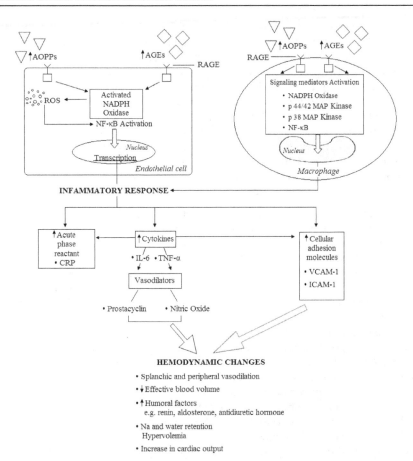

Fig. 4. Summary of the effects of AOPPs or AGEs and oxidative stress on markers of
inflammation and hemodynamic changes in cirrhotic patients. AOPPs or AGEs bind with
RAGE on the surface of endothelial cells lining blood vessels. AOPPS, AGEs ligands of
RAGE sustain stimulation of RAGE. One consequence of RAGE signaling is the activation of
NADPH oxidase and production of reactive oxygen species. Once formed, reactive oxygen
species activate key transcription factor such as NF-κB, which results in the transcriptional
activation of genes relevant for inflammation. Consequences include increased migration
and activation of RAGE-expressing macrophages. This results in release of the pro-
inflammatory cytokines. In this inflammatory environment, *via* interaction with RAGE on
the surface of macrophages, AOPPs or AGEs magnify activation of NF-κB and other factors,
thereby amplifying cellular stress and hepatic damage. In the aggregate, these processes
may contribute to propagation of inflammation and vascular perturbation in liver cirrhosis.
AGEs, advanced glycation endproducts; AOPPs, advanced oxidation protein products; CRP,
C-reactive protein; ICAM-1, intercellular adhesion molecule 1; IL, interleukin; MAP,
mitogen activated protein; NADPH, nicotinamide adenine dinucleotide phosphate (reduced
form); NF-κB, nuclear factor κB; RAGE, receptor for AGEs; ROS, reactive oxygen species;
TNF-α, tumour necrosis factor α; VCAM-1, vascular cell adhesion molecule.

These results suggested that this association of AOPPs with hemodynamic disturbances is dependent of the severity of cirrhosis. Additionally, AOPPs levels could be elevated as a result of insufficient renal elimination. However, the precise mechanism by which AOPPs is cleared from plasma is currently unknown. In addition, we found no correlation between AOPPs-albumin and serum creatinine levels. In any case, the values of AOPPs-albumin in patients with a low Child-Pugh score and absence of ascites suggests that AOPPs might have a role in the late stages of cirrhosis by aggravating the already initiated vasodilatation. Indeed, the presence of ascites, one of the major complications of cirrhosis and closely related to the hemodynamic disturbances of cirrhotic patients, was found in patients with higher levels of AOPPs. Finally, the extremely weak correlation between AOPPs-albumin levels and mean arterial pressure may suggest an indirect contribution of AOPPs to arterial vasodilatation through other mediators.

Structure and function of albumin are impaired in advanced liver disease by different mechanisms: plasma levels are decreased due to reduced synthesis and albumin is oxidatively modified. In this context, AOPPs-albumin may shows altered binding capacities for several substances. Decreased bilirubin binding was reported for *in vitro* oxidized albumin (Oettl & Stauber, 2007). As bilirubin is preferentially bound by the fully reduced form of albumin, impaired binding of bilirubin and other ligands (e.g. nitric oxide) is likely to occur in liver cirrhosis. Theoretically, increased circulating AOPPs-albumin may indirectly lead to elevation of nitric oxide which can, in turn, contribute to oxidative stress in cells (La Villa & Gentilini, 2008). Finally, nitric oxide plays a major key role in the development of hyperdynamic circulation and portal hypertension in cirrhosis (Iwakiri & Groszmann, 2007). Infusion of albumin (Garcovich *et al.*, 2009) as well as albumin dialysis has been shown to improve the circulatory dysfunction as evidenced by an increase in mean blood pressure and systemic vascular resistance (Mitzner *et al.*, 2001). This improvement in systemic hemodynamics might be due to a reduction in vasodilation following removal of nitric oxide which results in deactivation of the neurohormonal systems and a decrease in plasma levels of renin, aldosterone, norepinephrine and vasopressin (Chen *et al.*, 2009). However, other findings of the present study imply that AOPPs are not only the factors responsible for the hemodynamic changes observed in cirrhotic patients. The AOPPs-albumin levels did not correlate significantly with the parameters that accompany important hemodynamic alterations in cirrhotic patients, such as plasma renin activity and aldosterone. Finally, when cirrhotic patients in our study were divided according to high and low mean arterial pressure, we found similar AOPPs-albumin levels in both groups.

Portal hypertension and cirrhosis can increase gut permeability to endotoxin and impair reticuloendothelial function of the liver that may result in increased serum endotoxin concentrations (Cariello *et al.*, 2010). This may be a stimulus for the production of pro-inflammatory cytokines, resulting in the increased production of acute phase proteins such as C-reactive protein. In turn, CRP is capable of stimulating IL-6 and TNF-α production by monocytes (Ballou & Lozanski, 1992) and reactive oxygen species formation (Wang *et al.*, 2003). Advanced oxidation protein products, as pro-inflammatory factors, accumulated in cirrhotic patients (Zuwala−Jagiello *et al.*, 2006, 2009, 2011) and played an important role in the occurrence and progression of complications such as dysfunction of endothelial cells (Witko-Sarsat *et al.*, 1998). AOPPs correlate well with certain cytokines (Kalousová *et al.*, 2005) as well as with some markers of inflammation, including fibrinogen, orosomucoid. Even if the correlation between AOPPs-albumin and hs-CRP were poor, other studies demonstrated fair associations with other markers of oxidative stress, such as lipid peroxidation products and F2-isoprostanes. Furthermore, results from a recent study demonstrate that glycosylated and oxidized proteins indirectly up-regulate CRP expression in hepatocytes by stimulating monocytes to produce IL-6 (Li et al., 2007). It seem, based on our study, even though there was functional loss of hepatocytes in patients with hepatic cirrhosis, the serum CRP level was still maintained in high level and dependent of AOPPs-albumin level with its significant correlation. The remaining viable hepatocytes may still contribute to this result. IL-6 is the main stimulant for hepatic production of CRP but also has other important roles leading to increased endothelial cell adhesiveness by up-regulating ICAM-1, and VCAM-1 and releasing inflammatory mediators, including IL-6 itself (Szmitko et al., 2003). Finally, the significant correlation between the levels of AOPPs and IL-6 supports the existence of a link between AOPPs and hemodynamic changes present in cirrhotic patients (Fig. 4).

Portal hypertension is characterized by intrahepatic vascular resistance causing an increase of portal vein pressure, and leads to the development of ascite (Møller et al., 2008). IL-6 levels in our cirrhotic patients were different from those in controls, increased with the severity of liver disease, were independently associated to the presence of ascites. IL-6 increased in serum of patients with ascites compared to compensated (Child-Pugh class A) patients without ascites, and similar results were obtained in a study where serum IL-6 was also analyzed (Zhang et al., 2002). In a simplistic view, portal hypertension leads to the formation of portosystemic collateral veins in liver cirrhosis and the resulting shunting (Cichoz-Lach et al., 2008) also contributes to impaired hepatic uptake of IL-6. It has been convincingly shown that hepatic uptake of IL-6 is significantly impaired in patients with liver cirrhosis, and this may at least in part explain elevated serum levels in these patients (Soresi et al., 2006). In our study, IL-6 levels increase significantly in association with the severity of liver cirrhosis according to the MELD score. Moreover, IL-6 levels in all cirrhotic patients were independently associated to the presence of low mean arterial pressure, and showed significant correlations with parameters related to hemodynamic abnormalities. The mechanism by which IL-6 could cause vasodilatation is unknown. However, the effect appears to be independent of nitric oxide, possibly due to an important role of prostacyclin synthesis (Dagher & Moore, 2001). It is then possible that IL-6 would produces vasodilatation by inducing prostacyclin synthesis; the effect of IL-6 would be potentiated by AOPPs stimulation (Li et al., 2007).

Further, investigators have tried to find more noninvasive biomarkers for cirrhotic patients for years (Schuppan & Afdhal, 2008). We observed good abilities of plasma AOPPs-albumin, TNF-α and IL-6 levels to distinguish cirrhotic patients from healthy controls, with good sensitivities and specificities by ROC analysis. Additionally, these parameters also were found to be elevated in concordance with the severity of cirrhosis. Thereby, it is possible that plasma levels of AOPPs-albumin, TNF-α and IL-6 levels could be evaluated as candidate biomarkers for initial and long-term assessment of liver cirrhosis.

5. Conclusion

In conclusion, there are differences between advanced oxidation protein products modified–albumin (AOPPs-albumin), which act as a pure oxidative stress marker, and Nε (carboxymethyl)lysine modified-albumin (CML-albumin; as prototype of the advanced glycation endproducts-AGEs), which behave as both oxidative and carbonyl stress markers. AOPPs-albumin shows a closer relationship to inflammation than CML-albumin. Because of the relationship of AOPPs-albumin with formation of inflammation, their relationship with certain inflammatory markers and their changes during progression of chronic liver disease, AOPPs may provide a marker of chronic long-lasting liver damage. Oxidative stress not only contributes to the derangement of hemodynamic chronic liver failure, but also gradually increases with its progression to end-stage liver disease.

6. References

Alderman, C.J., Shah, S., Foreman, J.C., Chain, B.M. & Katz, D.R. (2002) The role of advanced oxidation protein products in regulation of dendritic cell function. *Free Radical Biology & Medicine*, Vol.32, No.5, (March 2002), pp. 377-385, ISSN:0891-5849

Ballou, S.P. & Lozanski, G. (1992) Induction of inflammatory cytokine release from cultured human monocytes by C-reactive protein. *Cytokine* Vol.4, No.5, (September 1992), pp. 361-368, ISSN 1043-4666

Bandara, P., George, J., McCaughan, G., Naidoo, D., Lux O., Salonikas, C., Kench, J., Byth, K. & Farrell, G.C. (2005) Antioxidant levels in peripheral blood, disease activity and fibrotic stage in chronic hepatitis C. *Liver International*, Vol.2, No.3, (June 2005), pp. 518-526, ISSN 1478-3223

Baskol, G., Demir, H., Baskol, M., Kilic, E., Ates, F., Karakukcu, C. & Ustdal, M. (2006) Investigation of protein oxidation and lipid peroxidation in patients with rheumatoid arthritis. *Cell Biochemistry and Function*, Vol.24, No.4, (July 2006), pp. 307-311, ISSN 263-6484

Baynes, J.W. & Thorpe, S.R. (1999) Role of oxidative stress in diabetic complications: a new perspective on an old paradigm. *Diabetes*, Vol.48, No.1, (January 1999), pp. 1-9, ISSN 0012-1797

Bierhaus, A., Hofmann, M.A., Ziegler, R. & Nawroth, P.P. (1998) AGEs and their interaction with AGE-receptors in vascular disease and diabetes mellitus. I. The AGE concept. *Cardiovascular Research*, Vol.37, No.3, (March 1998), pp. 586-600, ISSN 0008-6363

Bierhaus, A., Stern, D.M. & Nawroth, P.P. (2006) RAGE in inflammation: a new therapeutic target? *Current Opinion in Investigational Drugs*, Vol.7, No.11, (November 2006), pp. 985-991, ISSN 1472-4472

Cariello, R., Federico, A., Sapone, A., Tuccillo, C., Scialdone, VR., Tiso, A., Miranda, A., Portincasa, P., Carbonara, V., Palasciano, G., Martorelli, L., Esposito, P., Cartenì, M., Del Vecchio Blanco, C. & Loguercio, C. (2010) Intestinal permeability in patients with chronic liver diseases: Its relationship with the aetiology and the entity of liver damage. *Digestive and Liver Disease*, Vol.42, No.3, (March 2010), pp. 200-204, ISSN 1590-8658

Chen, T.A., Tsao, Y.C., Chen, A., Lo, G.H., Lin, C.K., Yu, H.C., Cheng, L.C., Hsu, P.I. & Tsai, W.L. (2009) Effect of intravenous albumin on endotoxin removal., cytokines., and nitric oxide production in patients with cirrhosis and spontaneous bacterial peritonitis. *Scandinavian Journal of Gastroenterology*, Vol.44, No.5, pp. 619-625, ISSN 0036-5521

Cichoz-Lach, H., Celiński, K., Slomka, M. & Kasztelan-Szczerbińska, B. (2008) Pathophysiology of portal hypertension. *Journal of Physiology and Pharmacology Supplement*, Vol.59, Suppl.2, (August 2008), pp. 231-238. ISSN 0867-5910

Cybulsky, M.I., Fries, J.W., Williams, A.J., Sultan, P., Eddy, R., Byers, M., Shows, T., Gimbrone, M.A.Jr. & Collins, T. (1991) Gene structure., chromosomal location., and basis for alternative mRNA splicing of the human VCAM1 gene. *Proceedings of the National Academy of Sciences of the United States of America*, Vol.88, No.17, (September 1991), pp. 7859-7863, ISSN 0027-8424

Dagher, L. & Moore, K. (2001) The hepatorenal syndrome. *Gut*, Vol.49, No.5, (November 2001), pp. 729-737, ISSN 0017-5749

Dalle-Donne, I., Rossi, R., Giustarini, D., Milzani, A. & Colombo, R. (2003) Protein carbonyl groups as biomarkers of oxidative stress. *Clinica Chimica Acta*, Vol.329, No.1-2, (March 2003), pp. 23-38, ISSN 0009-8981

Esposito, C., Gerlach, H., Brett, J., Stern, D. & Vlassara, H. (1989) Endothelial receptor-mediated binding of glucose-modified albumin is associated with increased monolayer permeability and modulation of cell surface coagulant properties. *The Journal of Experimental Medicine*, Vol.170, No.4, (October 1989), pp. 1387-1407, ISSN 0022-1007

Fialová, L., Malbohan, I., Kalousová, M., Soukupová, J., Krofta, L., Stípek, S. & Zima, T. (2006) Oxidative stress and inflammation in pregnancy. *Scandinavian Journal of Clinical and Laboratory Investigation*, Vol.66, No.2, pp. 121-127, ISSN 0036-5513

Friedman, S.L. Mechanisms of hepatic fibrogenesis. (2008) *Gastroenterology*, Vol.134, No.6, (May 2008), pp. 1655-1669, ISSN 0016-5085

Garcia-Tsao, G., Angulo, P., Garcia, JC., Groszmann, RJ. & Cadelina, GW. (1998) The diagnostic and predictive value of ascites nitric oxide levels in patients with spontaneous bacterial peritonitis. *Hepatology*, Vol.28, No.1, (July 1998), pp. 17-21, ISSN 0270-9139

Garcovich, M., Zocco, M.A. & Gasbarrini, A. (2009) Clinical use of albumin in hepatology. *Blood Transfusion*, Vol.7, No.4, (October 2009), pp. 268-77, ISSN 1723-2007

Gatta, A., Bolognesi, M. & Merkel, C. (2008) Vasoactive factors and hemodynamic mechanisms in the pathophysiology of portal hypertension in cirrhosis. *Molecular Aspects of Medicine*, Vol.29, No.1-2, (February-April 2008), pp. 119-129, ISSN 0098-2997

Genesca, J., González, A., Mujal, A., Cereto, F. & Segura, R. (1999) Vascular endothelia growth factor levels in liver cirrhosis. *Digestive Diseases and Sciences*, Vol.44, No.6, (June 1999), pp. 1261-1262, ISSN 0163-2116

Giron-González, JA., Martínez-Sierra, C., Rodriguez-Ramos, C., Macías, M.A., Rendon, P., Díaz, F., Fernández-Gutiérrez, C. & Martín-Herrera, L. (2004) Implication of inflammation-related cytokines in the natural history of liver cirrhosis. *Liver International*, Vol.24, No.5, (October 2004), pp. 437-445, ISSN 1478-3223

Gorka, J., Zuwala-Jagiello, J., Pazgan-Simon, M., Simon, K. & Warwas, M. (2008) [Fluorescence of age in serum in detecting liver cirrhosis and hepatocellular carcinoma among patients with anti-HCV antibodies]. *Przeglad Epidemiologiczny*, Vol.62, No.2, pp. 393-400. Polish. ISSN 0033-210

Grangé, JD. & Amiot, X. (2004) Nitric oxide and renal function in cirrhotic patients with ascites, from physiopathology to practice. *European Journal of Gastroenterology & Hepatology*, Vol.16, No.6, (June 2004), pp. 567-570, ISSN 0954-691X

Guo, ZJ., Niu, H.X., Hou, F.F., Zhang, L., Fu, N., Nagai, R., Lu, X., Chen, B.H., Shan, Y.X., Tian, J.W., Nagaraj, R.H., Xie, D. & Zhang, X. (2008) Advanced oxidation protein products activate vascular endothelial cells via a RAGE-mediated signaling pathway. *Antioxidants & Redox Signaling*, Vol.10, No.10, (October 2008), pp. 1699-1712, ISSN 1523-0864

Halliwell, B. (2007) Oxidative stress and cancer: have we moved forward? *The Biochemical Journal*, Vol.401, No1, (January 2007), pp. 1-11, ISSN 0264-6021

Hanck, C., Glatzel, M., Singer, M.V. & Rossol, S. (2000) Gene expression of TNF-receptors in peripheral blood mononuclear cells of patients with alcoholic cirrhosis. *Journal of Hepatology*, Vol.32, No.1, (January 2000), pp. 51-57, ISSN 0168-8278

Huo, T.I., Lin, H.C., Wu, J.C., Lee, F.Y., Hou, M.C., Lee, P.C., Chang, F.Y. & Lee, S.D. (2006) Proposal of a modified Child-Turcotte-Pugh scoring system and comparison with the model for end-stage liver disease for outcome prediction in patients with cirrhosis. *Liver Transplantation*, Vol.12, No.1, (January 2006), pp. 65-71, ISSN 1527-6465

Hyogo, H. & Yamagishi, S. (2008) Advanced glycation end products (AGEs) and their involvement in liver disease. *Current Pharmaceutical Design*, Vol.14, No.10,(October 2008), pp. 969-972, ISSN 1381-6128

Iwakiri, Y. & Groszmann, R.J. (2007) Vascular endothelial dysfunction in cirrhosis. *Journal of Hepatology*, Vol.46, No.5, (May 2007), pp. 927-934, ISSN 0168-8278

Jain, S.K., Pemberton, P.W., Smith, A., McMahon, R.F., Burrows, P.C., Aboutwerat, A. & Warnes, T.W. (2002) Oxidative stress in chronic hepatitis C, not just a feature of late stage disease. *Journal of Hepatology*, Vol.36, No.6, (June 2002), pp. 805-811, ISSN 0168-8278

Kalousová, M., Sulková, S., Fialová, L., Soukupová, J., Malbohan, IM., Spacek, P., Braun, M., Mikulíková, L., Fortová, M., Horejsí, M., Tesar, V. & Zima, T. (2003) Glycoxidation and inflammation in chronic haemodialysis patients. *Nephrology, Dialysis, Transplantation*, Vol.18, No.12, (December 2003), pp. 2577-2581, ISSN 0931-0509

Kalousová, M., Zima, T., Tesar, V., Dusilová-Sulková, S. & Skrha, J. (2005) Advanced glycoxidation end products in chronic diseases-clinical chemistry and genetic background. *Mutation Research*, Vol.579, No.1-2, (November 2005), pp. 37-46, ISSN 0027-5107

Kim, I., Moon, S.O., Kim, S.H., Kim, H.J., Koh, Y.S. & Koh, G.Y. (2001) Vascular endothelial growth factor expression of intercellular adhesion molecule 1 (ICAM-1)., vascular cell adhesion molecule 1 (VCAM-1)., and E-selectin through nuclear factor-kappa B activation in endothelial cells. *The Journal of Biological Chemistry*, Vol.276, No.10, (March 2001), pp. 7614-7620, ISSN 0021-9258

Kirkham, P. (2007) Oxidative stress and macrophage function, a failure to resolve the inflammatory response. *Biochemical Society Transactions*, Vol.35, No.Pt2 (April 2007), pp. 284-287, ISSN 0300-5127

La Villa, G. & Gentilini, P. (2008) Hemodynamic alterations in liver cirrhosis. *Molecular Aspects of Medicine*, Vol.29, No.1-2, (February-April 2008), pp. 112-118, ISSN 0098-2997

Li, J.T., Hou, F.F., Guo, Z.J., Shan, Y.X., Zhang, X. & Liu, Z.Q. (2007) Advanced glycation end products upregulate C-reactive protein synthesis by human hepatocytes through stimulation of monocyte IL-6 and IL-1 beta production. *Scandinavian Journal of Immunology*, Vol.66, No.5, (November 2007), pp. 555-562, ISSN 0300-9475

Mitzner, S.R., Klammt, S., Peszynski, P., Hickstein, H., Korten, G., Stange, J. & Schmidt, R.
(2001) Improvement of multiple organ functions in hepatorenal syndrome during
albumin dialysis with the molecular adsorbent recirculating system. *Therapeutic
Apheresis*, Vol.5, No.5, (October 2001), pp. 417-422, ISSN 1091-6660

Miyata, T., Inagi, R., Iida, Y., Sato, M., Yamada, N., Oda, O., Maeda, K. & Seo, H. (1994)
Involvement of beta 2-microglobulin modified with advanced glycation end
products in the pathogenesis of hemodialysis-associated amyloidosis. Induction of
human monocyte chemotaxis and macrophage secretion of tumor necrosis factor-
alpha and interleukin-1. *The Journal of Clinical Investigation*, Vol.94, No.2, (February
1994), pp. 521-528, ISSN 0021-9738

Møller, S., Henriksen, J.H. & Bendtsen, F. (2008) Pathogenetic background for treatment of
ascites and hepatorenal syndrome. *Hepatology International*, Vol.2, No.4, (December
2008), pp. 416-428, ISSN 1936-0533

Nagata, K., Suzuki, H. & Sakaguchi, S. (2007) Common pathogenic mechanism in
development progression of liver injury caused by non-alcoholic or alcoholic
steatohepatitis. *The Journal of Toxicological Sciences*, Vol.32, No.5, (December 2007),
pp. 453-468, ISSN 0388-1350

Nakhjavani, M., Mashayekh, A., Khalilzadeh, O., Asgarani, F., Morteza, A., Omidi, M. &
Froutan, H. (2011) Oxidized low-density lipoprotein is associated with viral load
and disease activity in patients with chronic hepatitis C. *Clinics and Research in
Hepatology & Gastroenterology*, Vol.35, No.2, (February 2011), pp. 111-116, ISSN
2210-7401

Oettl, K. & Stauber, R.E. (2007) Physiological and pathological changes in the redox state of
human serum albumin critically influence its binding properties. *British Journal of
Pharmacology*, Vol.151, No.5, (July 2007), pp. 580-590, ISSN 0007-1188

Oettl, K., Stadlbauer, V., Petter, F., Greilberger, J., Putz-Bankuti, C., Hallström, S., Lackner,
C. & Stauber, R.E. (2008) Oxidative damage of albumin in advanced liver disease.
Biochimica et Biophysica Acta, Vol.1782, No.7-8, (July-August 2008), pp. 469-473, ISSN
0006-3002

Porcel, J.M. (2002) Unilateral pleural effusion secondary to brachiocephalic venous
thrombosis, a rare complication of central vein catheterization. *Respiration*, Vol.69,
No.6, pp. 569, ISSN 0025-7931

Raj, D.S., Choudhury, D., Welbourne, T.C. & Levi, M. (2000) Advanced glycation end
products, a Nephrologist's perspective. *American Journal of Kidney Diseases*, Vol.35,
No.3, (March 2000), pp. 365-380, ISSN 0272-6386

Reddy, S., Bichler, J., Wells-Knecht, K.J., Thorpe, S.R. & Baynes, J.W. (1995) Nepsilon-
(carboxymethyl)lysine is a dominant advanced glycation end product (AGE)
antigen in tissue proteins. *Biochemistry*, Vol.34, No.34, (August 1995), pp. 10872-
10878, ISSN 0006-2960

Rhee, S.G., Chang, T.S., Bae, Y.S., Lee, S.R. & Kang SW. (2003) Cellular regulation by
hydrogen peroxide. Journal of the American Society of Nephrology : JASN, Vol.14,
No.8 Suppl 3, (August 2003), pp. S211-S215, ISSN 1046-6673

Riordan, S.M., Skinner, N., Nagree, A., McCallum, H., McIver, C.J., Kurtovic, J., Hamilton,
J.A., Bengmark, S., Williams, R. & Visvanathan, K. (2003) Peripheral blood
mononuclear cell expression of toll-like receptors and relation to cytokine levels in
cirrhosis. *Hepatology*, Vol.37, No.5, (May 2003), pp. 1154-1164, ISSN 0270-9139

Saito, H. & Ishii, H. (2004) Recent understanding of immunological aspects in alcoholic hepatitis. *Hepatology Research*, Vol.30, No.4, (December 2004), pp. 193-198, ISSN 1386-6346

Schuppan, D. & Afdhal, N.H. (2008) Liver cirrhosis. *Lancet* Vol.371, No.9615, (March 2008), pp. 838-851, ISSN 0140-6736

Szmitko, P.E., Wang, C.H., Weisel, R.D., de Almeida, J.R., Anderson, T.J. & Verma S. (2003) New markers of inflammation and endothelial cell activation: Part I. *Circulation*, Vol.108, No.16, (October 2003), pp. 1917-1923. ISSN 0009-7322

Sebeková, K., Kupcová, V., Schinzel, R. & Heidland, A. (2002) Markedly elevated levels of plasma advanced glycation end products in patients with liver cirrhosis - amelioration by liver transplantation. *Journal of Hepatology*, Vol.36, No.1, (January 2002), pp. 66-71, ISSN 0168-8278

Serejo, F., Emerit, I., Filipe, P.M., Fernandes, A.C., Costa, M.A., Freitas, J.P. & de Moura, M.C. (2003) Oxidative stress in chronic hepatitis C, the effect of interferon therapy and correlation with pathological features. *Canadian Journal of Gastroenterology*, Vol.17, No.11, (November 2003), pp. 644-650, ISSN 0835-7900

Servettaz, A., Guilpain, P., Goulvestre, C., Chéreau, C., Hercend, C., Nicco, C., Guillevin, L., Weill, B., Mouthon, L. & Batteux, F. (2007) Radical oxygen species production induced by advanced oxidation protein products predicts clinical evolution and response to treatment in systemic sclerosis. *Annals of The Rheumatic Diseases*, Vol.66, No.9, (September 2007), pp. 1202-1209, ISSN 0003-4967

Soresi, M., Giannitrapani, L., D'Antona, F., Florena, A.M., La Spada, E., Terranova, A., Cervello, M., D'Alessandro, N. & Montalto, G. (2006) Interleukin-6 and its soluble receptor in patients with liver cirrhosis and hepatocellular carcinoma. *World journal of gastroenterology*, Vol.12, No.16, (April 2006), pp. 2563-2568, ISSN 1007-9327

Sparvero, L.J., Asafu-Adjei, D., Kang, R., Tang, D., Amin, N., Im, J., Rutledge, R., Lin, B., Amoscato, A.A., Zeh, H.J. & Lotze, M.T. (2009) RAGE (Receptor for Advanced Glycation Endproducts); RAGE ligands., and their role in cancer and inflammation. *Journal of Translational Medicine*, Vol.17, No.7 (March 2007), pp. 17, ISSN 1479-5876

Spitzer, J.A. (1994) Cytokine stimulation of nitric oxide formation and differential regulation in hepatocytes and nonparenchymal cells of endotoxemic rats. *Hepatology*, Vol.19, No.1, (January 1994), pp. 217-228, ISSN 0270-9139

Svistounov, D. & Smedsrød, B. (2004) Hepatic clearance of advanced glycation end products (AGEs)--myth or truth? *Journal of Hepatology*, Vol.41, No.6, (December 2004), pp. 1038-1040, ISSN 0168-8278

Thornalley, PJ., Battah, S., Ahmed, N., Karachalias, N., Agalou, S., Babaei-Jadidi, R. & Dawnay, A. (2003) Quantitative screening of advanced glycation endproducts in cellular and extracellular proteins by tandem mass spectrometry. *The Biochemical Journal*, Vol.375, No.Pt3, (November 2003), pp. 581-592, ISSN 0264-6021

Tilg, H., Wilmer, A., Vogel, W., Herold, M., Nölchen, B., Judmaier, G. & Huber, C. (1992) Serum levels of cytokines in chronic liver diseases. *Gastroenterology*, Vol.103, No.1, (July 1992), pp. 264-274, ISSN 0016-5085

Tsutsui, H., Adachi, K., Seki, E. & Nakanishi, K. (2003) Cytokine-induced inflammatory liver injuries. *Current Molecular Medicine*, Vol.3, No.6, (September 2003), pp. 545-559, ISSN 1566-5240

Valko, M., Leibfritz, D., Moncol, J., Cronin, M.T., Mazur, M. & Telser, J. (2007) Free radicals
and antioxidants in normal physiological functions and human disease. *The
International Journal of Biochemistry & Cell Biology*, Vol.39, No.1, pp. 44-84, ISSN
1357-2725

Wang, C.H., Li, S.H., Weisel, R.D., Fedak, P.W., Dumont, A.S., Szmitko, P., Li, R.K., Mickle,
D.A. & Verma, S. (2003) C-reactive protein upregulates angiotensin type 1 receptors
in vascular smooth muscle. *Circulation*, Vol.107, No.13, (April 2003), pp. 1783-1790,
ISSN 0009-7322

Wautier, M.P., Chappey, O., Corda, S., Stern, D.M., Schmidt, A.M. & Wautier, J.L. (2001)
Activation of NADPH oxidase by AGE links oxidant stress to altered gene
expression via RAGE. *American Journal of Physiology. Endocrinology and Metabolism*,
Vol.280, No.5, (May 2001), pp.E685-E694, ISSN 0193-1849

Wiest, R. & Groszmann, R.J. (2002) The paradox of nitric oxide in cirrhosis and portal
hypertension, too much: not enough. *Hepatology*, Vol.35, No.2, (February 2002), pp.
478-491, ISSN 0270-9139

Witko-Sarsat, V., Friedlander, M., Capeillère-Blandin, C., Nguyen-Khoa, T., Nguyen, A.T.,
Zingraff, J., Jungers, P. & Descamps-Latscha, B. (1996) Advanced oxidation protein
products as a novel marker of oxidative stress in uremia. *Kidney International*,
Vol.49, No.5, (May 1996), pp. 1304-1313, ISSN 0085-2538

Witko-Sarsat, V., Friedlander, M., Nguyen, Khoaj T., Capeillère-Blandin, C., Nguyen, A.T.,
Canteloup, S., Dayer, J.M., Jungers, P., Drüeke, T. & Descamps-Latscha, B. (1998)
Advanced oxidation protein products as novel mediators of inflammation and
monocyte activation in chronic renal failure. *Journal Immunology*, Vol.161, No.5,
(September 5), pp 2524-2532, ISSN 0022-1767

Witko-Sarsat, V., Gausson, V., Nguyen, A., Touam, M., Drüeke, T., Santangelo, F. &
Descamps-Latscha, B. (2003) AOPP-induced activation of human neutrophil and
monocyte oxidative metabolism, a potential target for N-acetylcysteine treatment in
dialysis patients. *Kidney International*, Vol.64, No.1, (July 2003), pp. 82-91, ISSN
0085-2538

Wright, E.Jr, Scism-Bacon, J.L. & Glass, L.C. (2006) Oxidative stress in type 2 diabetes: the
role of fasting and postprandial glycaemia. *International Journal of Clinical Practice*,
Vol.60, No.3, (March 2006), pp. 308-314. ISSN 1368-5031

Yan, S.F., Ramasamy, R. & Schmidt, A.M. (2010) The RAGE axis: a fundamental mechanism
signaling danger to the vulnerable vasculature. *Circulation Research*, Vol.106, No5, (
March 2010), pp. 842-853, ISSN 0009-7330

Yagmur, E., Tacke, F., Weiss, C., Lahme, B., Manns, MP., Kiefer, P., Trautwein, C. &
Gressner, A.M. (2006) Elevation of Nepsilon-(carboxymethyl)lysine-modified
advanced glycation end products in chronic liver disease is an indicator of liver
cirrhosis. *Clinical Biochemistry*, Vol.39, No.1, (Jananuary 2006), pp. 39-45, ISSN 0009-
9120

Yazici, C., Köse, K., Calis, M., Kuzugüden, S. & Kirnap, M. (2004) Protein oxidation status in
patients with ankylosing spondylitis. *Rheumatology (Oxford*, Vol.43, No.10, (October
2004), 1235-1239, ISSN 1462-0324

Zeni, F., Tardy, B., Vindimian, M., Comtet, C., Page, Y., Cusey, I. & Bertrand, J.C. (1993)
High levels of tumor necrosis factor-alpha and interleukin-6 in the ascitic fluid of

cirrhotic patients with spontaneous bacterial peritonitis. *Clinical Infectious Diseases*, Vol.17, No.2, (August 1993), pp. 218-223, ISSN 1058-4838

Zhang, W., Yue, B., Wang, G.Q. & Lu, S.L. (2002) Serum and ascites levels of macrophage migration inhibitory factor: TNF-alpha and IL-6 in patients with chronic virus hepatitis B and hepatitis cirrhosis. *Hepatobiliary & pancreatic diseases international*, Vol.1, No.4, (November 2002), pp 577-580, ISSN 1499-3872

Zuwala-Jagiello, J., Pazgan-Simon, M., Gorka., J., Simon, K., Milczarska, J. & Warwas, M. (2007) Serum advanced glycation end products and the development of hepatocellular carcinoma among HBV carriers. *Polish Journal of Environmental Studies*, Vol.16, No.5C, pp. 747-751, ISSN 1230-1485

Zuwala-Jagiello, J., Pazgan-Simon, M., Simon, K. & Warwas, M. (2009) Elevated advanced oxidation protein products levels in patients with liver cirrhosis. *Acta Biochimica Polonica* Vol.56(4), 679-85. ISSN, 0001-527X

Zuwala-Jagiello, J., Pazgan-Simon, M., Simon, K. & Warwas, M. (2011) Advanced oxidation protein products and inflammatory markers in liver cirrhosis, a comparison between alcohol-related and HCV-related cirrhosis. *Acta Biochimica Polonica*, Vol.58, No.1, pp. 59-65 ISSN 0001-527X

6

Oxidative Stress in Multiple Organ Damage in Hypertension, Diabetes and CKD, Mechanisms and New Therapeutic Possibilities

Tatsuo Shimosawa* et al.
Department of Clinical Laboratory, Faculty of Medicine, University of Tokyo
Japan

1. Introduction

Hypertension, diabetes, hypercholesterolemia and chronic kidney disease (CKD) lead to cardiovascular (CV) events and cardiovascular death consists of main cause in mortality of those diseases. Understanding of pathophysiology that links them and CV events has been vigorously studied and several factors are believed to play roles such as NO, renin-angiotensin system, and oxidative stress. It has been shown that those factors affect endothelial function and consequently organ circulation as well as function and viability of cells and organs. Despite overwhelming evidences in the consequences of experimental models of ROS-induced organ damage, large-scale clinical trials of former antioxidant therapies, such as vitamin C, vitamin E or β-carotene, could not demonstrate satisfactory benefit to patients and they seemed to be harmful in some cases (Hennekens et al., 1996; Omenn et al., 1996; Virtamo et al., 1998; Hercberg et al., 1999; Lee et al., 1999; Yusuf et al., 2000; de Gaetano, 2001; Heart, 2002; Vivekananthan et al., 2003; Hercberg et al., 2004; Kris-Etherton et al., 2004; Lonn et al., 2005). Several studies concluded that β-carotene supplementation increased the relative risk of death in patients with some types of cancer and had no benefit on patients with cardiovascular disease. Another study said vitamin E increased hemorrhagic stroke. Even antioxidant cocktails increased in all-cause motality (Omenn et al., 1996; Rosen et al., 2001a). So far, supplementation with vitamins C and E, either alone or in combination with each other or with other antioxidant vitamins, does not appear to be efficacious for the treatment of cardiovascular disease (Lonn et al., 2005). We investigated role of oxidative stress in consequences of multiple organ damages in mouse and possible new therapeutic agent.

---

* Tomoyo Kaneko[2], Xu Qingyou[1], Yusei Miyamoto[3], Mu Shengyu[2], Hong Wang[2], Sayoko Ogura[2], Rika Jimbo[2], Bohumil Majtan[2], Yuzaburo Uetake[2], Daigoro Hirohama[2], Fumiko Kawakami-Mori[2], Toshiro Fujita[2] and Yutaka Yatomi[1]
[1] *Department of Clinical Laboratory, Faculty of Medicine, University of Tokyo, Japan*
[2] *Department of Endocrinology and Nephrology, Faculty of Medicinem, Japan*
[3] *Department of Integrated Biosciences, Graduate School of Frontier Sciences, University of Tokyo, Japan*

2. Adrenomedullin

2.1 Cardiovascular effects

Adrenomedullin is a 52-amino-acid peptide and was originally isolated from pheochromocytoma cell but is also produced and secreted in endothelial cells and potent hypotensive peptide (Shimosawa & Fujita, 2005). We generated its deficient mice and proved that it is an endogenous antioxidants.

We examined the antioxidant properties of AM in both angiotensin II-salt (Shimosawa *et al.*, 2002) and pulmonary hypertension models (Matsui *et al.*, 2004). We showed that AM-knockout mice had higher oxidative stress through 3 ways. the urinary excretion of oxidative stress markers, such as 8-hydroxydeoxyguanosine (Fig. 1a) and isoprostane (Fig. 1b) by ELISA method, the immunostaining of 3-nitrotyrosine to localize oxidative stress (Fig. 1c), and real-time oxidant production measurement by the electron spin resonance (ESR) method (Fig. 1d). Pathological changes in the heart, such as periarterial fibrosis and coronary artery occlusion, were prominent in the knockout mice, and further investigation revealed that high oxidative stress caused coronary artery damage in this model.

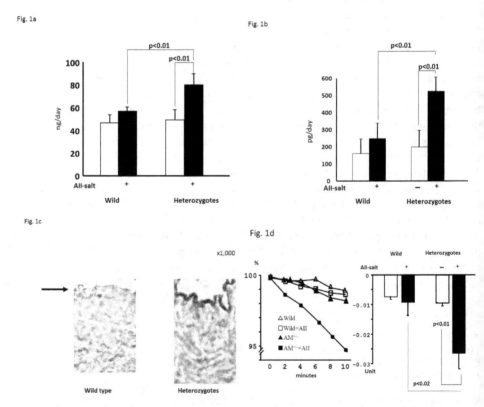

Fig. 1. Evaluation of oxidative stress in AM+/- mice after angiotensin II and salt loading.

We measured urinary excretion of 8-OHdG (a) and isoprostane (b) as footsteps of oxidative stress in whole body. As for confirming local oxidative stress, we immunostained coronary artery with 3-nitrotyrosine (c) that reflects NO-related oxidative stress and is a good marker for oxidative stress at the endothelial layer (arrow). For evaluation of real time production of oxidative stress, we measured ESR signal in the heart (d). On the left of panel d, time dependent of decay of ESR signal were plotted and on the right panel, the decay ratio of ESR signal that reflects production of oxidative stress was calculated. All these data suggest angiotensin II and salt-loaded $AM^{+/-}$ mice has higher oxidative stress than wild type mice. Modified from Shimosawa et al. 2002.

Hypoxia is known to cause pulmonary hypertension and pulmonary vascular remodeling. With this model, evidence is accumulating that ROS are the upstream signals of chronic hypoxia-induced pulmonary vasoconstriction and the development of vascular remodeling. It has been reported that ROS generated by hypoxia can induce calcium release from sarcoplasmic reticulum stores, which is followed by pulmonary vasoconstriction (Waypa et al., 2002). Moreover, ROS activated several growth factors such as vascular endothelial growth factor and platelet-activating factor (PAF) (Hartung et al., 1983; Ono et al., 1992; Chandel et al., 1998), which induce vascular remodeling. In the early phase of hypoxic conditions, ROS may act as a trigger of this signaling cascade in the process of vascular remodeling. In our experiments, 10% O_2 conditions caused higher ROS in AM-knockout mice compared with wild-type mice. Concomitant with increased oxidative stress in the pulmonary artery, AM-knockout mice showed marked vascular remodeling and higher mortality (Matsui et al., 2004), which were reversed by AM supplementation or by treatment with 4-Hydroxy-2,2,6,6-tetramethyl-piperidine-N-oxyl (hydroxyl-TEMPO), a mimetic of superoxide dismutase. Based on this basic research, translational research studies have been conducted in patients with pulmonary hypertension, and potent therapeutic effects of inhaling AM have been reported (Nagaya et al., 2004). As the mortality among patients with pulmonary hypertension is high even treatment with nitric oxide and prostaglandins, AM, or other agents may be novel and promising agents for the treatment of these patients.

2.2 Renal effects and possible role in CKD and diabetic kidney disease

Renal dysfunction or chronic kidney disease (CKD) draws high interests among nephrologists as well as cardiologists and other specialities. There are several factors implicated to connect renal dysfunction and cardiovascular events, endothelial dysfunction, renin-angiotensin-aldosterone axis, oxidative stress, and unknown toxic agents. In order to investigate if renin-angiotensin axis and oxidative stress can be therapeutical target in CKD, we used AM knockout mice. We found increased oxidative stress and local renin-angiotensin system in the ureter-obstructed kidney of AM knockout mice. Concordance with the increase of oxidative stress and renin-angiotensin system, severe interstitial fibrosis, and cell proliferation were prominent in AM-knockout mice. The interstitial fibrosis can be partly reversed by hydroxyl-TEMPO, angiotensin receptor blockers, or the systemic replacement of AM (unpublished observation). When dual treatment with angiotensin receptor blocker and hydroxyl-TEMPO were given, pathological changes are more effectively reversed compared with that by each agent alone. Angiotensin II is known to induce oxidative stress via activating NADPH oxidase and AM can block this pathway via c-Src (Liu et al., 2007). In the kidney, oxidative stress is generated not only by NADPH oxidase but from mitochondria and xanthine oxidase and thus AM or angiotensin II blocker alone therapy is not sufficient and together with hydroxyl-TEMPO that can totally block oxidative stress is effective.

Multiple lines of study have shown that diabetic patients have increased oxidative stress and the resultant organ damage (Rosen et al., 2001b). In turn, it is hypothesized that oxidative stress can induce diabetes by series of studies; oxidative stress impairs insulin internalization (Bertelsen et al., 2001), blocks insulin receptor substrate phosphorylation, impairs phosphoinositide-3 kinase activity (Najib & Sanchez-Margalet, 2001), induces protein glycation and as a consequence of advanced glycation end-products-receptor binding that leads to cytotoxicity in pancreatic beta cells and reduces the translocation of glucose transporter type-4 (Rudich et al., 1997; Rudich et al., 1998). An in vivo study also showed that the administration of oxidative stress aggravated diabetes in diabetes-prone obese-Zucker rats (Laight et al., 2000). Antioxidan supplementation studies have shown conflicting resuts; some reported beneficial effects in endothelial function, retinal blood flow and renal function outcomes (Blum et al.; Ziegler et al., 2004; Lopes de Jesus et al., 2008), in contrast, the recent metaanalysis of antioxidants, vitamin C and E, supplementation trials revealed ineffectiveness in glycemic control (Akbar et al., 2011). On the other hand, in clinical settings, diabetics have higher oxidative stress and AM levels (Hayashi et al., 1997), which led us to assume that AM is upregulated in order to antagonize oxidative stress. In fact, aged or angiotensin II-treated AM-knockout mice showed insulin resistance, and this was reversed by AM supplementation (Shimosawa et al., 2003; Xing et al., 2004) (Fig. 2). In this experiment, we showed that oxidative stress directly impaired insulin signaling by interfering with insulin receptor substrate 1 and 2 phosphorylations. By in vitro experiments, we found that oxidative stress not only impairs insulin signaling, but also reduces glucose transporter 4 transcription (unpublished observation).

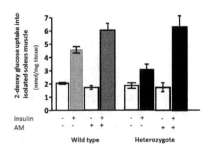

Fig. 2. Insulin resistance in skeletal muscle of aged AM[+/-] mice which was recovered by AM supplement.

Insulin resistance was measured by ex vivo experiments. Isolated soleus muscle was incubated with insulin and ^{12}C-deoxyglucose. Deoxyglucose uptake was evaluated as insulin sensitivity. Aged male AM[+/-] mice (55 weeks old) showed insulin resistance and AM supplementation for a month successfully reversed insulin resistant state. Modified from Shimosawa et al. 2003.

Recent clinical trial in diabetic CKD patients revealed bardoxolone methyl effectively improved in the estimated GFR independent from blood pressure which is the most important risk in CKD (Pergola *et al.*, 2011). Also, it improved insulin resistance by increasing glucose uptake in skeletal muscle (Saha *et al.*), which is compatible with our findings in AM knockout mice as described above. Bardoxolone methyl is an oral antioxidant inflammation modulator and activates the Keap1–Nrf2 pathway. Keap1-Nrf2 pathway regulates inflammation and oxidative stress (Dinkova-Kostova *et al.*, 2005) via increased expression of heme oxygenase 1 (Wu *et al.*, 2011). This compound is so far the only clinically available antioxidants that show organ protective effect, however, it requires further assessment if bardoxolone methyl can prevent hard end-point such as cardiovascular events or mortality in CKD patients.

3. Platinum nanoparticle

In diabetic-CKD, possible clinical usefulness of bardoxolone methyl was shown as mentioned above, so far few antioxidant agents are proved to be effective in cardiovascular protection. Although in pulmonary hypertension AM showed its clinical usefulness by inhaling, AM is a peptide and its clinical use is limited due to its short half life. We next investigated the possible therapeutic effect of platinum-nano-particle.

In recent years, a few reports about new drug delivery system using nanotechnology have been released (Muro *et al.*, 2006; Kajita *et al.*, 2007; Shimizu et al., 2010). Among them platinum nanoparticles as both superoxide dismutase (SOD) mimetic and catalase mimetic (Kajita *et al.*, 2007; Watanabe *et al.*, 2009). In vivo experiments shows its effectiveness in smoking-induced lung damage (Onizawa *et al.*, 2009) or stroke model mice (Takamiya et al.*et al.*, 2011). Platinum nanoparticles (PAA-Pt) was administered intranasally, which were then exposed to cigarette smoking for 3 days. Cigarette smoking induced NFkappaB activation, and neutrophilic inflammation in the lungs of mice, and intranasal administration of PAA-Pt inhibited these changes. Moreover in in vitro experiments, treatment of alveolar-type-II-like A549 cells with PAA-Pt inhibited cell death after exposure to a cigarette smoke extract. Transient middle cerebral artery occlusion (tMCAO) and reperfusion model was used in this study. PAA-Pt was administered intravenously. PAA-Pt dramatically reduced oxidative stress in the brain and significantly improved the motor function and greatly reduced the infarct volume, especially in the cerebral cortex with preserved collagen IV and a remarkable suppression of MMP-9. By these acute models the antioxidant effect and concomitant organ protection by PAA-Pt were established. We investigated the chronic effect of PAA-Pt and studied if PAA-Pt reverse cardiovascular damage in metabolic syndrome model by eliminating ROS.

Metabolic syndrome model was established by angiotensin II and high salt diet on diabetic 8-week-old male db/db mice. 10 µM platinum nanoparticles were given orally. The oxidative stress was measured in two ways. First, the urinary excretion of isoprostane reflects the state of systemic oxidative stress. Ang II and salt loading increased isoprostane excretion in db/db mice. PAA-Pt reduced isoprostane excretion, suggesting that PAA-Pt would be an effective antioxidant (Fig 3). Blood pressure was elevated by angiotensin II and high salt diet and PAA-Pt treatment did not have any effects on blood pressure and other metabolic biomarkers of mice.

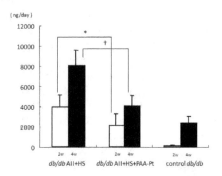

Fig. 3. PAA-Pt effect on oxidative stress

Db/db mice were treated with angiotensin II and high salt diet and urinary isoprostane was measured as a marker of oxidative stress. PAA-Pt effectively reduced oxidative stress.

Consistent with AM knockout mice, angiotensin II and high salt diet induced pericoronary fibrosis in db/db mice (Fig. 4a). The degree of deterioration of vascular damage was much severer in the db/db mice than wild type mice. PAA-Pt therapy reversed the vascular damage in angiotensin II/salt loaded db/db mice. The area of fibrous changes was calculated for each group (Fig 4b). The results of RT-PCR supported the aforesaid

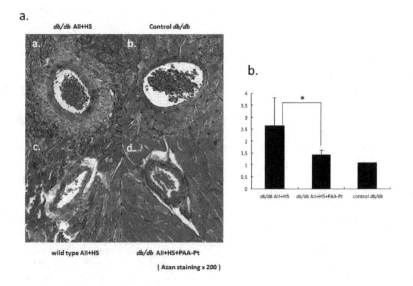

Fig. 4. Histopathological changes of db/db mice by angiotensin II/high salt diet and effect of PAA-Pt.

histopathological findings. We examined mRNA expression of ANP and procollagen type Iα in the heart as molecular markers of cardiac damage and we found that those molecular markers were apparently up-regulated by angiotensin II and salt loading (Fig 5). PAA-Pt reduced these markers expressions to the almost same extent with the control db/db mice and these results support our findings in the histopathological examinations.

Pericoronary fibrosis were evaluated by Azan staining. Fibrosis was robust by angiotensin II and high salt diet and was reversed by PAA-Pt.

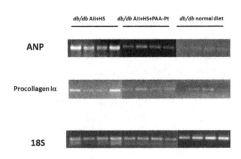

Fig. 5. Molecular markers for cardiac damage and fibrosis.

mRNA expression of ANP and procollagen type I were evaluated. Concomitant with figure 4, angiotensin II and high salt diet aggravated cardiac damage and it was reversed by PAA-Pt.

PAA-Pt behaves as an SOD mimetic and possesses a catalase activity, then its efficacy can last longer than vitamin C (Kajita et al., 2007). It is mainly because the catalytic property of PAA-Pt which is distinguished from vitamin C that is quickly consumed. Therefore we examined the effect of PAA-Pt in metabolic syndrome model mice to reduce oxidative stress and protect the cardiovascular system. Concordance with in vitro study, PAA-Pt effectually reduced ROS and prevented cardiovascular damage in hypertensive db/db mice.

We found the extension of cardiovascular damages was much severer and the level of systemic oxidative stress, which was evaluated by urinary excretion of isoprostane, was much higher in db/db mice. Then, it is indicated that the 'metabolic-syndrome-status' is equal to 'high-oxidative-stress-status' and this status seems to be the trigger of the vicious cycle of organ damage. Thus, the scavenging ROS would be a promising approach to preventing organ damage in metabolic syndrome.

PAA-Pt also has the possibility to solve the problems of former antioxidants as vitamin C, E or β-carotene. It is said the one of the reasons why those former antioxidants would have failed to be effective against organ damage that they are short-acting and could not eliminate intracellular ROS, including mitochondrial ROS. PAA-Pt possesses long-lasting activity (means catalytic effect) and dual properties as SOD and catalase mimetics.   The pharmacokinetics of PAA-Pt in detail have not been revealed yet that PAA-Pt should enter to systemic circulation from digestive apparatus, and be excreted from urine. Also we suggestively expect the novel antioxidant PAA-Pt enables to eliminate intracellular ROS. Further studies in pharmacokinetics and safety of PAA-Pt in chronic use are required. Specially, PAA-Pt contains platinum, which is sometimes toxic to the kidney, we must be careful about its nephrotoxicity before clinical applications. In site of some assignments such as its safety or full internal dynamism must be proven before clinical use, this novel agent will be certain way to open the door to interdisciplinary treatments between nanotechnology and medicine.

4. Perspectives

Piles of evidence accumulated that oxidative stress is a new therapeutic targets in preventing cardiovascular events and its risks. Among risks, hypertension, glucose metabolism and lipid accumulation in the vasculature are closely related with oxidative stress, however, some but not all antioxidant lower blood pressure or prevent atherosclerosis. Blood pressure is regulated by both circulating blood volume and vascular resistance and most of the data suggest that oxidative stress may affect vascular resistance but not blood volume, which can explain antioxidant therapy is not consistent in lowering blood pressure. For lipid disorders and consecutive atherosclerosis, oxidized LDL as well as its receptor is important. In mice LDL level are almost negligible however, oxidized LDL receptor such as LOX-1 overexpression leads to vascular damage (Inoue et al., 2005). It suggests that ligand regulation by targeting oxidative stress is not sufficient but we should consider receptor regulation as well. On the other hand, insulin resistance or glucose metabolism can be a good target of antioxidant therapy. Insulin signaling is impaired by oxidative stress and it is reversed by antioxidants. Our and others' basic trials as well as recent data by bardoxolone methyl are promising, although the optimal glucose regulation level to prevent cardiovascular complications in diabetes are still under debate. The 'lower the better' of blood glucose is controversial because of hypoglycemia induces poor prognosis in cardiovascular event, however, targeting oxidative stress recovers impaired-insulin resistance and does not stimulate insulin release, therefore, it would be safer than insulin replacement or insulin secretion stimulating agents such as sulphonylurea. Further clinical trials are required.

5. Conclusion

AM, PAA-PT or bardoxolone are possibly promising antioxidants in treating or reducing risks for chronic metabolic disease and organ damages. Other possible agents are also under investigations and novel technology will lead us to invent antioxidants with more efficacy and less side effect. Oxidative stress is indispensable for host defence and appropriate reduction of oxidative stress is required when considering antioxidants in clinical use.

6. References

Akbar S, Bellary S & Griffiths HR. (2011). Dietary Antioxidant Interventions in Type 2 Diabetes Patients A Meta-analysis. *Br J Diabetes Vasc Dis* 11, 62-68.

Bertelsen M, Anggard EE & Carrier MJ. (2001). Oxidative stress impairs insulin internalization in endothelial cells in vitro. *Diabetologia* 44, 605-613.

Blum S, Vardi M, Brown JB, Russell A, Milman U, Shapira C, Levy NS, Miller-Lotan R, Asleh R & Levy AP. Vitamin E reduces cardiovascular disease in individuals with diabetes mellitus and the haptoglobin 2-2 genotype. *Pharmacogenomics* 11, 675-684.

Chandel NS, Maltepe E, Goldwasser E, Mathieu CE, Simon MC & Schumacker PT. (1998). Mitochondrial reactive oxygen species trigger hypoxia-induced transcription. *Proc Natl Acad Sci U S A* 95, 11715-11720.

de Gaetano G. (2001). Low-dose aspirin and vitamin E in people at cardiovascular risk: a randomised trial in general practice. Collaborative Group of the Primary Prevention Project. *Lancet* 357, 89-95.

Dinkova-Kostova AT, Liby KT, Stephenson KK, Holtzclaw WD, Gao X, Suh N, Williams C, Risingsong R, Honda T, Gribble GW, Sporn MB & Talalay P. (2005). Extremely potent triterpenoid inducers of the phase 2 response: correlations of protection against oxidant and inflammatory stress. *Proc Natl Acad Sci U S A* 102, 4584-4589.

Hartung HP, Parnham MJ, Winkelmann J, Englberger W & Hadding U. (1983). Platelet activating factor (PAF) induces the oxidative burst in macrophages. *Int J Immunopharmacol* 5, 115-121.

Hayashi M, Shimosawa T, Isaka M, Yamada S, Fujita R & Fujita T. (1997). Plasma adrenomedullin in diabetes. *Lancet* 350, 1449-1450.

Heart P, Study, Collaborative, Group. (2002). MRC/BHF Heart Protection Study of antioxidant vitamin supplementation in 20,536 high-risk individuals: a randomised placebo-controlled trial. *Lancet* 360, 23-33.

Hennekens CH, Buring JE, Manson JE, Stampfer M, Rosner B, Cook NR, Belanger C, LaMotte F, Gaziano JM, Ridker PM, Willett W & Peto R. (1996). Lack of effect of long-term supplementation with beta carotene on the incidence of malignant neoplasms and cardiovascular disease. *N Engl J Med* 334, 1145-1149.

Hercberg S, Galan P, Preziosi P, Bertrais S, Mennen L, Malvy D, Roussel AM, Favier A & Briancon S. (2004). The SU.VI.MAX Study: a randomized, placebo-controlled trial of the health effects of antioxidant vitamins and minerals. *Arch Intern Med* 164, 2335-2342.

Hercberg S, Preziosi P, Galan P, Faure H, Arnaud J, Duport N, Malvy D, Roussel AM, Briancon S & Favier A. (1999). "The SU.VI.MAX Study": a primary prevention trial using nutritional doses of antioxidant vitamins and minerals in cardiovascular diseases and cancers. SUpplementation on VItamines et Mineraux AntioXydants. *Food Chem Toxicol* 37, 925-930.

Inoue K, Arai Y, Kurihara H, Kita T & Sawamura T. (2005). Overexpression of lectin-like oxidized low-density lipoprotein receptor-1 induces intramyocardial vasculopathy in apolipoprotein E-null mice. *Circ Res* 97, 176-184.

Kajita M, Hikosaka K, Iitsuka M, Kanayama A, Toshima N & Miyamoto Y. (2007). Platinum nanoparticle is a useful scavenger of superoxide anion and hydrogen peroxide. *Free Radic Res* 41, 615-626.

Kris-Etherton PM, Lichtenstein AH, Howard BV, Steinberg D & Witztum JL. (2004). Antioxidant vitamin supplements and cardiovascular disease. *Circulation* 110, 637-641.

Laight DW, Desai KM, Anggard EE & Carrier MJ. (2000). Endothelial dysfunction accompanies a pro-oxidant, pro-diabetic challenge in the insulin resistant, obese Zucker rat in vivo. *Eur J Pharmacol* 402, 95-99.

Lee IM, Cook NR, Manson JE, Buring JE & Hennekens CH. (1999). Beta-carotene supplementation and incidence of cancer and cardiovascular disease: the Women's Health Study. *J Natl Cancer Inst* 91, 2102-2106.

Liu J, Shimosawa T, Matsui H, Meng F, Supowit SC, Dipette DJ, Ando K & Fujita T. (2007). Adrenomedullin Inhibits Angiotensin II-Induced Oxidative Stress Via Csk-Mediated Inhibition of Src Activity. *Am J Physiol Heart Circ Physiol* 292, H1714-H1721.

Lonn E, Bosch J, Yusuf S, Sheridan P, Pogue J, Arnold JM, Ross C, Arnold A, Sleight P, Probstfield J & Dagenais GR. (2005). Effects of long-term vitamin E supplementation on cardiovascular events and cancer: a randomized controlled trial. *Jama* 293, 1338-1347.

Lopes de Jesus CC, Atallah AN, Valente O & Moca Trevisani VF. (2008). Vitamin C and superoxide dismutase (SOD) for diabetic retinopathy. *Cochrane Database Syst Rev*, CD006695.

Matsui H, Shimosawa T, Itakura K, Guanqun X, Ando K & Fujita T. (2004). Adrenomedullin Can Protect Against Pulmonary Vascular Remodeling Induced by Hypoxia. *Circulation* 109, 2246-2251.

Muro S, Mateescu M, Gajewski C, Robinson M, Muzykantov VR & Koval M. (2006). Control of intracellular trafficking of ICAM-1-targeted nanocarriers by endothelial Na+/H+ exchanger proteins. *Am J Physiol Lung Cell Mol Physiol* 290, L809-817.

Nagaya N, Kyotani S, Uematsu M, Ueno K, Oya H, Nakanishi N, Shirai M, Mori H, Miyatake K & Kangawa K. (2004). Effects of adrenomedullin inhalation on hemodynamics and exercise capacity in patients with idiopathic pulmonary arterial hypertension. *Circulation* 109, 351-356.

Najib S & Sanchez-Margalet V. (2001). Homocysteine thiolactone inhibits insulin signaling, and glutathione has a protective effect. *J Mol Endocrinol* 27, 85-91.

Omenn GS, Goodman GE, Thornquist MD, Balmes J, Cullen MR, Glass A, Keogh JP, Meyskens FL, Valanis B, Williams JH, Barnhart S & Hammar S. (1996). Effects of a combination of beta carotene and vitamin A on lung cancer and cardiovascular disease. *N Engl J Med* 334, 1150-1155.

Onizawa S, Aoshiba K, Kajita M, Miyamoto Y & Nagai A. (2009). Platinum nanoparticle antioxidants inhibit pulmonary inflammation in mice exposed to cigarette smoke. *Pulm Pharmacol Ther* 22, 340-349.

Ono S, Westcott JY & Voelkel NF. (1992). PAF antagonists inhibit pulmonary vascular remodeling induced by hypobaric hypoxia in rats. *J Appl Physiol* 73, 1084-1092.

Pergola PE, Raskin P, Toto RD, Meyer CJ, Huff JW, Grossman EB, Krauth M, Ruiz S, Audhya P, Christ-Schmidt H, Wittes J & Warnock DG. (2011). Bardoxolone methyl and kidney function in CKD with type 2 diabetes. *N Engl J Med* 365, 327-336.

Rosen P, Nawroth PP, King G, Moller W, Tritschler HJ & Packer L. (2001a). The role of oxidative stress in the onset and progression of diabetes and its complications: a summary of a Congress Series sponsored by UNESCO-MCBN, the American Diabetes Association and the German Diabetes Society. *Diabetes Metab Res Rev* 17, 189-212.

Rosen P, Nawroth PP, King G, Moller W, Tritschler HJ & Packer L. (2001b). The role of oxidative stress in the onset and progression of diabetes and its complications: a summary of a Congress Series sponsored by UNESCO-MCBN, the American Diabetes Association and the German Diabetes Society. *Diabetes Metab Res Rev* 17, 189-212.

Rudich A, Kozlovsky N, Potashnik R & Bashan N. (1997). Oxidant stress reduces insulin responsiveness in 3T3-L1 adipocytes. *Am J Physiol* 272, E935-940.

Rudich A, Tirosh A, Potashnik R, Hemi R, Kanety H & Bashan N. (1998). Prolonged oxidative stress impairs insulin-induced GLUT4 translocation in 3T3-L1 adipocytes. *Diabetes* 47, 1562-1569.

Saha PK, Reddy VT, Konopleva M, Andreeff M & Chan L. (2010). The triterpenoid 2-cyano-3,12-dioxooleana-1,9-dien-28-oic-acid methyl ester has potent anti-diabetic effects in diet-induced diabetic mice and Lepr(db/db) mice. *J Biol Chem* 285, 40581-40592.

Shimizu H, Hori Y, Kaname S, Yamada K, Nishiyama N, Matsumoto S, Miyata K, Oba M, Yamada A, Kataoka K & Fujita T. (2010). siRNA-based therapy ameliorates glomerulonephritis. *J Am Soc Nephrol* 21, 622-633.

Shimosawa T & Fujita T. (2005). Adrenomedullin and its related peptide. *Endocr J* 52, 1-10.

Shimosawa T, Ogihara T, Matsui H, Asano T, Ando K & Fujita T. (2003). Deficiency of adrenomedullin induces insulin resistance by increasing oxidative stress. *Hypertension* 41, 1080-1085.

Shimosawa T, Shibagaki Y, Ishibashi K, Kitamura K, Kangawa K, Kato S, Ando K & Fujita T. (2002). Adrenomedullin, an Endogenous Peptide, Counteracts Cardiovascular Damage. *Circulation* 105, 106-111.

Takamiya M, Miyamoto Y, Yamashita T, Deguchi K, Ohta Y, Ikeda Y, Matsuura T & Abe K. (2011). Neurological and pathological improvements of cerebral infarction in mice with platinum nanoparticles. *J Neurosci Res* 89, 1125-1133.

Virtamo J, Rapola JM, Ripatti S, Heinonen OP, Taylor PR, Albanes D & Huttunen JK. (1998). Effect of vitamin E and beta carotene on the incidence of primary nonfatal myocardial infarction and fatal coronary heart disease. *Arch Intern Med* 158, 668-675.

Vivekananthan DP, Penn MS, Sapp SK, Hsu A & Topol EJ. (2003). Use of antioxidant vitamins for the prevention of cardiovascular disease: meta-analysis of randomised trials. *Lancet* 361, 2017-2023.

Watanabe A, Kajita M, Kim J, Kanayama A, Takahashi K, Mashino T & Miyamoto Y. (2009). In vitro free radical scavenging activity of platinum nanoparticles. *Nanotechnology* 20, 455105.

Waypa GB, Marks JD, Mack MM, Boriboun C, Mungai PT & Schumacker PT. (2002). Mitochondrial reactive oxygen species trigger calcium increases during hypoxia in pulmonary arterial myocytes. *Circ Res* 91, 719-726.

Wu QQ, Wang Y, Senitko M, Meyer C, Wigley WC, Ferguson DA, Grossman E, Chen J, Zhou XJ, Hartono J, Winterberg P, Chen B, Agarwal A & Lu CY. (2011). Bardoxolone methyl (BARD) ameliorates ischemic AKI and increases expression of protective genes Nrf2, PPARgamma, and HO-1. *Am J Physiol Renal Physiol* 300, F1180-1192.

Xing G, Shimosawa T, Ogihara T, Matsui H, Itakura K, Qingyou X, Asano T, Ando K & Fujita T. (2004). Angiotensin II-induced insulin resistance is enhanced in adrenomedullin-deficient mice. *Endocrinology* 145, 3647-3651. Epub 2004 Apr 3622.

Yusuf S, Dagenais G, Pogue J, Bosch J & Sleight P. (2000). Vitamin E supplementation and cardiovascular events in high-risk patients. The Heart Outcomes Prevention Evaluation Study Investigators. *N Engl J Med* 342, 154-160.

Ziegler D, Nowak H, Kempler P, Vargha P & Low PA. (2004). Treatment of symptomatic diabetic polyneuropathy with the antioxidant alpha-lipoic acid: a meta-analysis. *Diabet Med* 21, 114-121.

7

The Relationship Between Thyroid States, Oxidative Stress and Cellular Damage

Cano-Europa, Blas-Valdivia Vanessa,
Franco-Colin Margarita and Ortiz-Butron Rocio
*Escuela Nacional de Ciencias
Biológicas del Instituto Politécnico Nacional,
México*

1. Introduction

The thyroid hormones play an important role in many physiological processes, such as differentiation, growth, development, and the physiology of all cells. One of the most studied effects of the thyroid hormone is the control of the basal metabolic rate. Modifications in its levels can produce several alterations including modifications in the ROS steady-state and the REDOX environment in the cells. There is much evidence that show both hyperthyroidism and hypothyroidism are related to oxidative stress and cellular damage. For hypothyroidism, there are other findings that point to its protective effects. In this chapter we show both findings and propose that hypothyroidism is a protective state against toxic agents.

2. Thyroid hormones

Thyroid hormones (THs) T_4 (thyroxine or 3′,5′,3,5-L-tetra-iodothyronine) and T_3 (3′,3,5-triiodothyronine) are synthesized in the thyroid gland located in the anterior part of the trachea, just below the larynx. It consists of two lobes joined in the middle by a narrow portion of the gland. The major thyroid-secretor cells, known as follicular cells, are arranged into hollow spheres, each of which it forms a functional unit called a follicle. On a microscopic section, rings of follicular cells enclosing an inner lumen filled with colloid form the follicles (figure 1).

The principal constituent of the colloid is a large protein molecule, thyroglobulin, where thyroid hormones are incorporated in their various stages of synthesis. The follicular cells produce the two iodine-containing hormones derived from the amino acid tyrosine; T_4 and T_3, the thyroid hormones. The mechanism involved in thyroid hormone syntheses and their release from thyrolobulin are shown in figure 2. Iodine, an essential element of the thyroid molecule, is actively transported by the Na^+-I^- symporter (NIS, encoded by the *SLC5A5* gene) at the basolateral membrane of the thyrocyte and it diffuses by an exchanger, known as pendrin (PDS, encoded by the *SLC26A4* gene) to the lumen at the apical membrane. At the extracellular apical membrane, thyroperoxidase (TPO, EC 1.11.1.8) with hydrogen peroxide (H_2O_2), generated by dual oxidase 2 (DUOX2, EC 1.6.3.1), oxidizes and binds iodine

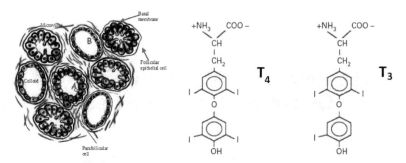

Fig. 1. Histological thyroid and the thyroid hormone structure. There are thyroid follicles with different activity. A: Thyroid follicle resting. B: Follicle with high activity.

covalently to tyrosyl residues, producing monoiodotyrosine (MIT) and diiodotyrosine (DIT) within the thyrogloblin macromolecule. The enzyme thyroperoxidase catalyzes the coupling of two iodotyrosine residues to produce the prohormone T_4 and smaller amounts of the active hormone T_3. After endocytosis, iodinated thyroglobulin is hydrolyzed in the lysosomes by cathepsins and the thyroid hormone is released from the thyroglobulin backbone. The released MIT and DIT are deiodinated by a specific iodotyrosine deiodinase (IYD, or DEHAL1, EC 1.22.1.1), and the released iodine is recycled within the cell. The mechanism involved in the last step in the process, the thyroid hormone secretion, remains unknown (Di Cosmo et al., 2010). About 90% of the secretory product released from the thyroid gland is in the form of T_4 though T_3 is about four times more potent in its biologic activity. Most of the secreted T_4 is converted into T_3 by a group of enzymes known as iodothyronine deiodinases (D1 and D2, EC 1.97.1.10), which also include an inactivating deiodinase, the type 3 deiodinase (D3), that inactivates both T_4 and T_3 (Bianco et al., 2002)(table 1).

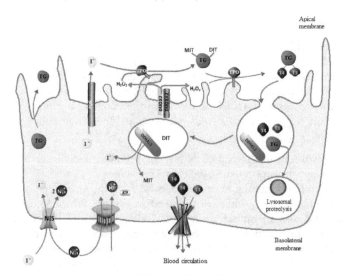

Fig. 2. Thyroid hormone syntheses in a follicular thyroid cell. Based on the model proposed by Di Cosmo (Di Cosmo et al., 2010)

The role played by the deiodinases is physiologically relevant. They have a role in various aspects of mammalian physiology, such as the maintenance of plasma T_3 concentration (Bianco et al., 2002), TSH and TRH feedback regulation (Christoffolete et al., 2006; Larsen, 1982), and the clearance of sulfated iodothyronines (Schneider et al., 2006).

Deiodinase	Tissue
D1	Liver, thyroid, kidney
D2	Brain, pituitary gland, BAT, thyroid, muscle
D3	Developing tissues and placenta, adult skin, brain

Table 1. Tissue distribution of deiodinases in humans (St.Germain et al., 2009).

2.1 Thyroid hormone regulation levels

Because the thyroid hormones have a crucial role in the function of every tissue in the body, their levels must be maintained relatively constant around an optimum level. This homeostatic-control mechanism primarily operates on the principle of negative feedback (figure 3). In this homeostatic-control mechanism, the hypothalamic thyrotropin-releasing hormone (TRH), the thyroid-stimulating hormone (TSH), and thyroid hormone all together form the hypothalamus-pituitary-thyroid axis. Thus, TRH in trophic fashion turns on the TSH secretion by the anterior pituitary, whereas thyroid hormone, in negative feedback fashion, turns off the TSH secretion. In the hypothalamus-pituitary-thyroid axis, inhibition is exerted primarily at the level of the anterior pituitary. As with other negative feedback loops, the one between thyroid hormone and TSH tends to maintain a stable thyroid hormone output (Hulbert, 2000).

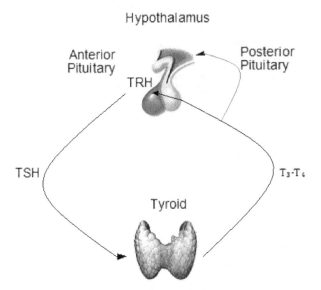

Fig. 3. Hypothalamus-pituitary-thyroid axes. In this image is shown the negative feedback exerted by a high T_3 concentration.

2.2 Mechanism of action of the thyroid hormones

Because thyroid hormones have been considered as lipophilic in their intracellular action, passive hormone diffusion through the lipid bilayer has been accepted. Because of the existence of nuclear receptors for thyroid hormones, it has long been believed that the THs caused effects only via genomic effects, however since the late 1980s it has been proposed that the THs may cause effects independently of genetic mechanisms, affecting membrane lipid composition or activation of enzymes (Lazar, 1993). Receptors for thyroid hormones (TRs) are proteins that act as transcription factors that belong to the superfamily of nuclear receptors, which includes steroids, vitamin D, retinoic acid, fatty acids, prostaglandins, and orphan receptors (Zhang & Lazar, 2000). The TRs have regions, the DNA binding domain (DBD), ligand binding domain (LBD), hinge region (HR), and amino terminal domain (A-B) (Sap et al., 1986). The THs cross the lipid membrane because of their hydrophobic nature. In the cytoplasm they can bind to newly synthesized TRs, however most THs bind to nuclear receptors. The TRs bound to the THs may regulate the transcription process by modifying the structure of chromatin, allowing other factors to exert their action on elements of the TH responses (figure 4). In addition, the THs interact, directly or indirectly, through bridge or coactivator molecules, with the transcriptional machinery of the process. A critical aspect in regulating the transcription process by the TRs is the conformational change that T_3 exerts on the receiver itself. The T_3 decreases the ability of the hydrophobic TRs and can modify the way in which the TRs, either as dimer or heterodimer, bind to DNA. The TRs bind to regulate transcription as a monomer, homodimer, heterodimer, or heteromultimer. The thyroid hormones binding to the TRs cause changes in the structure of these complexes thus modulating the interaction with other elements of the transcriptional apparatus to determine the type of response, enhancing or inhibiting (Cheng et al., 2010; Yen, 2001).

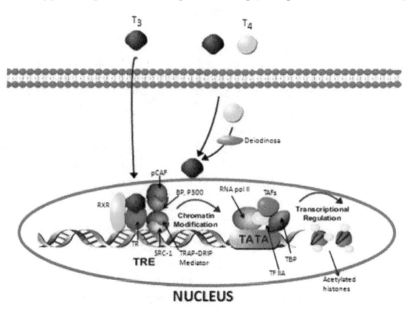

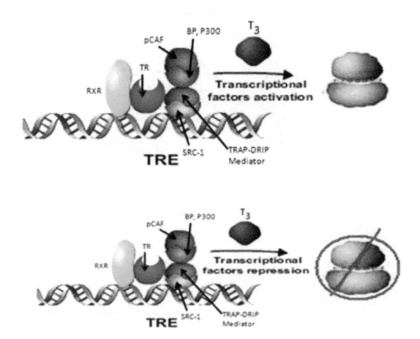

Fig. 4. Nuclear gene expressions by T_3 to the thyroid hormone receptors TRα and TRβ. Heterodimers must bind to specific thyroid response elements (TREs) in the promoters of T_3-target genes and activate or repress transcription in response to hormone.

The existence of cell-surface receptors for thyroid hormones has been acknowledged. The presence of binding sites for thyroid hormones on the cell surface has been known for many years in the red blood-cell membrane (Davis et al., 1983) and in the synaptosome (Giguere et al., 1996; Giguere et al., 1992). The identity of the proteins involved in membrane binding of hormones was not established in these studies and there has been a reluctance to believe that integrin avβ3, containing a binding site for the thyroid hormones, is an initiation site for complex hormone-directed cellular events, such as cell division and angiogenesis (Davis et al., 2005).

Integrins are ubiquitous heterodimeric structural proteins of the cell membrane that convey signals from the cell interior to the extracellular matrix (ECM) (inside-out) and from the ECM to the cell (outside-in). The integrin purified from the plasma membrane bound radiolabeled thyroid hormones and with high affinity. This integrin contains a binding site for thyroid hormones caused by the functional consequences of the binding activation of MAPK (figure 5). The receptor has been located at the Arg-Gly-Asp (RGD) recognition site on the integrin that is important to the binding of a number of extracellular-matrix proteins and growth factors. From this site, the thyroid hormone signals are transduced by MAPK (ERK1-2) in angiogenesis in endothelial cells and the cell proliferation of tumor cell lines. The T_4 in concentrations that are physiological (10^{-10} M free T_4) and T_3 in supraphysiological concentrations cause ERK-dependent cell proliferation. It is now clear that the hormone receptor domain on the integrin is more complex than initially thought. There is a T_3-specific

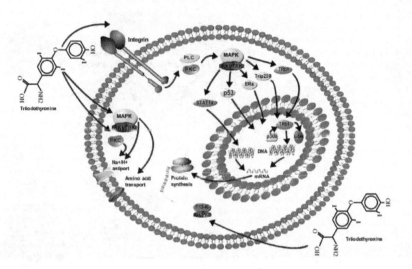

Fig. 5. Nongenomic mechanisms of action of thyroid hormones mediated by integrine-type membranal receptors.

site in the domain and a site at which both T_4 and T_3 may act. The T_3-specific site activates PI3K and is linked not to cell proliferation, but to trafficking of certain intracellular proteins such as shuttling of TRs from the cytoplasm to nucleus and to the transcription of specific genes, such as hypoxia-inducible factor-1 (HIF-1). T_4 is unable to activate PI3K (Cheng et al., 2010).

2.3 Thyroid hormone actions

The T_3 is recognized as a key metabolic hormone of the body. It has many physiological actions and it modulates all metabolic pathways through alterations in oxygen consumption and changes in protein, lipid, carbohydrate, and vitamin metabolism. Through its direct manipulation of protein expression associated with such pathways, T_3 affects the synthesis and degradation of many other hormones and growth factors and indirectly influences additional endocrine signaling. To see more details of the thyroid hormone action, check the review of Hulbert (Hulbert, 2000).

2.4 Abnormalities of the thyroid function

Abnormalities of the thyroid function are among the most common of all endocrine disorders. They fall into two categories, hypothyroidism and hyperthyroidism, reflecting deficient and excess thyroid hormone secretion. There are many causes that generate these conditions. Whatever the cause, the consequences of too little or too much thyroid hormone secretion are largely predictable, given the knowledge of the functions of the thyroid hormones.

3. Relationship between alterations of the thyroid hormones and the ROS-steady state

It has been proposed that the thyroid hormones influence the ROS steady-state and REDOX environment in the cell. The most common idea is that hyperthyroidism enhances the ROS production that perturbs the ROS steady-state and changes the REDOX environment to facilitate cell damage. A hypometabolic state caused by hypothyroidism could be a protective state. In the next part of this chapter we show the evidence for this.

3.1 Hyperthyroidism and the ROS-steady state

Because thyroid hormones modulate many functions, if thyroid hormone levels change, many cellular processes could be altered, including modifications in the REDOX environment. Related to this issue we can ask what happens in a hyperthyroid condition?. Because one of the most studied effects of the thyroid hormone is the control of the basal metabolic rate, a hypermetabolic state produces a modification of the REDOX environment (Venditti & Di Meo, 2006). It is well-known that a higher T_3 level, a hypermetabolic state, causes calorigenesis in two ways. The first is a short-term signaling mechanism with the allosteric activation of cytochrome-C oxidase and the second is a long-term pathway producing nuclear and mitochondrial gene transcription through T_3 signaling, thus stimulating basal thermogenesis (Oppenheimer et al., 1994). This last mechanism causes the synthesis of the enzymes involved in energy metabolism and the components of the respiratory-chain apparatus, leading to a higher capacity of oxidative phosphorylation (Videla, 2000; Soboll, 1993). These short- and long-term pathways are mainly responsible for the increased cellular respiration caused by the hyperthyroid state. Other processes may also play a role, namely 1) energy expenditure caused by a higher active cation transport, 2) loss of energy from futile cycles caused by increases in catabolic and anabolic pathways of intermediary metabolism, 3) higher activity of membrane-bound enzymes associated with electron transfer and metabolite carriers caused by changes in the lipid composition of mitochondrial membranes (Soboll, 1993), and 4) O_2 equivalents related to oxidative stress (Videla, 2000), a REDOX imbalance that leads to various pathological events in several organs as the liver (Jaeschke et al., 2002). In these pathologies, the cellular damage occurs when the balance between oxidant and antioxidants is disturbed and the antioxidant system does not balance the oxidants, thus altering the ROS steady-state level (Lushchak, 2011). An enhanced ROS causes lipid peroxidation, enhancement of reactive oxygen species, nitration, carbonylation, or glutathionylation of proteins, and fragmentation of DNA.

Fernandez et al. found that thyroid calorigenesis is a hormonal stimulus for the REDOX activation of NFκB, a response that is triggered in Kupffer cells having higher respiratory-burst activities (Fernandez et al., 2006). These findings are in agreement with studies showing that the NFκB activation can be achieved by physiological levels of the ROS, which are produced during the respiratory burst after stimulation of isolated or cultured macrophages (Kaul & Forman, 1996) and in carbon-stimulated Kupffer cells in the isolated, perfused rat liver (Romanque et al., 2003). This may be caused by the damage produced by the oxidative stress generated by an excess of thyroid hormones. There are data indicating that excess thyroid hormones act at multiple levels to cause apoptosis, because this higher level enhances the expression of several death receptors and their ligands, such as TNF-α,

FasL, proNGF, and proBDNF, resulting in activation of apical caspase-8, which is further amplified through the activation of the p75NRT-mediated pathways (Kumar et al., 2007). Hyperthyroid animals appear to have a shorter lifespan and, at an advanced age, have a myelin deficit (Carageorgiou et al., 2005). It is known that hyperthyroidism increases hepatic protein oxidation, as evidenced by a significant 88% increase in the content of protein hydrazone derivatives 3 days after a T_3 treatment. This effect may be caused by the increased generation of ROS generated by T_3 (Fernández et al., 1985; Fernandez & Videla, 1993) leading to the formation of carbonyl derivatives mainly occurring at the arginyl, prolyl, lysyl, and histidyl residues in proteins (Stadtman, 1990; Reznick & Packer, 1994). The T_3-caused ROS formation can cause the conversion of cysteinyl residues to protein-protein disulfide conjugates or to mixed-disulfide derivatives (Stadtman, 1990), whereas T_3-caused NO generation (Fernandez et al., 1997) may lead to protein oxidation or nitration through peroxynitrite formation (Alvarez & Radi, 2001). The biological significance of the oxidative modification of proteins in T_3-caused liver oxidative stress can be visualized on two levels; 1) loss of protein function and 2) increased protein degradation. Under high rates of ROS and RNS input, the oxidative modification of enzymes can occur with the consequent reduction in enzyme activity (Stadtman, 1990; Lissi et al., 1991). The inactivation of hepatic antioxidant enzymes has been described in several conditions in vivo involving oxidative stress in the tissue, including hyperthyroidism (Lissi et al., 1991), which determines a decrease in the activity of superoxide dismutase and catalase (Fernandez et al., 1988) and in the content of cytochrome P450 (Fernández et al., 1985). In agreement with this contention, inactivation of superoxide dismutase by H_2O_2 (Bray et al., 1974) and of catalase by O_2^{-} (Kono & Fridovich, 1982) has been reported in conditions in vitro. In addition to enzyme inactivation, thyrotoxicosis in mammals results in the stimulation of both synthesis and degradation of protein, with a predominance of degradation, as shown by the increase in protein catabolism, negative nitrogen balance, and the loss of protein from muscle and other body stores (Loeb, 1996). These findings are in accordance with those of Tapia et al. who found a higher oxidation of the liver protein and an increase in lipid peroxidation levels in hyperthyroid rats (Tapia et al., 2010).

Interestingly, the activities of both GPx-1 and GR are decreased in hyperthyroid rats. It is remarkable that both enzymatic activities are strongly GSH-dependent. The GPx-1 catalyzes the reduction of H_2O_2 and lipid hydroperoxides coupled with oxidation of GSH into GSSG whereas the GR replenishes the GSH pool with the help of NADPH principally provided by the pentose-phosphate pathway. The intracellular GSH status appears to be a sensitive indicator of the cellular ability to resist ROS. Furthermore, it has been found that total GSH equivalents and the GSH and GSSG pools were increasingly depleted by T_3 over time (Chattopadhyay et al., 2007). The liver is especially rich in GST that metabolizes xenobiotics by conjugating with GSH. In fact, the GST-catalyzed conjugation of GSH with exogenous compounds and endogenous metabolites such as 4-hydroxynonenal is regarded as a major cellular-defense mechanism against toxicity (Cheng et al., 2001). The activities of GST were considerably impaired with the progression of the T_3 treatment (Chattopadhyay et al., 2007). Because recycling of oxidized glutathione consumes NADPH, the cellular levels of NADPH and its synthesis represent the rate limiting factors of H_2O_2 consumption by catalase-deficient tissues (Ho et al., 2004). Moreover, prolonged hyperthyroidism diminishes GR but elevated G6PD activity indicates that severe hyperthyroidism may compromise the cellular ability to maintain the redox state. Lombardi et al. have demonstrated that injection of T_3 into hypothyroid rats caused an increase in both enzyme activity and mRNA expression of G6PD in the liver. Nevertheless, the reduced activities of GPx, GR, and GST in the hyperthyroid liver prevent optimum GSH use and recycling. Accumulation of GSSG can lead to protein modifications because of interactions with –SH groups (Reed, 1990). Though the T_3 exerted a positive stimulatory effect on the NADPH supply, it was not sufficient to compensate for the massive GSH depletion and this probably explains the negative regulatory impact of T_3 on activities of GSH-dependent enzymes such as GPx and GR in the rat liver. Under such conditions, the cellular redox-status is disturbed, as reflected in the high oxidative-stress index of hyperthyroid rats.

In brief, thyroid calorigenesis resulting from acceleration of energy metabolism and secondary electron-transfer processes lead to a higher generation of ROS in the target tissue. This prooxidant condition enhances the oxidative-stress status of the organs when the decrease in the antioxidant potential is not adequately compensated for, leading to

a) substantial oxidative deterioration of biomolecules, with loss of their functions that may compromise cell viability, b) activities of GPx, GR, GST, catalase, and superoxide dismutase are considerably impaired, c) total GSH equivalents and GSH and GSSG pools were increasingly depleted, d) a higher susceptibility of the liver to toxic stimuli that exacerbate liver injury, e) upregulation of gene expression, f) apoptosis, g) shorter lifespan, and h) myelin deficit.

3.2 Hypothyrodism and the ROS-steady state

Hypothyroidism has been related to some diseases because it causes a hypometabolic state. This condition can be beneficial. Why can we make this assertion? There are many findings to support this suggestion. It is well-known that a deficiency of the thyroid hormones results in decreased metabolism and lowering of the basal metabolic rate (BMR). There is evidence that supports the lower cell stress in the hypothyroidism condition. Tenorio-Velázquez et al. have demonstrated that hypothyroidism attenuates oxidative stress and renal injury caused by ischemia-reperfusion, produced by an increase in the ROS and reactive nitrogen species (Tenorio-Velásquez et al., 2005). Most research has been done in the kidney and liver models of ischemia (Swaroop & Ramasarma, 1985; Paller, 1986). The postulated mechanism in such organs has been either a decrease in the general metabolic rate or a reduced free radical scavenging response after ischemia. The lipid peroxidation in hypothyroid animals with renal ischemia was decreased (Paller, 1986). The content of malondialdehyde, which is an indirect measure of the generation of oxygen free radicals, was decreased and the cortical content of glutathione, a free radical scavenger, was increased in the hypothyroid, ischemic animals. Similarly, in the liver model of hypothyroid-ischemic injury, lipid peroxidation and free-radical generation were decreased in the hypothyroid animals (Swaroop & Ramasarma, 1985). These investigators have shown a significant decrease in hydrogen peroxide, a measure of the oxygen free-radical status, in the liver mitochondrion in the hypothyroid animals. Hypothyroidism attenuates not only renal but also cardiac damage caused by ischemia and reperfusion. Bobadilla et al. have shown that hypothyroidism conferred protection against reperfusion arrhythmias and the cardiac release of creatine kinase and aspartate amino transferase and preserved the normal structure of the myocardial tissue (Bobadilla et al., 2002). It has been proposed that hypothyroidism protects against pore opening and heart reperfusion (Chávez et al., 1998). This may be relevant to the protective effect of hypothyroidism in ischemia and reperfusion because it has been recognized that the mitochondria play a key role in cell-death pathways by activating the mitochondrial-permeability transition pore and causing the release of cytochrome C, proapoptotic factors, and the Ca^{2+} overload that causes a nonselective permeability of the inner membrane. The prolonged opening of the membrane-permeability transition pore during the first few minutes of reperfusion is a critical determinant of cell death, and pharmacological inhibition of the pore at the time of reperfusion protects the cell (Halestrap et al., 2004). It has been found that there is a decreased glutamate release during hypothyroidism and this is correlated to a protection in cerebral ischemia (Shuaib et al., 1994). The reason why hypothyroidism results in a decreased release of glutamate is as yet unknown. It is possible that the hypothyroidism affects the release mechanisms in the presynaptic receptors. It is also possible that the hypothyroid state results in an increase in the reuptake mechanism for glutamate.

We have noted the protector effect of hypothyroidism, but many investigators use methimazole to cause it. There are some indications that antithyroid-caused hypothyroidism can produce cellular damage. Although, some results indicate that this drug causes cellular protection because of its chemical structure (Bruck et al., 2007; Tutuncu et al., 2007). In addition, there is evidence of extrathyroidal effects of antithyroid drugs, such as thionamides, in humans and animals (Bandyopadhyay et al., 2002). One of the effects of thionamides is the contribution to oxidative stress and cellular damage. These effects can produce an increase of oxidant species that causes lipid peroxidation, nitration, carbonylation, or glutathionylation of proteins, and fragmentation of DNA (Halliwell & Gutteridge, 2007; Valko et al., 2007). Because of this, we determined if methimazole or hypothyroidism causes cellular damage in several organs. After producing a hypothyroid animal caused by thyroidectomy or methimazole administration, the spleen, heart, liver, lung, and kidney were obtained. A portion of these tissues was processed for histological study and another portion was used for the biochemical assay for determining oxidative stress. Histologically, we demonstrated that only methimazole-caused hypothyroidism causes cellular damage in the kidney, lung, liver, heart, and spleen. Animals with methimazole and with T_4 supplementation showed cellular damage in the lung, spleen, and renal medulla with lesser damage in the liver, renal cortex, and heart. Hypothyroidism did not produce cellular damage in any organs except the lung. The thyroidectomy group showed no other tissue alterations (Cano-Europa et al., 2011). These results are in accordance with what others have observed in animals and humans. Five percent of patients with hyperthyroidism treated with antithyroid drugs, including methimazole, are reported to have liver (Casallo Blanco et al., 2007; Woeber, 2002), lung (Tsai et al., 2001) and kidney damage (Calañas-Continente et al., 2005). The methimazole-caused hypothyroidism in animals has tumorigenic effects (Jemec, 1977) and modifies the pulmonary function (Liu & Ng, 1991). No tissue damage was seen in a model of hypothyroidism caused by a thyroidectomy (Tenorio-Velásquez et al., 2005). We also compared, over a time-course, markers of oxidative stress, the REDOX environment, and the antioxidant enzymatic system in the liver and the spleen of rats with methimazole- or thyroidectomy-caused hypothyroidism. We found that the cell damage was related with an increase of oxidative stress markers (ROS and lipid peroxidation) that were not compensated for by the antioxidant system. The catalase activity is reduced in hepatic tissue and this allows H_2O_2-caused hepatic damage (Cano-Europa et al., 2010). The increase of the glutathione-cycle enzymes was insufficient to prevent oxidative-stress markers (Ortiz-Butron et al., 2011). All these findings together pointed out that methimazole and not the hypothyroidism is responsible for the cell damage. The tissues evaluated, especially the kidney and liver, have a high metabolic activity that generates ROS. Under physiological conditions the presence of antioxidant enzymes, in particular peroxidases and dismutases, prevent oxidative stress and tissue damage (Halliwell & Gutteridge, 2007; Angermuller et al., 2009). Some drugs, such as methimazole, disturb the physiological steady state. Methimazole alters the intracellular REDOX environment and causes cellular damage because of oxidant generation and ROS, and consequently the lipid peroxidation is not completely neutralized by the antioxidant system. We suggest that the central mechanism of the methimazole-caused cell damage is based on the reduction of catalase activity caused by a methimazole-inactivated catalytic center (Bandyopadhyay et al., 1995; Bandyopadhyay et al., 2002).

Other investigators, like Bergman and Brittebo, have demonstrated this anthytiroid-caused damage in other models, i.e. an olfactory mucosa model. They found that this drug covalently binds to the tissue, and pretreatment with the cytochrome-P450 inhibitor metyrapone prevented both the covalent binding and the toxicity of methimazole in this tissue. They suggest a cytochrome P450-dependent metabolic activation of methimazole to a reactive and toxic intermediate at this site (Bergman & Brittebo, 1999). The pretreatment with thyroxin did not protect against the methimazole-caused necrosis, suggesting that this lesion is not related to a transient decrease in thyroid hormone levels. The covalent binding shown by methimazole in this tissue has been found in other tissues, such as the bronchial epithelium and the centrilobular parts of the liver. It is possible that methimazole suffers activation at these sites. Further, this drug is metabolized stepwise to the corresponding sulfenic and sulfinic acids with a concurrent formation of reactive intermediates (Poulsen et al., 1974). It is known that methimazole produces a decrease of P450 at the hepatic level (Decker & Doerge, 1992). In rodents given the methimazole analogs 1-methy-imidazole, 4-methylimidazol, or methyl pyrrole, which are devoid of a thiol group, no morphological changes were observed in the olfactory mucosa (Brittebo, 1995). The thiol group in methimazole seems to be important for the methimazole-caused toxicity, suggesting that enzyme-catalyzed changes of the thiol group will give rise to an intermediate toxin in the tissue.

Other methimazole-caused damage mechanisms are associated with its chemical structure and its biotransformation. Some investigators suggest that this drug binds covalently to the hepatocytes, mainly those next to the hepatic triad (Decker & Doerge, 1992; Lee & Neal, 1978). For biotransformation, methimazole may be oxidized by the P450 enzymes to form the 4,5-epoxide. The enzymatic or nonenzymatic hydrolysis of the epoxide formed would produce an unstable hemiketal-like intermediate, which it is expected to undergo spontaneous ring cleavage to form glyoxal and N-methylthiourea. The metabolism of N-methylthiourea is complex, but it is believed that sulfur oxidation, mediated mainly by flavin-monooxigenase (FMO, EC.EC 1.14.13.8), proceeds primarily to the sulfenic acids and then possibly to the sulfinic acids. It is known that this step is necessary in the bioactivation of thioureas resulting in protein binding, enzyme inactivation, and organ toxicity (Mizutani et al., 1994; Neal & Halpert, 1982).

The thyroidectomy group examinee showed no other tissue alterations, except for the lung. There is some evidence that demonstrates molecular mechanisms by which hypothyroidism itself may produce a protected state of the tissues, such as reducing the enzyme activity associated with the mitochondrial-respiratory chain (Paradies et al., 1994), the decrease in adenine nucleotide translocase (Schonfeld et al., 1997), reduced activity of cytochrome-C oxidase (Paradies et al., 1997), and the resistance to forming the permeability transition-pore formation of the inner mitochondrial membrane (Chávez et al., 1998).

With all this evidence it is important to develop other therapies or antithyroid drugs with fewer side effects. We suggest that hypothyroidism is a protective state against toxic agents and it is related to an increase of reduced glutathione or γ-L-glutamyl- cysteinyl-glycine (GSH) synthesis and a mild immunosuppression.

3.3 Enhanced GSH synthesis in the hypothyroid state, a mechanism of cell protection

Before we show evidence of the relationship between hypothyroidism and high γ –L-glutamyl- cysteinyl-glycine (GSH) concentration, we need to know more about GSH. The synthesis of reduced glutathione or GSH involves two ATP-dependent enzymatic steps made in the cell cytoplasm. Figure 6 shows the cycle of GSH.

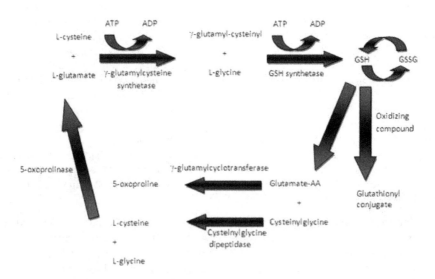

Fig. 6. GSH cycle. GR is glutathione reductase; GST glutathione S-transferase, and GSSG oxidized glutathione.

The synthesis of GSH starts with the entry into the cells of its amino acid precursors: glutamate, cysteine, and glycine. Glutamate and glycine can enter the cell by secondary active transport. Some of the glutamate cotransport carriers transfer cysteine. For the cysteine, entry into the cell may also be caused by the transporters of the neutral amino acid system. It is believed that cysteine is the limiting amino acid for the synthesis of GSH because it occurs in lower concentrations in plasma and has a lower Km (Aoyama et al., 2008). Once the amino acids have entered the cell, the g-glutamylcysteine synthetase (γ-GCS, EC 6.3.2.2) forms the g-glutamylcysteine. The formation of the product involves two steps. The first is the interaction between glutamate and ATP in the presence of Mg^{+2} to form g - glutamylphosphate and the second involves the interaction of this intermediate with the cysteine and with ADP release (Griffith & Mulcahy, 1999; Griffith, 1999). This first step is the most important in the formation of GSH because γ-GCS is the limiting enzyme in the synthesis of GSH. The γ-GCS is an heterodimeric enzyme composed of a catalytic subunit known as the heavy subunit (γ-GCS$_H$ Mr, $\approx$ 73 kDa) and a regulatory or light subunit (γ-GCS$_L$ Mr, $\approx$ 31 kDa). The γ-GCS activity depends primarily on the substrates and is inhibited by GSH. The γ-GCS$_L$ activity is under the control of kinases such as protein kinase A (PKA) and PKC (Griffith, 1999)

Two processes can occur thermodynamically once γ-glutamyl-cysteinyl is formed. The compound may be used by the GSH synthetase (GS, EC 6.3.2.3) to form GSH when conjugated with glycine or it may interact with g -glutamyl cyclotransferase to form 5-oxo-L-proline and L-cysteine. The pathway that prevails depends of the Km of each enzyme. Under physiological conditions the Km of GS is 12 times greater than the γ-glutamyl cyclotransferase so it favors the formation of GSH in more than 95% (Weber, 1999). Once GSH has been synthesized there are different processes in which it participates;

1. In the hydrolysis of plasma GSH to synthesize GSH de novo for another cell. For example, if a hepatocyte secretes GSH, another cell can hydrolyze such a compound into its precursors (cysteinylglycine and glutamate) by the g-glutamyl transpeptidase (γ-GT, EC 2.3.2.2) expressed on the outside of the plasmatic membrane. The cysteinylglycine compounds or their S-conjugates can be hydrolyzed by dipeptidases to yield free amino acids that can be introduced into the cell and start the formation of GSH (Weber, 1999).

2. In the detoxification of electrophiles by conjugating these with a-carbonyls and by b-unsaturation by glutathion-S transferase (GST, EC. 2.5.1.18). This reaction results in the elimination of the electrophile by the consequent metabolism of the glutathione S-conjugate by the γ-GT enzymes and the cysteinylglycine dypeptidase. This process is not always in the favor for the cell, because it can sometimes create more toxic species (Weber, 1999).

3. In the detoxification of hydrogen peroxide by the action of the glutathione peroxidase enzymes (GPX, EC 1.11.1.19) (Beckett & Arthur, 2005).

4. In maintaining ascorbic acid and vitamin E (Van Acker et al., 1993).

5. In intracellular communication processes as a modulator of diverse signaling pathways (Cruz et al., 2003).

6. In the modulation of membrane receptors as for NMDA receptors in the central nervous system (Oja et al., 2000).

7. In the transport of metals such as Cu^{+2}, Hg^{+2}, Pb^{+2}, and Zn^{+2} (Filomeni et al., 2002).

The mitochondrial concentration of GSH is approximately 11-15 mM. The entry of GSH into the mitochondria depends on the electroneutral transporters, such as the tricarboxylic or dicarboxylic acids (Lash, 2006). In general, the ratio GSH/GSSG is greater than 10 for the cells and organelles, such as mitochondria and nucleus, whereas the endoplasmic reticulum has the lowest GSH/GSSG ratio of 1 to 3. The best indicator of the REDOX environment is the $GSH^2/GSSG$ ratio because the REDOX environment involves the transfer of electrons, for which the theoretical model of Schafer and Buettner uses the Nernst equation (Schafer & Buettner, 2001). These authors proposed that other REDOX couples can participate in the REDOX environment maintaining the ratios of $NADPH/NADP^+$, reduced thioredoxin/oxidized thioredoxin ($TrxSH_2/TrxSS$), and $GSH^2/GSSG$. These REDOX couples could participate in the maintaining of the REDOX environment because their pKas are above the physiological pH and the ratio of the reduced pair to its oxidized counterpart is 1:100, 1:1000, or greater. The $GSH^2/GSSG$ ratio is the most important couple in the REDOX environment because their chemical structures are not susceptible to any peptidase and their use is in the cell, particularly for cell antioxidant protection, and not in essential biosynthetic pathways. Also, the $GSH^2/GSSG$ ratio has the highest concentration of the three REDOX ratios mentioned, and this one best buffers the REDOX potential changes between -300 and - 100 mV, despite varying the concentration of the GSH. The change in the half-cell reduction potential of this REDOX couple is related to the processes such as cell proliferation, differentiation, apoptosis, and necrosis in biological experiments (Cai & Jones, 1998; Cai et al., 2000; Hwang et al., 1992; Jones et al., 1995; Kirlin et al., 1999).

In our group we are studying the effect of the hypothyroid state and the GSH synthesis in various organs, with special interest in the liver and kidney. For that we used thyroidectomyzed rats with a parathyroid gland reimplant (only to affect thyroid hormone system). Two weeks postsurgery we determined the GSH content by a fluorometric method and the γ-GCS by a spectophotometric method as described (Cano-Europa et al., 2010; Ortiz-Butron et al., 2011). Figure 7 shows that hypothyroid animals have a higher GSH content than euthyroid animals because they have an enhanced γ -GCS activity.

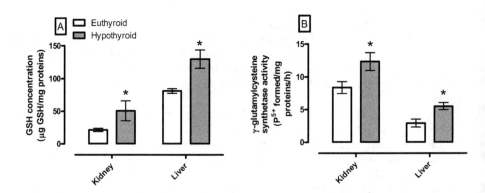

Fig. 7. Effect of hypothyroidism on GSH content (A) and γ-GCS activity in liver and kidney. Values are the mean ± SE. (*) $P < 0.05$ vs. euthyroid ($n = 5$ for each group).

It is possible that some intracellular signals modify the γ-GCS activity. It is also probable that the thyroid hormone-receptor complex acts as a negative regulator for γ-GCS because the putative thyroid-hormone response-element sequences in the promoter regions of both the catalytic and modulator subunit of γ-GCS genes have been seen. However, we need to do further experiments to demonstrate this. If this does occur then hypothyroidism could be a protective state against chemical- or physical-caused oxidative stress and cell damage. At present, we are evaluated if the hypothyroid state protects against ethylene glycol-caused oxidative stress and renal damage. At this respect, the results are in accordance with the idea that the hypothyroidism-enhanced REDOX environment (unpublished data).

3.4 Thyroid hormone alteration levels and the immune response

There is evidence of the thyroid hormones and immune systems and their development and function from amphibious animals to mammals (Rollins-Smith & Blair, 1990; Lam et al., 2005; Watanabe et al., 1995; Nakamura et al., 2007).

In zebra fish the thyroid state participates in thymus development and lymphopoiesis (Lam et al., 2005). In humans and other mammals, clinical hyperthyroidism increased the size and cellularity of the thymus, particularly a larger number of thymus nurse cells, Thy1+ thymocytes, and the CD4-CD8- and CD44-positive cells (Villa-Verde et al., 1993; Scheiff et al., 1977). Hyperthyroidism increases T cells in the spleen and thymus with high levels of NK cells only in the spleen (Watanabe et al., 1995). Hypothyroidism reduces the cellularity in the spleen and thymus (Bendyug et al., 2003). In neonatal hypothyroidism, it has been observed that the NK cells and regulatory T cells (CD4+CD25+) are enhanced in thymus, spleen, and peripheral blood. The dendritic cells integrate signals from several pathways and receptors, including those arising from engagement of uptake and pattern recognition receptors, proinflammatory and antiinflammatory cytokines, chemokines, and hormones like THs. The T_3 promotes the dendritic-cell maturation and Th1-type cytokine secretion (Mascanfroni et al., 2008). The dendritic cells are modulated by the THs because the T_3-TRβ1 causes Akt signaling-pathway activation and NFκB-dependence, but a PI3K-independent pathway (Mascanfroni et al., 2010).

There are reports that hypothyroidism decreases immune system activity and increases infection in humans (Schoenfeld et al., 1995; Amadi et al., 2008).

All this evidence suggests the proposal that the hypothyroid condition decreases the immune response. This could be protective in the case of toxicant-caused oxidative stress and cell damage caused by the immune system activation by a substance like aniline. Aniline is a toxic, aromatic amine and it is an extensively used industrial chemical. Exposure to aniline is known to cause toxicity to the hematopoietic system. Aniline toxicity is generally characterized by methemoglobinemia, hemolysis, and hemolytic anemia and by the development of splenic hyperplasia, fibrosis, and a variety of primary sarcomas after chronic exposure in rats. The immunological system participates actively in aniline-caused oxidative stress and spleen damage (Wang et al., 2011; Wang et al., 2010; Wang et al., 2008).

In our laboratory, we evaluated the participation of hypothyroidism and aniline-caused oxidative stress and spleen damage. We used male Wistar rats weighing 240 to 260 g divided into four groups; 1) euthyroid, 2) euthyroid + aniline, 3) hypothyroid, and 4) hypothyroid + aniline. The hypothyroidism was produced by thyroidectomy with implantation of the parathyroid gland. Two weeks after surgery, the animals were treated with 1 mmol/kg/d ig aniline for five days. On the fifth day, the animals were killed, the blood obtained to determine the lymphocyte count and the spleen was dissected to assess lipid peroxidation and the quantification of reactive oxygen species as preliminary results. In figure 8 are the results. It was shown that hypothyroid rats had decreased ROS concentration, lipid peroxidation, and the lymphocyte counts in aniline-treated rats compared with euthyroid rats.

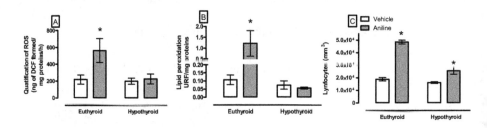

Fig. 8. Effect of hypothyroidism on ROS concentration (A), lipid peroxidation (B), and lymphocyte count (C) in the spleen of rats treated with aniline. Values are the mean ± SE. (*) $P < 0.05$ vs. euthyroid ($n = 5$ for each group).

The hypothyroid state can be protective in this situation because it decreased the ROS production, increased the GSH, and decreased the lymphocyte count. Although we need to do more experiments to demonstrate the idea of a mild immunosuppression participating in cell protection, these preliminary results suggest it.

4. Final remarks

In earlier years it was believed that the hypo- and hyperthyroid conditions modifying the ROS steady state caused cell damage. Presently, there is much evidence that only hyperthyroidism does this. Hypothyroidism is believed to be a protective state because scientists did not believe the drug-caused hypothyroidism modified the ROS steady state. On this point, by studying hypothyroidism in animal models that modify only TH concentrations, such as thyroidectomy, we can observe cell protection against chemical- and physical-caused oxidative stress. Right now, we can describe two different pathways by which the hypothyroid state can protect: GSH synthesis and mild immunosuppression. This is now an open field in which to study these possibilities.

5. Acknowledgement

This study was partially supported by SIP-IPN 20110283 y 20110336. Thanks to Dr. Ellis Glazier for editing this English-language text.

6. References

Alvarez,B. & Radi,R. (2001). Peroxynitrite decay in the presence of hydrogen peroxide, mannitol and ethanol: a reappraisal. *Free Radical Research* Vol. 34, No. 5, pp. 467-475 (2001) ISSN 1071-5762

Amadi,K., Sabo,A.M., Ogunkeye,O.O. & Oluwole,F.S. (2008). Thyroid hormone: a "prime suspect" in human immunodeficiency virus (HIV/AIDS) patients? *Physiological Society of Nigeria* Vol. 23, No. 1-2, pp 61-66 (Jun-Dec, 2008) ISSN 0794-859X

Angermuller,S., Islinger,M. & Volkl,A. (2009). Peroxisomes and reactive oxygen species, a lasting challenge. *Histochemistry and Cell Biology* Vol. 131, No. 4, pp 459-463 (April, 2009) ISSN

Aoyama,K., Watabe,M. & Nakaki,T. (2008). Regulation of neuronal glutathione synthesis. *Journal of Pharmacology Sciences* Vol. 108, No. 3, pp 227-238 (November, 2008) ISSN 1347-8648

Bandyopadhyay,U., Bhattacharyya,D.K., Chatterjee,R. & Banerjee,R.K. (1995). Irreversible inactivation of lactoperoxidase by mercaptomethylimidazole through generation of a thiyl radical: its use as a probe to study the active site. *Biochemical Journal* Vol. 306 (Pt 3), pp 751-757 (March, 2005) ISSN 0264-6021

Bandyopadhyay,U., Biswas,K. & Banerjee,R.K. (2002). Extrathyroidal actions of antithyroid thionamides. *Toxicology Letters* Vol. 128, No. 1-3, pp 117-127 (March, 2008) ISSN 0378-4274

Beckett,G.J. & Arthur,J.R. (2005). Selenium and endocrine systems. *Journal of Endocrinology* Vol. 184, No. 3, pp 455-465 (March, 2005) ISSN 0022-0795

Bendyug,G.D., Grinevich,Y.A., Khranovskaya,N.N., Fil'chakov,F.V., Yugrinova,L.G. & Kad'kalenko,A.G. (2003). The state of the immune system in thyroidectomized rats. *Bulletin of Experimental Biology and* Medicine Vol. 135, No. 2, pp 154-157 (February, 2003) ISSN 0007-4888

Bergman,U. & Brittebo,E.B. (1999). Methimazole toxicity in rodents: covalent binding in the olfactory mucosa and detection of glial fibrillary acidic protein in the olfactory bulb. *Toxicology and Applied Pharmacology* Vol. 155, No. 2, pp 190-200 (March, 1999) ISSN 0041-008X

Bianco,A.C., Salvatore,D., Gereben,B., Berry,M.J. & Larsen,P.R. (2002). Biochemistry, cellular and molecular biology, and physiological roles of the iodothyronine selenodeiodinases. *Endocrine Reviews* Vol. 23, No. 1, *pp* 38-89 (February, 2002) ISSN 0163-769X

Bobadilla,I., Franco,M., Cruz,D., Zamora,J., Robles,S.G. & Chavez,E. (2002). Hypothyroidism protects the myocardium against ischemia-reperfusion injury. Archivos de cardiología de México Vol. 72 Suppl 1, pp S27-S30 (January, 2002) ISSN 1405-9940

Bray,R.C., Cockle,S.A., Fielden,E.M., Roberts,P.B., Rotilio,G. & Calabrese,L. (1974). Reduction and inactivation of superoxide dismutase by hydrogen peroxide. *Biochemical Journal* Vol. 139, No.1, pp 43-48 (April, 1974) ISSN 0264-6021

Brittebo,E.B. (1995). Metabolism-dependent toxicity of methimazole in the olfactory nasal mucosa. *Pharmacology & Toxicology* Vol. 76, No. 1, pp 76-79 (January, 1995) ISSN 0901-9928

Bruck,R., Weiss,S., Traister,A., Zvibel,I., Aeed,H., Halpern,Z. & Oren,R. (2007). Induced hypothyroidism accelerates the regression of liver fibrosis in rats. *Journal of Gastroenterology and Hepatology* Vol. 22, No. 12, pp 2189-2194 (December, 2007) ISSN 1440-1746

Cai,J. & Jones,D.P. (1998). Superoxide in Apoptosis. Mitocondrial generation triggered by cytocrome c loss. *Journal of Biological Chemistry* Vol. 273, No. 19, pp 11401-11404 (May, 1998) ISSN 1083-351X

Cai,J., Wallace,D.C., Zhivotovsky,B. & Jones,D.P. (2000). Separation of cytochrome c-
    dependent caspase activation from thiol-disulfide redox change in cells lacking
    mitochondrial DNA. *Free Radical Biology & Medicine* Vol. 29, No. (3-4), pp 334-342
    (Ausgut, 2000) ISSN 0891-5849
Calañas-Continente,A., Espinosa,M., Manzano-García,G., Santamaría,R., López-Rubio,F. &
    Aljama,P. (2005). Necrotizing glomerulonephritis and pulmonary hemorrhage
    associated with carbimazole therapy. *Thyroid* Vol. 15, No. 3, pp 285-288 (March,
    2005) ISSN 1050-7256
Cano-Europa,E., Blas-Valdivia,V., Franco-Colin,M., Gallardo-Casas,C.A. & Ortiz-Butron,R.
    (2011). Methimazole-induced hypothyroidism causes cellular damage in the spleen,
    heart, liver, lung and kidney. *Acta Histochemica* Vol. 113, No. 1, pp 1-5 (January,
    2011) ISSN 0044-5991
Cano-Europa,E., Blas-Valdivia,V., Lopez-Galindo,G.E., Franco-Colin,M., Pineda-
    Reynoso,M., Hernandez-Garcia,A. & Ortiz-Butron,R. (2010). Methimazole-induced
    hypothyroidism causes alteration of the REDOX environment, oxidative stress, and
    hepatic damage; events not caused by hypothyroidism itself. *Annals of hepatology*
    Vol. 9, No.1, pp 80-88 (January, 2010) ISSN 1665-2681
Carageorgiou,H., Pantos,C., Zarros,A., Mourouzis,I., Varonos,D., Cokkinos,D. & Tsakiris,S.
    (2005). Changes in antioxidant status, protein concentration, acetylcholinesterase,
    (Na$^+$,K$^+$)-, and Mg^{2+}-ATPase activities in the brain of hyper- and hypothyroid adult
    rats. *Metabolic Brain Disease* Vol. 20, No. 2, pp 129-139 (June, 2005) ISSN 0885-7490
Casallo Blanco,S., Valero,M.A., Marcos Sanchez,F., de Matias Salces,L., Blanco Gonzalez,J.J.
    & Martin Barranco,M.J. (2007). [Methimazole and propylthiouracil induced acute
    toxic hepatitis]. *Journal of Gastroenterology and Hepatology* Vol. 30,No. 5, pp 268-270
    (May, 2007) ISSN 1440-1746
Chattopadhyay,S., Sahoo,D.K., Subudhi,U. & Chainy,G.B. (2007). Differential expression
    profiles of antioxidant enzymes and glutathione redox status in hyperthyroid rats:
    a temporal analysis. *Comparative Biochemistry and Physiology - Part C: Toxicology &
    Pharmacology* Vol. 146, No. 3,pp 383-391 (September, 2007) ISSN 0742-8413
Chávez,E., Franco,M., Reyes-Vivas,H., Zazueta,C., Ramirez,J. & Carrillo,R. (1998).
    Hypothyroidism renders liver mitochondria resistant to the opening of membrane
    permeability transition pore. *Biochimica et biophysica acta* Vol. 1407, No. 3, pp 243-
    248 (September, 1998) ISSN 0006-3002
Cheng,J.Z., Singhal,S.S., Sharma,A., Saini,M., Yang,Y., Awasthi,S., Zimniak,P. & Awasthi,Y.C.
    (2001). Transfection of mGSTA4 in HL-60 cells protects against 4-hydroxynonenal-
    induced apoptosis by inhibiting JNK-mediated signaling. *Archives of Biochemistry and
    Biophysics* Vol. 392, No. 2, pp 197-207 (August, 2001) ISSN 0003-9861
Cheng,S.Y., Leonard,J.L. & Davis,P.J. (2010). Molecular aspects of thyroid hormone actions.
    Endocrine Reviews Vol. 31, No.2, pp 139-170 (April, 2010) ISSN 0163-769X
Christoffolete,M.A., Ribeiro,R., Singru,P., Fekete,C., da Silva,W.S., Gordon,D.F., Huang,S.A.,
    Crescenzi,A., Harney,J.W., Ridgway,E.C., Larsen,P.R., Lechan,R.M. & Bianco,A.C.
    (2006). Atypical expression of type 2 iodothyronine deiodinase in thyrotrophs
    explains the thyroxine-mediated pituitary thyrotropin feedback mechanism.
    Endocrinology Vol. 147, No. 4, pp 1735-1743 (April, 2006) ISSN 0013-7227

Cruz,R., Almaguer-Melian,J.A. & Bergado-Rosado,J.A. (2003). El glutatión en la función cognitiva y la neurodegeneración. Revista de Neurología Vol. 36, No. 9, pp 877-886 (May, 2003) ISSN 0210-0010

Davis,F.B., Davis,P.J. & Blas,S.D. (1983). Role of calmodulin in thyroid hormone stimulation in vitro of human erythrocyte Ca^{2+}-ATPase activity. The Journal of Clinical Investigation Vol. 71, No. 3, pp 579-586 (March, 1983) ISSN 0021-9738

Davis,P.J., Davis,F.B. & Cody,V. (2005). Membrane receptors mediating thyroid hormone action. Trends in Endocrinology and Metabolism Vol. 16, No. 9, pp 429-435 (November, 2005) ISSN 1043-2760

Decker,C.J. & Doerge,D.R. (1992). Covalent binding of 14C- and 35S-labeled thiocarbamides in rat hepatic microsomes. Biochemical pharmacology Vol. 43, No. 4, pp 881-888 (February, 1992) ISSN 0006-2952

Di Cosmo,C., Liao,X.H., Dumitrescu,A.M., Philp,N.J., Weiss,R.E. & Refetoff,S. (2010). Mice deficient in MCT8 reveal a mechanism regulating thyroid hormone secretion. The Journal of Clinical Investigation Vol. 120, No. 9, pp 3377-3388 (September, 2010) ISSN 0021-9738

Fernandez,V., Cornejo,P., Tapia,G. & Videla,L.A. (1997). Influence of hyperthyroidism on the activity of liver nitric oxide synthase in the rat. Nitric Oxide Vol. 1, No. 6, pp 463-468 (1997) ISSN 1089-8603

Fernandez,V., Llesuy,S., Solari,L., Kipreos,K., Videla,L.A. & Boveris,A. (1988). Chemiluminescent and respiratory responses related to thyroid hormone-induced liver oxidative stress. Free Radical Research Communications Vol. 5, No.2, pp 77-84 (1988) ISSN 8755-0199

Fernandez,V., Tapia,G., Varela,P., Romanque,P., Cartier-Ugarte,D. & Videla,L.A. (2006). Thyroid hormone-induced oxidative stress in rodents and humans: a comparative view and relation to redox regulation of gene expression. Comparative Biochemistry and Physiology. Toxicology & Pharmacology Vol. 142, No. 4-3, pp 231-239 (March-April, 2006) ISSN 1532-0456

Fernandez,V. & Videla,L.A. (1993). Influence of hyperthyroidism on superoxide radical and hydrogen peroxide production by rat liver submitochondrial particles. Free Radical Research Communications Vol. 18, No. 6, pp 329-335 (1993) ISSN 8755-0199

Fernández,V.I.R.G., BARRIENTOS,X.I.M.E., KIPREOS,K.A.T.I., VALENZUELA,A.L.F.O. & Videla,L.A. (1985). Superoxide Radical Generation, NADPH Oxidase Activity, and Cytochrome P-450 Content of Rat Liver Microsomal Fractions in an Experimental Hyperthyroid State: Relation to Lipid Peroxidation. Endocrinology Vol. 117, No. 2, pp 496-501 (August, 1985) ISSN 0013-7227

Filomeni,G., Rotilio,G. & Ciriolo,M.R. (2002). Cell signalling and the glutathione redox system. Biochemical Pharmacology Vol. 64, No. 5-6, pp 1057-1064 (September, 2002) ISSN 0006-2952

Giguere,A., Fortier,S., Beaudry,C., Gallo-Payet,N. & Bellabarba,D. (1996). Effect of thyroid hormones on G proteins in synaptosomes of chick embryo. Endocrinology Vol. 137, No. 6, pp 2558-2564 (June, 1996) ISSN 0013-7227

Giguere,A., Lehoux,J.G., Gallo-Payet,N. & Bellabarba,D. (1992). 3,5,3'-Triiodothyronine binding sites in synaptosomes from brain of chick embryo. Properties and ontogeny. Brain Research. Developmental Brain Research Vol. 66, No. 2,pp 221-227 (April, 1992) ISSN 0165-3806

Griffith,O.W. (1999). Biologic and pharmacologic regulation of mammalian glutathione synthesis. *Free Radical Biology & Medicine* Vol. 27, No.9-10, pp 922-935 (November, 1999) ISSN 0891-5849

Griffith,O.W. & Mulcahy,R.T. (1999). The enzymes of glutathione synthesis: gamma-glutamylcysteine synthetase. *Advances in Enzymology and Related Areas of Molecular Biology* Vol. 73, pp 209-67, xii (1999) ISSN 0065-258X

Halestrap,A.P., Clarke,S.J. & Javadov,S.A. (2004). Mitochondrial permeability transition pore opening during myocardial reperfusionGÇöa target for cardioprotection. *Cardiovascular Research* Vol. 61, No.3, pp 372-385 (February, 2004) ISSN 0008-6363

Halliwell,B. & Gutteridge,J.M.C. (2007). Free radicals in biology and medicine. UK: Oxford University Press.

Ho,Y.S., Xiong,Y., Ma,W., Spector,A. & Ho,D.S. (2004). Mice Lacking Catalase Develop Normally but Show Differential Sensitivity to Oxidant Tissue Injury. *Journal of Biological Chemistry* Vol. 279, No. 31, pp 32804-32812 (Juliy, 2000) ISSN 0021-9258

Hulbert,A. (2000). Thyroid hormones and their effects: a new perspective. *Biological Reviews of the Cambridge Philosophical* Society Vol. 75, No. 4, pp 519-621 (November, 2000) ISSN 1464-7931

Hwang,C., Sinskey,A.J. & Lodish,H.F. (1992). Oxidized redox state of glutathione in the endoplasmic reticulum. *Science* Vol. 257, No. 5076, pp 1496-1502 (September, 1992) ISSN 0036-8075

Jaeschke,H., Gores,G.J., Cederbaum,A.I., Hinson,J.A., Pessayre,D. & Lemasters,J.J. (2002). Mechanisms of Hepatotoxicity. *Toxicological Sciences* Vol. 65, No. 2, pp 166-176 (February, 2002) ISSN 1096-6080

Jemec,B. (1977). Studies of the tumorigenic effect of two goitrogens. *Cancer* Vol. 40, No. 5, pp 2188-2202 (November, 1977) ISSN 0008-543X

Jones,D.P., Maellaro,E., Jiang,S., Slater,A.F. & Orrenius,S. (1995). Effects of N-acetyl-L-cysteine on T-cell apoptosis are not mediated by increased cellular glutathione. *Immunology Letters* Vol. 45, No. 3, pp 205-209 (March, 1995) ISSN 0165-2478

Kaul,N. & Forman,H.J. (1996). Activation of NF kappa B by the respiratory burst of macrophages. *Free Radical Biology & Medicine* Vol. 21, No. 3, pp 401-405 (1996) ISSN 0891-5849

Kirlin,W.G., Cai,J., Thompson,S.A., Diaz,D., Kavanagh,T.J. & Jones,D.P. (1999). Glutathione redox potential in response to differentiation and enzyme inducers. *Free Radical Biology & Medicine* Vol. 27, No. 11-12, pp 1208-1218 (December, 1999) ISSN 0891-5849

Kono,Y. & Fridovich,I. (1982). Superoxide radical inhibits catalase. *Journal of Biological Chemistry* Vol. 257, No. 10, pp 5751-5754 (May, 1982) ISSN 0021-9258

Kumar,A., Sinha,R.A., Tiwari,M., Singh,R., Koji,T., Manhas,N., Rastogi,L., Pal,L., Shrivastava,A., Sahu,R.P. & Godbole,M.M. (2007). Hyperthyroidism induces apoptosis in rat liver through activation of death receptor-mediated pathways. *Journal of Hepatology* Vol. 46, No. 5, pp 888-898 (May, 2007) ISSN 0168-8278

Lam,S.H., Sin,Y.M., Gong,Z. & Lam,T.J. (2005). Effects of thyroid hormone on the development of immune system in zebrafish. *General and Comparative Endocrinology* Vol. 142, No. 33, pp 325-335 (July, 1995) ISSN 0016-6480

Larsen,P.R. (1982). Thyroid-pituitary interaction: feedback regulation of thyrotropin secretion by thyroid hormones. *The New England Journal of Medicine* Vol. 306, No. 1, pp 23-32 (Janu, 1982) ISSN 0028-4793

Lazar,M.A. (1993). Thyroid hormone receptors: multiple forms, multiple possibilities. *Endocrine Reviews* Vol. 14, No. 2, pp 184-193 (April, 1993) ISSN 0163-769X

Lee,P.W. & Neal,R.A. (1978). Metabolism of methimazole by rat liver cytochrome P-450-containing monoxygenases. *Drug Metabolism and Disposition: the biological fate of chemicals* Vol. 6, No. 5, pp 591-600 (October, 1978) ISSN 0090-9556

Lissi,E.A., Salim-Hanna,M., Faure,M. & Videla,L.A. (1991). 2,2'-Azo-bis-amidinopropane as a radical source for lipid peroxidation and enzyme inactivation studies. *Xenobiotica* Vol. 21, No. 8, pp 995-1001 (August, 1991) ISSN 0049-8254

Liu,W.K. & Ng,T.B. (1991). Effect of methimazole-induced hypothyroidism on alveolar macrophages. *Virchows Archiv. B, Cell Pathology Including Molecular Pathology* Vol. 60, No. 1, pp 21-26 (1991) ISSN 0340-6075

Lushchak,V.I. (2011). Adaptive response to oxidative stress: Bacteria, fungi, plants and animals. *Comparative Biochemistry and Physiology. Toxicology & Pharmacology* Vol. 153, No. 2, pp 175-190 (March, 2011) ISSN 1532-0456

Mascanfroni,I.D., del Mar Montesinos,M., Alamino,V.A., Susperreguy,S., Nicola,J.P., Ilarregui,J.M., Masini-Repiso,A.M., Rabinovich,G.A. & Pellizas,C.G. (2010). Nuclear factor (NF)-kB-dependent thyroid hormone receptor b1 expression controls dendritic cell function via Akt signaling. *Journal of Biological Chemistry* Vol. 285, No.13, pp 9569-9582 (March, 2010) ISSN 0021-9258

Mascanfroni,I., Montesinos,M.d.M., Susperreguy,S., Cervi,L., Ilarregui,J.M., Ramseyer,V.D., Masini-Repiso,A.M., Targovnik,H.M., Rabinovich,G.A. & Pellizas,C.G. (2008). Control of dendritic cell maturation and function by triiodothyronine. *The FASEB journal : official publication of the Federation of American Societies for Experimental Biology* Vol. 22, No. 4, pp 1032-1042 (April, 2008) ISSN 0892-6638

Mizutani,T., Yoshida,K. & Kawazoe,S. (1994). Formation of toxic metabolites from thiabendazole and other thiazoles in mice. Identification of thioamides as ring cleavage products. *Drug Metabolism and Disposition: the biological fate of chemicals* Vol. 22, No. 5, pp 750-755 (September-October, 1994) ISSN 0090-9556

Nakamura,R., Teshima,R., Hachisuka,A., Sato,Y., Takagi,K., Nakamura,R., Woo,G.H., Shibutani,M. & Sawada,J. (2007). Effects of developmental hypothyroidism induced by maternal administration of methimazole or propylthiouracil on the immune system of rats. *International Immunopharmacology* Vol.7, No.13 (December 2007), pp.1630-1638, ISSN 1567-5769

Neal,R.A. & Halpert,J. (1982). Toxicology of thiono-sulfur compounds. *Annual Review of Pharmacology and Toxicology* Vol.22, pp.321-339, ISSN 0362-1642

Oja,S.S., Janaky,R., Varga,V. & Saransaari,P. (2000). Modulation of glutamate receptor functions by glutathione. *Neurochemstry International* Vol.37, No. 2-3 (August 2000) pp.299-306, ISSN 0197-0186

Oppenheimer,J.H., Schwartz,H.L. & Strait,K.A. (1994). Thyroid hormone action 1994: the plot thickens. *European Journal of Endocrinology* Vol.130, No.1 (January 1994) Pp.15-24, ISSN 0804-4643

Ortiz-Butron,R., Blas-Valdivia,V., Franco-Colin,M., Pineda-Reynoso,M. & Cano-Europa,E. (2011). An increase of oxidative stress markers and the alteration of the antioxidant enzymatic system are associated with spleen damage caused by methimazole-induced hypothyroidism. *Drug and Chemical Toxicology* Vol.34, No.2 (April 2011), pp.180-188, ISSN 0148-0545

Paller,M.S. (1986). Hypothyroidism protects against free radical damage in ischemic acute renal failure. *Kidney International* Vol.29, No.6 (June 1986), pp.1162-1166, ISSN 0085-2538

Paradies,G., Petrosillo,G. & Ruggiero,F.M. (1997). Cardiolipin-dependent decrease of cytochrome c oxidase activity in heart mitochondria from hypothyroid rats. *Biochimica et Biophysica Acta* Vol.1319, No.1 (March 1997), pp.5-8, ISSN 0006-3002

Paradies,G., Ruggiero,F.M., Petrosillo,G. & Quagliariello,E. (1994). Enhanced cytochrome oxidase activity and modification of lipids in heart mitochondria from hyperthyroid rats. *Biochimica et Biophysica Acta* Vol.1225, No. 2(January 1994), pp.165-170, ISSN 0006-3002

Poulsen,L.L., Hyslop,R.M. & Ziegler,D.M. (1974). S-oxidation of thioureylenes catalyzed by a microsomal flavoprotein mixed-function oxidase. *Biochemical Pharmacology* Vol.23, No. 24 (December 1974), pp.3431-3440, ISSN 006-2952

Reed,D.J. (1990). Glutathione: toxicological implications. *Annual Review Of Pharmacology and Toxicology* Vol.30, pp.603-631, ISSN 0362-1642

Reznick,A.Z. & Packer,L. (1994). Oxidative damage to proteins: spectrophotometric method for carbonyl assay. *Methods in Enzymology* Vol.233, No.233, pp.357-363, ISSN 0076-6879

Rollins-Smith,L.A. & Blair,P. (1990). Expression of class II major histocompatibility complex antigens on adult T cells in *Xenopus* is metamorphosis-dependent. *Developmental and comparative Immunology* Vol.1, No.2, pp.97-104, ISSN

Romanque,P., Tapia,G. & Videla,L.A. (2003). Kupffer cell stimulation in the isolated perfused rat liver triggers nuclear factor-kappaB DNA binding activity. *Redox Report* Vol.8, No.6 (December 2003) pp.341-346, ISSN 1351-0002

Sap,J., Munoz,A., Damm,K., Goldberg,Y., Ghysdael,J., Leutz,A., Beug,H. & Vennstrom,B. (1986). The c-erb-A protein is a high-affinity receptor for thyroid hormone. *Nature* Vol.324, No.6098 (December 1986), pp.635-640, ISSN 0028-0836

Schafer,F.Q. & Buettner,G.R. (2001). Redox environment of the cell as viewed through the redox state of the glutathione disulfide/glutathione couple. *Free Radical Biology and Medicine* Vol.30, No.11 (June 2001), pp.1191-1212, ISSN 0891-5849

Scheiff,J.M., Cordier,A.C. & Haumont,S. (1977). Epithelial cell proliferation in thymic hyperplasia induced by triiodothyronine. *Cliical and Experimental Immunology* Vol.27, No. 3 (March 1977), pp.516-521, ISSN 1365-2249

Schneider,M.J., Fiering,S.N., Thai,B., Wu,S.Y., St.Germain,E., Parlow,A.F., St.Germain,D.L. & Galton,V.A. (2006). Targeted disruption of the type 1 selenodeiodinase gene (Dio1) results in marked changes in thyroid hormone economy in mice. *Endocrinology* Vol.147, No.1 (January 2006), pp.580-589, ISSN 0013-7227

Schoenfeld,P.S., Myers,J.W., Myers,L. & LaRocque,J.C. (1995). Suppression of cell-mediated immunity in hypothyroidism. *Southern Medical Journal* Vol.88, No.3 (March 1995), pp.347-349, ISSN 0038-4348

Schonfeld,P., Wieckowski,M.R. & Wojtczak,L. (1997). Thyroid hormone-induced expression of the ADP/ATP carrier and its effect on fatty acid-induced uncoupling of oxidative phosphorylation. *FEBS Letters* Vol.416, No.1 (October 1997), pp.19-22, ISSN 0014-5793

Shuaib,A., Ijaz,S., Hemmings,S., Galazka,P., Ishaqzay,R., Liu,L., Ravindran,J. & Miyashita,H. (1994). Decreased glutamate release during hypothyroidism may contribute to protection in cerebral ischemia. *Experimental Neurology* Vol.128, No.2 (August 1994), pp.260-265, ISSN 0014-4886

Soboll,S. (1993). Thyroid hormone action on mitochondrial energy transfer. *Biochimica et Biophysica Acta* Vol.1144, No.1 (August 1993), pp.1-16, ISSN 0006-3002

St.Germain,D.L., Galton,V.A. & Hernandez,A. (2009). Defining the roles of the iodothyronine deiodinases: current concepts and challenges. *Endocrinology* Vol.150, No.3 (March 2009), pp.1097-1107, ISSN 0013-7227

Stadtman,E.R. (1990). Metal ion-catalyzed oxidation of proteins: biochemical mechanism and biological consequences. *Free Radical Biology and Medicine* Vol.9, No.4, pp.315-325, ISSN 0891-5849

Swaroop,A. & Ramasarma,T. (1985). Heat exposure and hypothyroid conditions decrease hydrogen peroxide generation in liver mitochondria. *Biochemical Journal* Vol.226, No.2 (March 1985), pp.403-408, ISSN 0264-6021

Tapia,G., Santibanez,C., Farias,J., Fuenzalida,G., Varela,P., Videla,L.A. & Fernandez,V. (2010). Kupffer-cell activity is essential for thyroid hormone rat liver preconditioning. *Mollecular and Cellular Endocrinology* Vol.323, No.2 (July 2010), pp.292-297, ISSN 0303-7207

Tenorio-Velásquez,V.M., Barrera,D., Franco,M., Tapia,E., Hernández-Pando,R., Medina-Campos,O.N. & Pedraza-Chaverri,J. (2005). Hypothyroidism attenuates protein tyrosine nitration, oxidative stress and renal damage induced by ischemia and reperfusion: effect unrelated to antioxidant enzymes activities. *BMC Nephrology* Vol.7, pp.6-12, ISSN 1471-2369

Tsai,M.H., Chang,Y.L., Wu,V.C., Chang,C.C. & Huang,T.S. (2001). Methimazole-induced pulmonary hemorrhage associated with antimyeloperoxidase-antineutrophil cytoplasmic antibody: a case report. *Journal of the Formosan Medical Association* Vol.100, No.11 (November 2001), pp.772-775, ISSN 0929-6646

Tutuncu,T., Demirci,C., Gozalan,U., Yuksek,Y.N., Bilgihan,A. & Kama,N.A. (2007). Methimazole protects lungs during hepatic ischemia-reperfusion injury in rats: an effect not induced by hypothyroidism. *Journal of Gastroenterology and Hepatology* Vol.22, No.5 (May 2007), pp.704-709, ISSN 1440-1746

Valko,M., Leibfritz,D., Moncol,J., Cronin,M.T., Mazur,M. & Telser,J. (2007). Free radicals and antioxidants in normal physiological functions and human disease. *International Journal of Biochemestry and Cell Biology* Vol.39, No.1 (January 2007), pp.44-84, ISSN 1357-2725

Van Acker,S.A.B.E., Koymans,L.M.C. & Bast,A. (1993). Molecular pharmacology of vitamin E: structural aspects of antioxidant activity. *Free Radical Biology and Medicine* Vol.15, No.3 (September 1993), pp.311-328, ISSN 0891-5849

Venditti,P. & Di Meo,S. (2006). Thyroid hormone-induced oxidative stress. *Cellular and Molecular Life Sciences* Vol.63, No.4 (February 2006), pp.414-434, ISSN1420-682X

Videla,L.A. (2000). Energy metabolism, thyroid calorigenesis, and oxidative stress: functional and cytotoxic consequences. *Redox Report* Vol.5, No.5 (October 2000), pp.65-275, ISSN 1351-0002

Villa-Verde,D.M., de Mello-Coelho,V., Farias-de-Oliveira,D.A., Dardenne,M. & Savino,W. (1993). Pleiotropic influence of triiodothyronine on thymus physiology. *Endocrinology* Vol.133, No.2 (August 1993) 867-875, ISSN 0013-7227

Wang,J., Ma,H., Boor,P.J., Ramanujam,V.M., Ansari,G.A. & Khan,M.F. (2010). Up-regulation of heme oxygenase-1 in rat spleen after aniline exposure. *Free Radical Biology and Medicine* Vol.48, No.4 (February 2010), pp.513-518, ISSN 0891-5849

Wang,J., Wang,G., Ansari,G.A. & Khan,M.F. (2008). Activation of oxidative stress-responsive signaling pathways in early splenotoxic response of aniline. *Toxicology and Applied Pharmacology* Vol.230, No.2 (July 2008), pp.227-234, ISSN 0041-008X

Wang,J., Wang,G., Ma,H. & Khan,M.F. (2011). Enhanced expression of cyclins and cyclin-dependent kinases in aniline-induced cell proliferation in rat spleen. *Toxicology and Applied Pharmacology* Vol.250, No.2 (January 2011), pp.213-220, ISSN 0041-008X

Watanabe,K., Iwatani,Y., Hidaka,Y., Watanabe,M. & Amino,N. (1995). Long-term effects of thyroid hormone on lymphocyte subsets in spleens and thymuses of mice. *Endocrine Journal* Vol.42, No.5 (October 1995), pp.661-668, ISSN 1348-4540

Weber,G.F. (1999). Final common pathways in neurodegenerative diseases: regulatory role of the glutathione cycle. *Neuroscience and Biobehavioral Reviews* 23, No.8 (December 1999), pp.1079-1086, ISSN 0149-7634

Woeber,K.A. (2002). Methimazole-induced hepatotoxicity. *Endocrine Practice* Vol.8, No.3 (May 2002), pp.222-224, ISSN 1530-891X

Yen,P.M. (2001). Physiological and molecular basis of thyroid hormone action. *Physiological Reviews* Vol. 81, No.3 (July 2001) , pp.1097-1142, ISSN 0031-9333

Zhang,J. & Lazar,M.A. (2000). The mechanism of action of thyroid hormones. *Annual Review of Physiology* Vol.62, No.62 (March 2000), pp.439-466, ISSN 0066-4278

8

Retinal Vein Occlusion Induced by a MEK Inhibitor – Impact of Oxidative Stress on the Blood-Retinal Barrier

Amy H. Yang and Wenhu Huang
Drug Safety Research & Development,
Pfizer Inc., La Jolla Laboratories,
USA

1. Introduction

The retina is a highly specialized sensory organ that transduces light energy into neural signal. It also has high energy requirement and an extensive vascular network. Reactive oxygen species (ROS) generated via light exposure, normal energy production, phagocytosis of spent photoreceptor membranes by retinal pigment epithelium (RPE) cells, and circulating toxins render retina at an increased risk for oxidative stress (Hardy et al., 2005; Siu et al., 2008). To cope with the high oxidant load, the retina is equipped with various antioxidant defense mechanisms, such as the expression of glutathione peroxidase and superoxide dismutase, the production of glutathione by Müller cells, high levels of vitamins C and E, and the presence of free radical scavenger melanin in RPE cells (Siu et al., 2008). However, when the redox balance is disrupted, retinal pathologies could result, and one of the consequences is impairment of the blood retinal barrier (BRB). Indeed, several retinal diseases have been shown or postulated to be linked to a state of oxidative stress and resulting BRB dysfunction.

Previously, we investigated the molecular mechanisms towards the development of retinal vein occlusion (RVO) in cancer patients treated with a mitogen-activated protein kinase kinase (MEK) inhibitor, PD0325901 (LoRusso et al., 2010). Through gene expression profiling analysis, we identified several mechanisms relevant to the development of RVO, including oxidative stress response, acute phase and inflammatory response, blood-retinal barrier (BRB) breakdown, leukostasis, and coagulation cascade (Huang et al., 2009).

This chapter aims to provide an overview of BRB structures and functions, the role of oxidative stress in BRB disruption and development of retinal pathologies, a detailed overview of RVO, and finally, a description of proposed mechanisms of PD0325901-induced RVO, highlighting several important cellular and molecular processes relevant to this pathology.

2. Blood-retinal barrier

The BRB has an endothelial and an epithelial component, namely the tight junctions between the endothelial cells of the inner retinal vessels, and those between cells of the RPE;

these cell types comprise integral components of the inner and outer BRB, respectively (Siu et al., 2008; Fig. 1A). The BRB regulates the transport of fluid and molecules between the retinal tissue and vasculature, hence playing an important role in maintaining the homeostatsis of the retinal microenvironment (Kaur et al., 2008; Siu et al., 2008).

2.1 Inner blood-retinal barrier

The inner BRB is composed of endothelial cells, astrocytes, Müller cells, and pericytes (Fig. 1A). Tight junctions between capillary endothelial cells form the basis of the inner BRB (Fig. 1B). Astrocytes, Müller cells and pericytes, all closely associated with the endothelial cells of the inner BRB, contribute to proper BRB functions.

The endothelial cells of inner retinal capillaries are not fenestrated, contributing to their low permeability. Solutes traverse the retinal endothelium via both the transcellular and paracellular pathways: the former involves vesicle-mediated transport of macromolecules, and the latter, passage through minute intercellular space safeguarded by junctional proteins (Vandenbroucke et al., 2008; Fig. 1B). Tight junctions consist of occludins, claudins, and junctional adhesion molecules (JAMs), all of which form complexes between adjacent endothelial cells. Zonula occludens (ZO) proteins link occludins and claudins to the endothelial actin cytoskeleton via cingulin. In addition to tight junctions, adherens junctions (AJ) also contribute to the endothelial barrier, where vascular endothelial (VE) cadherins on adjacent endothelial cells form a homophilic complex (Garrido-Urbani et al., 2008; Vandenbroucke et al., 2008). The C-terminal domain of VE-cadherin binds β-catenins and α-catenins, linking the AJs to the actin cytoskeleton. Several lines of evidence show that the regulation of actin cytoskeletal dynamics is central to the proper functioning of the endothelial barrier (Houle and Huot, 2006; Houle et al., 2003; Huot et al., 1998; Lum and Roebuck, 2001). It has also been reported that retinal endothelial cells are more susceptible to oxidative damage, leading to increased permeability, than endothelial cells at other sites. Indeed, ROS are also known to induce the expression of vascular endothelial growth factor (VEGF), a well-known endothelial mitogen and permeability factor, which contributes to the breakdown of BRB in experimental diabetes models (Chua et al., 1998; El-Remessy et al., 2003).

Pericytes line the outer surface of endothelial cells (Fig. 1A) and are contractile in nature, expressing actin, myosin, and tropomyosin (Kaur et al., 2008). They contract in response to signals such as hypoxia, endothelin-1, and angiontensin II, and relax on exposure to carbon dioxide, nitric oxide and adenosine. Therefore, pericytes regulate the vascular tone and blood flow. Under normoxia, they maintain the integrity of the inner BRB by inducing mRNA and protein expression of occludin and ZO-1, and by partially reversing the occludin decrease under hypoxia. Loss of pericytes and disruption of inner BRB are early events in diabetes.

Müller cells are the principal glial cells of the retina, and a functional link between neurons and vessels (Reichenbach et al., 2007). They span the inner and outer limiting membranes of the retina, with their foot processes in close contact with the retinal endothelial cells (Fig. 1A). Under physiological conditions, Müller cells contribute to the integrity of the BRB; however, when exposed to cellular stress they impair the barrier function. Under normoxia, Müller cells secrete pigment epithelium-derived factor (PEDF),

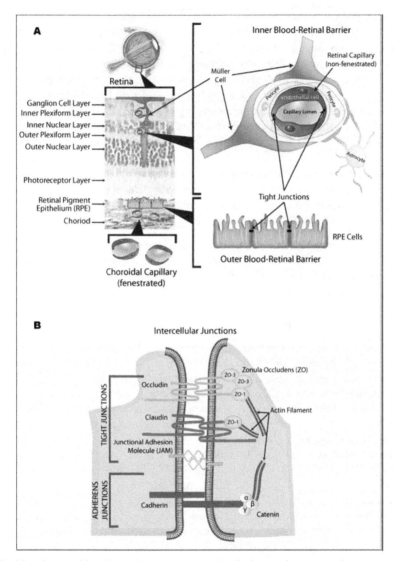

Fig. 1. The blood-retinal barrier (A) The retina is a multi-layered tissue in the posterior segment of the eye, and is shown by the H&E stained micrograph on the left. The cell types comprising the inner BRB (endothelial cells, pericytes, astrocytes, and Müller cells) and outer BRB (retinal pigment epithelial cells) are overlaid on the retinal micrograph, and are magnified on the right. Tight junctions between retinal capillary endothelial cells and retinal pigment epithelial cells form the basis of the inner and outer BRB, respectively. The endothelial cells of inner retinal capillaries are not fenestrated, whereas those of the choroidal capillaries are (depicted below the retinal micrograph). (B) Protein components of the intercellular junctions. Sources: Kaur et al., 2008; Niessen, 2007; http://www.landesbioscience.com/curie/images/chapters/Hosoya1.jpg

which downregulates VEGF expression and decreases vascular permeability (Yafai et al., 2007). Under hypoxia (Kaur et al., 2008) and oxidative stress (Yoshida et al., 2009), PEDF expression is decreased in Müller cells, thus favoring the secretion of VEGF and breakdown of the inner BRB. In addition, Müller cells are a source of matrix metalloproteinases, which proteolytically degrade the tight junction protein occludin, impairing the barrier function of retinal endothelial cells under cellular stress conditions. Müller cells also play a vital role in maintaining the retinal fluid balance (Reichenbach et al., 2007). Under physiological conditions, Müller cells carry out transcellular water transport from the retinal interstitial space into the blood, thus preventing excess fluid buildup within the retina. The transcellular water transport is osmotically coupled to the transport of potassium ions. When exposed to oxidative stress and inflammation, Müller cells have been shown to contribute to retinal edema through a disturbed intracellular fluid transport. Finally, Müller cells also respond to oxidative stress by increasing their production of the antioxidant glutathione (Siu et al., 2008).

Similar to pericytes and Müller cells, astrocytes are closely associated with the retinal vessels (Fig. 1A). They help maintain the BRB integrity by increasing the expression of the tight junction protein ZO-1 and modifying endothelial morphology (Kaur et al., 2008). Dysfunction of astrocytes has been linked to inner BRB breakdown and vasogenic edema.

2.2 Outer blood-retinal barrier

The outer BRB consists of tight junctions between RPE cells (Fig. 1A). The RPE is a monolayer of cells between the neuroretina and the choroid, and regulates access of blood components to the retina. Similar to the endothelium, movement across RPE is both transcellular and paracellular. The RPE cells exhibit a polarized morphology, with apical microvilli in contact with photoreceptor outer segments, and basal infoldings adjacent to the Bruch's membrane that separates the retina from choroidal capillaries called choriocapillaris (Kaur et al., 2008). Unlike the capillaries of the inner retina, the choroidal capillaries are fenestrated (Fig. 1A) and therefore do not contribute to the outer BRB (Kaur et al., 2008; Siu et al., 2008). Na+,K+-ATPase and aquaporin 1 (AQP1) expressed on the apical surface regulate movement of sodium, potassium, and water molecules across the RPE. Tight junctions, located at the apical side of the lateral membrane of the RPE cells, restrict paracellular movement between neighboring RPE cells. In addition to its function to regulate molecular transport integral to the outer BRB, RPE is responsible for phagocytizing photoreceptor outer segment membranes, which are digested by an extensive lysosome system, whose waste products are removed by the adjacent choriocapillaris (Burke, 2008; Siu et al., 2008). In pigmented animals, RPE cells also express melanin, a free radical scavenger that is also capable of absorbing stray light, and is thought to contribute to the retinal antioxidant mechanisms (Siu et al., 2008).

3. Oxidative stress, BRB dysfunction and ocular diseases

3.1 Sources of ROS in the Retina

In addition to the mitochondria, cellular sources of ROS in retina include endothelial cell xanthine oxidase, NAD(P)H oxidase, cyclooxygenase (COX), nitric oxide synthase (NOS), and lipoxygenase pathways (Frey and Antonetti, 2011; Hardy et al., 2005; Kunsch and Medford, 1999). ROS can also act as intracellular second messengers and activate various signaling transduction pathways.

3.2 Ocular pathologies associated with oxidative stress and BRB dysfunction

3.2.1 Diabetic retinopathy

Diabetic retinopathy is a significant cause of blindness. Tissue hypoxia and hyperglycemia are generally regarded as contributors to diabetic retinopathy, but how these lead to the disease state is unclear. Current hypotheses, which are not mutually exclusive, for pathogenic mechanisms leading to diabetic retinopathy include oxidative stress, hemodynamic changes, inflammation and the activation of microglia, and increased leukocyte adhesion to the endothelial cells and entrapment (leukostasis) (Chibber et al., 2007). Many of these processes in fact have an association with excessive production of ROS (Yang et al., 2010b). For instance, growing evidence supports an important role for leukostasis in the development of diabetic retinopathy, with downstream consequences including capillary occlusion and localized production of ROS, resulting in endothelial cell damage, BRB breakdown, and increased vascular permeability (Chibber et al., 2007). Inflammatory cytokines interact with cell surface receptors in various cell types to activate signaling pathways that mediate responses of cell adhesion, permeability and apoptosis; they also increase the production of ROS by mitochondria (Busik et al., 2008; Sprague and Khalil, 2009). There are several models that recapitulate various aspects of diabetic retinopathy, including streptozotocin-induced diabetes and ischemia-reperfusion injury.

3.2.2 Retinopathy of prematurity

Retinopathy of prematurity (ROP) is a vasoproliferative disease that often develops when premature infants are given supplemental oxygen, and is a leading cause of blindness in children (Hardy et al., 2005; Uno et al., 2010). The developing eye is at an increased risk for oxidative injury from hyperoxia, as the retinal vasculature in premature infants lacks fully-developed mechanisms to auto-regulate oxygen tension (Hardy et al., 2005). ROP develops in two phases (Hardy et al., 2005). Hyperoxia in the retina leads to cessation of vascular development, resulting in endothelial cell death, vaso-obliteration and consequently, ischemia. To re-establish retinal perfusion, the retina mounts an exaggerated intravitreal preretinal neovascularization, which may ultimately result in retinal detachment and vision loss. Many features of ROP are recapitulated in oxygen induced retinopathy (OIR) (Brafman et al., 2004; Gu et al., 2002; Uno et al., 2010), in which neonatal animals are exposed to hyperoxia, leading to the generation of ROS, which have been postulated to be causal for vaso-obliteration, death of endothelial cells, and consequently, impairment of the BRB.

3.2.3 Age-related macular degeneration

One of the major causes of blindness in the elderly population, age-related macular degeneration (AMD) is characterized by regional degeneration of photoreceptors and the RPE, lipofuscin accumulation in RPE cells, chronic inflammation, and drusen formation. Chronic oxidative stress has also been suggested to be an important factor to the pathogenesis of AMD. As alluded to earlier, RPE may be at a high risk for oxidative stress due to its location and function. The RPE is in an oxygen-rich environment adjacent to the choriocapillaris, is continuously exposed to light, sometimes at phototoxic wavelengths, and is responsible for the renewal of photoreceptor outer segments via phagocytosis (Burke, 2008). The high content of polyunsaturated fatty acids of these membrane segments make them susceptible to lipid peroxidation and subsequent free radical formation. Experimental evidence supporting a role of oxidative stress in AMD showed that supplementation of antioxidants in AMD patients has a protective effect, and that cigarette smoking, known to be a source of exogenous free radicals, is a risk factor for AMD (Burke, 2008). Given the central role RPE plays in AMD pathogenesis, a commonly used experimental model to study the link between oxidative stress and AMD involves the use of cultured human RPE (ARPE-19) cells. Common endpoints include cell survival, morphology, activation of signaling cascades, and cytokine production (Chan et al., 2008; Dong et al., 2011; Glotin et al., 2006; Jiang et al., 2009; Klettner and Roider, 2009; Qin et al., 2006; Tsao et al., 2006; Wang et al., 1998; Wu et al., 2010).

4. Retinal vein occlusion

Retinal vein occlusion is a vascular disorder of the retina that occurs when one or more of the retinal veins are blocked and the circulation of retinal blood becomes obstructed. This ocular pathology can be a primary lesion or secondary to other retinal diseases. With the blockage, poor venous drainage and increased retinal capillary pressure and permeability ultimately lead to retinal ischemia and edema. Retinal ischemia could lead to the generation of ROS, impacting the integrity of the BRB. Diabetic retinopathy and RVO are the two most common causes of inner BRB breakdown. Among complications found in the clinical examination are hemorrhages, edema, ischemia, neovascularization of the retina as well as increased intraocular pressure. Depending on the location and severity, loss of visual acuity can range from very mild to severe. While some patients with RVO may not have any symptoms, some patients may complain of blurred vision or visual field defects. In severe cases, RVO can lead to vision loss in the affected eye. The most common cause of decreased vision is macular degeneration secondary to RVO, which occurs when leakage within macula leads to macular edema and ischemia. Neovascularization and neovascular glaucoma are the other vision-threatening complications that are devastating for patients with RVO and that should be promptly diagnosed and treated. Vein occlusion is commonly diagnosed by examining the fundus with ophthalmoscope for characteristic morphological changes such as venous tortuosity, cotton-wool spots, dot and flame hemorrhage, and edematous optic nerves, and by fluorescein angiography for vasculature blood flow obstruction, leakage in the retina, retinal ischemia, aneurysm, neovascularization, and macular edema. Sometimes optical coherence tomography (OCT) is used to measure retinal thickness for the determination of the presence of macular edema. Central and peripheral visual disturbance should be evaluated by functional tests in the physical examination.

4.1 Classification of RVO

Retinal vein occlusion is primarily classified into central retinal vein occlusion (CRVO) and branch retinal vein occlusion (BRVO) based on the location of obstruction. In CRVO, the occlusion of the central retinal vein can slow or stop blood from leaving the retina and therefore most of the retina is affected. In BRVO, when macular venules are occluded, a vision decrease can occur depending on the amount of ischemia and edema. When one of the vein's two trunks is blocked and half of the retina is affected, it is called hemi-central retinal vein occlusion (HRVO). According to several RVO epidemiology studies, the prevalence of both CRVO and BRVO increases significantly with age, more in middle-aged and elderly populations and uncommon in young adults under the age of 40. Most patients with CRVO are male and over 65 years of age, but there seems to be no gender difference for BRVO. Most CRVO cases are unilateral and painless and only 6-14% of cases are found to be bilateral (Cheung et al., 2008; Klein et al., 2008; Lim et al., 2008; Marcucci et al., 2011; Xu et al., 2007). A recent combined world-wide data pool, containing 68,751 individuals with ages ranging from 30 to 101 years, suggested that approximately 16 million people are affected by RVO with 5.2 per 1000 for any RVO, 4.42 per 1000 for BRVO and 0.8 per 1000 for CRVO. The incidence of CRVO was lower than that of BRVO in all ethnic populations (Rogers et al., 2010). However, CRVO is the most clinically relevant RVO as it is associated with severe vision loss, especially for the ischemic (non-perfused or hemorrhagic retinopathy) RVO. Among the complications of RVO, the devastating neovascular glaucoma resulting from anterior segment neovascularization is seen only in ischemic CRVO. Fortunately, most cases (81%) (Hayreh et al., 1994) are of the non-ischemic type that rarely develops blindness-causing complications.

4.2 Risk factors of RVO

CRVO and BRVO have different symptoms, risk factors, pathogenesis, and therefore treatment. The pathogenesis for RVO is multifactorial and still under investigation. Anatomical positions of retinal veins play an important role in the pathogenesis of RVO (Fraenkl et al., 2010). The central retinal artery and vein share a common adventitial sheath in the optic nerve head. In CRVO, the tract of central retinal vein passing through the narrowing lamina cribrosa is the most frequent site of occlusion. In BRVO, vein occlusions occur at the junction of retinal vein and artery crossings in the retina. The mechanical compression of the veins at the narrowing passage or arteriovenous crossings predispose retinal veins to thrombus formation by various factors, including slowed or disturbed blood flow, endothelial damage in the vessel wall, changes in the blood viscosity, perivascular changes such as in lamina cribrosa (Albon et al., 1995), and sclerotic changes in the retinal arteries. Ocular risk factors associated with RVO include glaucoma or ocular hypertension. In glaucoma, increased intraocular pressure causes mechanical compression of retinal veins, which may induce RVO.

RVO has often been associated with a variety of systemic vascular disorders including arterial hypertension, arteriosclerosis, diabetes mellitus, dyslipidemia, and systemic vasculitis (The Eye Disease Case-Control Study Group, 1993, 1996; Koizumi et al., 2007; Mitchell et al., 1996; Sperduto et al., 1998). The increased rigidity of arterial wall affiliated with these diseases may result in compression of retinal veins.

Abnormal blood viscosity, platelets, and coagulation have also been suggested to be involved in RVO pathogenesis (Trope et al., 1983). Hematological dysfunction, such as increased plasma fibrinogen and disruption of the thrombosis-fibrinolysis balance, have been implicated in the development of RVO (Rehak and Rehak, 2008). Increased fibrinogen has been associated with RVO in several clinical reports (Lip et al., 1998; Patrassi et al., 1987; Peduzzi et al., 1986). An increasing number of studies have sought to establish an association between RVO and thrombophilic abnormalities. Thrombophilic risk factors related to RVO include hyperhomocysteinemia, methylenetetrahydrofolate reductase (MTHFR) gene mutation, factor V Leiden mutation, protein C and S deficiency, antithrombin deficiency, prothrombin gene mutation, anticardiolipin antibodies and lupus anticoagulant. High levels of circulating homocysteine may damage the vascular endothelium by releasing free radicals, creating a hypercoagulable environment (Angayarkanni et al., 2008). It appears that there is an association between RVO and hyperhomocysteinaemia and anti-phospholipid antibodies. However, for the other thrombophilic risk factors, there is a lack of consistency among the studies and the association with RVO is inconclusive (Fegan, 2002; Janssen et al., 2005; Rehak and Rehak, 2008). More recently, elevated levels of soluble endothelial protein C receptor (sEPCR) emerged an important candidate risk factor especially for CRVO (Gumus et al., 2006).

Significantly increased concentrations of growth factors, cytokines, and chemokines such as VEGF, interleukin (IL)-6, IL-8, interferon-inducible 10-kDa protein (IP-10), monocytochemotactic protein-1 (MCP-1), and platelet-derived growth factor (PDGF)-AA were observed in vitreous or aqueous humor samples of patients with RVO (Funk et al., 2009; Noma et al., 2009; Yoshimura et al., 2009). Excessive production of VEGF and inflammatory cytokines can be induced by ischemic conditions. The levels of VEGF and inflammatory cytokines are correlated with severity of retinal ischemia and macular edema (Noma et al., 2006), as well as neovascularization. A close correlation between aqueous VEGF levels and iris neovascularization and vascular permeability in CRVO patients has been found (Boyd et al., 2002).

4.3 Oxidative stress and RVO

Retinal ischemia that occurs in some cases of RVO could lead to the generation of ROS, and compromise the integrity of the BRB. In fact, RVO is a common complication of diabetic retinopathy, in which hypoxia-ischemia is thought to play a role in its pathogenesis. Many of the risk factors for RVO described above, such as alterations in blood flow, systemic vascular disorders, hypercoagulability, and elevated levels of pro-inflammatory cytokines, may also be associated with a state of oxidative stress (Simoncini et al., 2005). Indeed, in a case-control prospective study in young adult CRVO patients, serum levels of paraoxonase-1 arylesterase (PON1-ARE) activity, reported to have antioxidant potential, were found to be negatively correlated with hyperhomocysteinemia and lipid peroxidation, an indicator of oxidative stress (Angayarkanni et al., 2008). Decreased levels of PON1-ARE activity as well as increased levels of the lipid peroxidation marker were shown to be risk factors for CRVO. In another case study, an individual with glucose-6-phosphate dehydrogenase(G6PD) deficiency was exposed to an oxidative stressor, and later developed CRVO (Kotwal et al., 2009). G6PD deficiency is known to increase erythrocyte vulnerability to oxidative stress, which may precipitate hemolysis, increased erythrocyte aggregation and erythrocyte-endothelium interaction, leading to thrombosis (Kotwal et al., 2009). Anti-phospholipid antibodies have been associated with the development of RVO, and shown to induce oxidative stress in endothelial cells (Simoncini et al., 2005). Taken together, these lines of evidence suggest that a state of oxidative stress may predispose individuals to RVO.

4.4 Therapeutics associated with the clinical presentation of RVO

In addition to RVO that arises due to pathophysiological causes described above, this ocular disorder can also develop as an adverse event from treatment with certain therapeutics.

4.4.1 Interferon-α

Interferon-α (IFN-α) is used for the treatment of many cancers and chronic hepatitis C. Interferon-associated retinopathy has been documented since the 1990s, most commonly characterized by hemorrhage and cotton-wool spots, and sometimes by macular edema, retinal vascular occlusion, and retinal ischemia. The RVO could involve either the vein or artery, or both, and in most cases is reversible. The exact pathophysiological mechanism of interferon-induced retinopathy is unknown, although there are similarities with early stages of diabetic retinopathy (Bajaire et al., 2011; Esmaeli et al., 2001). Several risk factors have been suggested, including hypertension, hyperlipidemia, a hypercoagulable state, and diabetes (Nadir et al., 2000). In addition, IFN-α is known to cause the development of autoantibodies in 10% of the patients receiving treatment, and to exacerbate certain systemic autoimmune diseases. It is speculated that IFN-α therapy might cause deposition of immune complexes in retinal vasculature, with sequelae of retinal ischemia, hemorrhages and cotton wool spots.

4.4.2 Tumor necrosis factor

Tumor necrosis factor (TNF) is a proinflammatory cytokine that has been implicated in various diseases, including autoimmune diseases, diabetes, and cancer. In a phase II trial of recombinant TNF in patients with advanced colon cancer, TNF was administered by i.v. infusions twice daily for 5 consecutive days every other week for 8 weeks (Kemeny et al., 1990). Two out of 16 patients developed retinal vein thrombosis several weeks following completion of therapy. This finding is consistent with the known role of TNF in vascular leakage and blood-retinal barrier breakdown in diabetic retinopathy (Frey and Antonetti, 2011). In support of this, a patient with macular edema secondary to BRVO saw an improvement in visual acuity and cessation of macular edema during treatment with infliximab, a TNF-α antibody, administered for rheumatoid arthritis (Kachi et al., 2010). Paradoxically, infliximab therapy has also been linked in several case studies to the development of retinal vein thrombosis/occlusion in a patient being treated for ulcerative colitis (Veerappan et al., 2008), psoriasis (Vergou et al., 2010), or Crohn's disease (Puli and Benage, 2003). The temporal relationship between infliximab infusion and retinopathy suggested the two may be causally related. In two of the three cases, a medical history of myocardial infarction or hyperlipidemia was noted, both of which considered risk factors for RVO. Moreover, all three of these diseases are inflammatory in nature, and may predispose patients to weakened BRB.

4.4.3 MEK inhibitor PD0325901

PD0325901 is a potent and selective MEK inhibitor, developed for the treatment of advanced cancer. MEK is a key molecule in the Ras-mitogen-activated protein kinase (MAPK) pathway, which has roles including cellular proliferation and survival, and its only known substrate is the extracellular signal-regulated kinase (ERK), which in turn phosphorylates and activates downstream molecules in the pathway (Fig. 2). In the phase I dose escalation clinical trial of PD0325901, dose-limiting RVO was observed, characterized by the presence of cotton wool spots, hemorrhages, and vein occlusion. RVO developed in 2 patients after 3.5-4 months of 10 or 15 mg BID continuous treatment schedule, and in 1 patient after 9 months of 10 mg BID on a 5 days on/2 days off schedule, and was reversible upon treatment discontinuation (LoRusso et al., 2010). It was noted that at doses >= 4 mg BID, the systemic exposure of PD0325901 was equivalent to that in animal models that resulted in 90% phosphorylated ERK (pERK) suppression (LoRusso et al., 2010). Therefore, the ocular lesions could be related to the prolonged and/or significant levels of pERK suppression.

Two other MEK inhibitors, CI-1040 and selumetinib (AZD6244), also progressed to the clinic, but did not cause RVO. CI-1040 is a structural analogue of PD0325901. Insufficient clinical efficacy was reported due to poor bioavailability and metabolic instability (Rinehart et al., 2004). Selumetinib caused blurred vision in 12% of patients at >=100 mg BID in a Phase I trial (Adjei et al., 2008); this finding was not reported in subsequent Phase II trials at the 100 mg BID dose (Bekaii-Saab et al., 2011; Bodoky et al., 2011). Compared to PD0325901, selumetinib is approximately 10-fold less potent, and has a relatively poor bioavailability. Taken together, even though it is at present unclear whether PD0325901 caused RVO due to its deep inhibition of pERK, or to its chemotype, the above evidence suggests that the incidence of ocular lesions correlates with the efficacy of MEK inhibition.

5. Molecular mechanisms of MEK inhibitor PD0325901-Induced RVO

To develop an animal model of RVO to investigate mechanisms of toxicity, an in-life study was performed in rabbits, in which PD0325901 was administered by intravitreal injection at doses of 0.5 and 1 mg/eye, with an observation period of 2 weeks (Huang et al., 2009). The high dose was extrapolated to be a potentially toxic dose, while the low dose was chosen as a subtoxic dose, based on in vitro cytotoxicity data (Huang et al., 2009). As early as 1 day after treatment, the high dose produced hemorrhages and vascular leakage with branch occlusion. These lesions progressed to retinal detachment, edema, abnormal kinetic blood flow, and retinal vessel occlusion after 7 days. At the low dose, retinal vascular leakage was observed without vascular occlusion. Therefore, the rabbit model provided evidence that PD0325901 at sufficient ocular concentrations could lead to similar retinal lesions seen in the clinic.

The retinal vascular toxicity was not observed in preclinical safety studies where PD0325901 was administered orally in rats and dogs for up to 13 weeks (Huang et al., 2009). The difference in the level of ocular toxicity between rabbits and rats/dogs could be attributed to ocular drug concentration differences arising from local vs. systemic routes of administration. Since molecular events could precede overt signs of tissue injury, an investigative study was conducted in which rats were dosed orally for 3 or 5 days at 45 mg/kg/day, estimated to be at 70% maximal tolerated dose (Huang et al., 2009). No retinal toxicity was observed by ophthalmic examinations or fundus fluorescein angiography. Despite the absence of overt injury, global gene expression profiling on vehicle and PD0325901-treated retinas revealed several mechanisms relevant to the development of RVO, including oxidative stress response, acute phase and inflammatory response, BRB

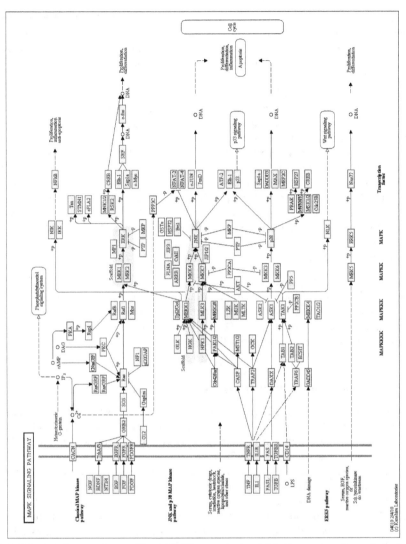

Fig. 2. MAPK signaling pathway The MAPK cascade is a highly conserved module that is involved in various cellular functions, including cell proliferation, differentiation and migration. Mammals express at least four distinctly regulated groups of MAPKs, ERK1/2, JNK1/2/3, p38alpha/beta/gamma/delta and ERK5, that are activated by specific MAPKKs: MEK1/2 for ERK1/2, MKK3/6 for the p38, MKK4/7 (JNKK1/2) for the JNKs, and MEK5 for ERK5. Each MAPKK, however, can be activated by more than one MAPKKK, increasing the complexity and diversity of MAPK signalling. Presumably each MAPKKK confers responsiveness to distinct stimuli. For example, activation of ERK1/2 by growth factors depends on the MAPKKK c-Raf, but other MAPKKKs may activate ERK1/2 in response to pro-inflammatory stimuli. Source: KEGG (http://www.genome.jp/dbget-bin/www_bget?map04010) (Kanehisa, 2000, 2012)

breakdown, leukostasis, and activation of coagulation cascade (Huang et al., 2009). Progressive induction of oxidative stress response genes was observed over time, suggesting the tissue was mounting a response against ongoing oxidative stress. The induced genes encoding for antioxidant proteins include heat shock protein 27 (HSP27), αβ-crystalline, and those involved in glutathione synthesis/metabolism (GCLM, GSS, GSTs), and adhesion molecules; the only repressed genes were glutaredoxin 2 and peroxiredoxins. Of the oxidative stress response genes induced in this study, HSP27 and αβ-crystalline show some of the greatest magnitude of induction. These are small heat shock proteins that have diverse cytoprotective functions, including modulation of ubiquitin-proteosome pathway, inhibition of apoptosis, and increased resistance to oxidative stress and inflammation (Arrigo et al., 2007). Notably, the ubiquitin-proteosome pathway was significantly perturbed on both day 3 and day 5, which could be a response to misfolded proteins arising from oxidative stress. Antigen presentation by retinal cells is also a cited response to oxidative stress (Tezel et al., 2007; Zhang et al., 2005), consistent with the induction of β-2-microglobulin gene of the major histocompatibility complex (MHC) class I on both day 3 and day 5. MHC class I molecules are normally expressed in the vascular endothelium and RPE (Zhang et al., 1997).

Multiple lines of evidence demonstrate that ROS mediate activation of the MAPK signaling pathway, which in turn modulates inflammation, intercellular junction assembly, actin cytoskeleton reorganization, and water transport, all of which are of critical importance to the maintenance of the BRB integrity. Disruption of one of the key MAPK pathways by PD0325901 could contribute to impaired BRB integrity, ultimately leading to retinal edema and RVO.

5.1 MAPK activation and oxidative stress response in the retina

Three main MAPK groups have been identified – ERKs, p38 MAPKs, and c-Jun N-terminal kinases (JNK). Typically, activation of ERKs are associated with growth-related signals, whereas p38 MAPKs and JNKs become activated in response to stress stimuli, including inflammation and oxidative stress (Fig. 2). However, the exact responses of these kinases in different cell types and tissue microenvironment under various experimental stimuli have proven to be more dynamic and less dichotomous than characterized above. Some degree of cross-talk also exist between these pathways (Houle and Huot, 2006). In the retina, the dynamic balance and cross-talk of these MAPK signaling pathways in cell types comprising the BRB, chiefly RPE, endothelial, and Müller cells, has been shown in experimental systems to be critical for modulating the integrity of the BRB.

The role of the MAPK pathway has been extensively investigated in cultured RPE cells (ARPE-19) following experimentally induced oxidative stress. In response to oxidants such as hydrogen peroxide or tert-butyl hydroperoxide, MAPK (most notably ERKs and p38 MAPK) activation has been shown to either protect against or exacerbate oxidative injury, differentiated by the amount of RPE cell death (Chan et al., 2008; Dong et al., 2011; Glotin et al., 2006; Jiang et al., 2009; Klettner and Roider, 2009; Qin et al., 2006; Tsao et al., 2006; Wang et al., 1998; Wu et al., 2010). These findings also raise questions as to the impact of oxidative stress on the outer BRB when the MAPK pathway is modulated pharmacologically by a MEK inhibitor. In endothelial cells, ROS are known to modulate the expression of redox-sensitive signaling pathways, including the MAPK cascades (Kunsch and Medford, 1999;

Ushio-Fukai and Alexander, 2004), and inhibition of the ERK1/2 has been shown to lead to endothelial apoptosis (Huot et al., 1998). In a porcine model of retinal ischemia-reperfusion, Müller cells exhibited increased levels of glial fibrillary acidic protein (GFAP) and phosphorylated ERK proteins, implicating ERK in the process of glial activation in response to oxidative/ischemic stress (Wurm et al., 2011).

The following sections highlight specific MAPK-mediated molecular and cellular events in response to oxidative stress that are important for the maintenance of BRB function. The published data at times reveal contradictory findings with regards to whether the MAPKs are protective or disruptive in modulating these cellular processes following oxidative stress, underscoring the complexicity of these signaling pathways. In each section, literature review is followed by a discussion of relevant gene expression profiling data from the rat investigative study in which PD0325901 was administered orally for 3 or 5 days at 45 mg/kg/day.

5.1.1 MAPK activation and Inflammation in the retina

Inflammation is a non-specific response to injury, and involves a plethora of cellular and molecular mediators. Chronic inflammatory processes are also an important source of ROS in the retina, and have been implicated in ocular diseases such as diabetic retinopathy and AMD. Both oxidative stress and inflammation negatively impact the integrity of the BRB. MAPKs mediate some of the downstream effects of proinflammatory cytokines such as IL-1, IL-6 and TNF-α (Du et al., 2010; Larrayoz et al., 2010; Wang et al., 2010).

In the 5-day rat investigative study involving PD0325901, hematology analysis revealed an increased number of phagocytes (neutrophils and monocytes) and increased plasma fibrinogen levels, indicating a mild inflammation in the compound treated animals. Consistently, the gene expression of many acute phase response proteins, such as lipocalin 2, fibronectin, fibrinogen, ferritin light chain, complement proteins, and coagulation factors, are significantly induced on day 5. In addition to being an acute phase response protein, fibrinogen is also a key player in the coagulation cascade; as alluded in Section 4.2, it has been associated with the development of RVO in several clinical reports. Notably, some studies show that fibrinogen may increase endothelial permeability and mediate vasoconstriction through activation of ERK1/2 (Sen et al., 2009; Tyagi et al., 2008). Though the expression levels of IL-1, IL-6, IL-8, and TNF-α genes were not affected, induction of downstream genes within these signaling pathways, including TRAF6 (TNF receptor associated factor 6), TNF receptor, IκB kinase, signal transducer and activator of transcription (STAT) 3, c-Jun, collagen type I, intracellular adhesion molecule (ICAM-1), vascular cell adhesion molecule (VCAM)-1, and cyclooxygenase (Cox)-2, suggests pathway activation downstream of these cytokines. TNF-α has been shown to increase vascular permeability via modulation of tight junction proteins in diabetic retinopathy (Aveleira et al., 2010). IL-1β may have a role in mediating retinal capillary degradation in diabetic retinopathy (Frey and Antonetti, 2011). Intravitreal levels of IL-6 are correlated with macular edema in branch retinal vein occlusion (Noma et al., 2006). ICAM-1, VCAM-1 and Cox-2 are downstream effectors of NF-κB in the IL-8 signaling pathway. ICAM-1 and VCAM-1 are adhesions molecules expressed on vascular endothelial cells and their induction play a critical role in leukostasis and inflammation. Cox-2 mediates the production of proinflammatory prostaglandins. NF-κB, c-Jun and STAT3 are important regulators of many genes mediating mammalian inflammatory and immune responses. In addition, the expression of GFAP was induced on day 5, suggesting activation of Müller cells in response to retinal stress such as inflammation.

Conversely, glucocorticoids, which have anti-inflammatory properties, have been shown to have a positive impact on promoting barrier integrity. In a porcine model of RVO, triamcinolone treatment, a widely used glucocorticoid in ocular applications, reduced VEGF and increased tight junction occludin levels in the retina, suggesting increased BRB integrity (McAllister et al., 2009). A study conducted to examine the impact of Streptococcus suis infection on blood-CSF (cerebral spinal fluid) barrier showed that the glucocorticoid dexamethasone improved the barrier function by preventing tight junction protein reorganization and degradation, and attenuated ERK activation and matrix metalloproteinase 3 (MMP3) expression (Tenenbaum et al., 2008). Paradoxically, dexamethasone has also been shown to activate ERK and JNK, which in turn induce the expression of the anti-inflammatory mediator MAPK phosphatase 1 (MKP-1) in human umbilical vein endothelial cells (Furst et al., 2008). Activated MKP-1 is then able to terminate the activity of activated MAPKs in a negative feedback loop (Furst et al., 2008). These data further demonstrate the complex spatiotemporal signaling network in which the MAPKs participate.

In our gene expression analysis, the glucocorticoid receptor signaling pathway was one of the few pathways significantly perturbed on day 3 following PD0325901 administration at 45 mg/kg/day, and many genes within the pathway were induced. On day 5, many of the same pathway genes became repressed. This could indicate an adaptive response to ongoing inflammation in the retina. As alluded to earlier, HSP27 and αβ-crystallin levels were also induced following PD0325901 administration. Their gene products have the ability to interfere with inflammatory signaling, such as attenuation of TNF-α, NF-κB signaling pathways, and may represent another cellular response to inflammation.

5.1.2 MAPK-mediated modulation of intercellular junctions in response to oxidative stress

Intercellular junctions are of critical importance to the integrity of the BRB function (Fig. 1). Oxidative stress is known to disrupt the structure and function of tight junctions and adherens junctions through MAPK activation, in both endothelial cells (Niwa et al., 2001; Simoncini et al., 2005; Usatyuk and Natarajan, 2004; Usatyuk et al., 2006; Yuan, 2002) and epithelial cells (Basuroy et al., 2006; Gonzalez et al., 2009), and these adverse effects on the junctional complexes could be ameliorated with the application of specific MAPK inhibitors. The presence of ROS could also induce the expression of the vascular permeability factor VEGF in endothelial cells (Chua et al., 1998; El-Remessy et al., 2003), often associated with downstream MAPK activation (Yang et al., 2010a; Zheng et al., 2010). In Müller cells, oxidative stress leads to decreased PEDF expression (Yoshida et al., 2009), thus relieving its antagonistic effect on VEGF action and subsequent MAPK activation in endothelial cells (Yafai et al., 2007), contributing to increased vascular permeability and breakdown of the inner BRB.

In our 5-day rat investigative study, the expression of myosin light chain kinase (MLCK) was significantly increased as a result of NF-κB signaling. Phosphorylation of myosin light chain by MLCK leads to actin-mediated endothelial cell contraction and increases permeability of endothelial junctional barrier. Activation of phosphatidylinositol 3-kinase (PI3K), whose expression was induced on both day 3 and day 5, has also been shown to increase vascular permeability (Abid et al., 2004; Lee et al., 2006; Serban et al., 2008). In addition, induction of MMP14 was observed on day 5. Under inflammatory conditions, Müller cells are a source of matrix metalloproteinases which impair the barrier function of retinal endothelial cells by degradation of the tight junction protein occludin (Reichenbach et al., 2007). Taken together, these data suggest increased retinal endothelial cell permeability and impaired BRB function as a result of PD0325901 administration. On the other hand, our data set also uncovered induction of genes important for maintaining the vascular endothelial barrier function, presumably as feedback mechanism to counteract permeability increases, such as repression of RhoA and induction of Rac1 to facilitate reannealing of adherens junctions (Vandenbroucke et al., 2008); induction of tight junction components claudin 11 and JAM-2; and induction of PEDF, likely in Müller cells, which represses expression of VEGF and decreases vascular permeability (Reichenbach et al., 2007).

5.1.3 Regulation of actin dynamics by MAPK in response to oxidative stress

The remodeling of actin cytoskeleton is an important response in endothelial cells exposed to oxidative stress, and contributes to increased permeability of the endothelial barrier (Houle and Huot, 2006; Lum and Roebuck, 2001). Under physiological and pathological stress conditions, endothelial cells undergo cell shape change, intercellular gap formation, and remodeling of the actin cytoskeleton, characterized by stress fiber formation and reduced cortical actin band. The formation of stress fibers is dependent on actin polymerization, and increases the endothelial cells' capacity to resist stress. On the other hand, stress fibers also pull apart intercellular junctions, likely contributing to their disruption and impaired endothelial barrier integrity.

ROS-induced MAPK activation plays an important role in actin remodeling. ERKs, p38 and JNK have all been shown to regulate actin dynamics induced by oxidative stress in endothelial cells (El-Remessy et al., 2011; Houle and Huot, 2006; Houle et al., 2003; Huot et al., 1998; Schweitzer et al., 2011; Usatyuk and Natarajan, 2004). p38 MAPK activation leads to phosphorylation of HSP27, which promotes actin polymerization. ERK activation results in phosphorylation of tropomyosin-1, which contributes to focal adhesion assembly and stress fiber formation, and modulates cell contractility. Inhibition of ERK activity by the MEK inhibitor PD098059 led to misassembly of focal adhesions and membrane blebbing, ultimately resulting in apoptosis (Huot et al. 1998). Physio-pathological consequence of surface blebbing of endothelial cells includes narrowing of vascular lumen associated with increased vascular resistance. Bleb shedding may also contribute to obstruction of blood vessels. Consistent with this interplay of MAPKs and actin dynamics, treatment with PD0325901 in the 5-day rat investigative study led to induction of genes in actin cytoskeleton and focal adhesion signaling pathways on study day 5, supporting perturbation of actin dynamics, likely downstream of oxidative stress. These data also raise the possibility of membrane blebbing in retinal vasculature following PD0325901 treatment, contributing to the development of RVO.

5.1.4 MAPK pathway and water permeability in response to oxidative stress

Macular edema was observed in our animal model administered with PD0325901. Two factors contribute to the development of chronic edema in the retina: increased vascular permeability leading to excessive fluid buildup, and reduced fluid absorption from the retina back into the blood. Müller and RPE cells play an integral role in transcellular fluid equilibrium (Reichenbach et al., 2007). Aquaporin 4 (AQP4) expressed on Müller cells and AQP1 expressed on RPE cells facilitate bidirectional water movements to maintain the osmotic and hydrostatic equilibrium in the retina. The Müller cell-specific AQP4 is co-localized with the inwardly rectifying potassium channel Kir4.1. Together they mediate the co-transport of water and potassium ions from the retinal tissue into the blood under normal conditions. In various animal models of retinopathy, characterized by inflammatory or oxidative stress conditions, Kir4.1 channel becomes mislocalized, and its expression is decreased in some cases (Reichenbach et al., 2007). This may lead to an intracellular potassium overload, increased osmotic pressure, and consequently, Müller cell swelling. In a study employing a rat model of RVO, downregulation of AQP1, AQP4, and Kir4.1 were observed, in addition to an altered distribution of Kir4.1 protein. Consequently, Müller cells displayed a decrease in potassium currents and increased in size (Rehak et al., 2008).

Application triamcinolone, a glucocorticoid frequently used for diabetic macular edema due to its anti-inflammatory properties (Felinski and Antonetti, 2005), reduced Müller cell swelling in animal models of ischemia-reperfusion and diabetic retinopathy (Reichenbach et al., 2007). In a porcine model of RVO, triamcinolone treatment reduced the glial activation marker GFAP expression in Müller cells, and also increased BRB integrity, as evidenced by reduced VEGF and increased tight junction occludin levels, potentially contributing to the resolution of edema in the retina (McAllister et al., 2009).

MAPKs are known to play an important role in cellular osmotic stress regulation (Cowan and Storey, 2003; de Nadal et al., 2002). In RPE cells, ultraviolet radiation (UVB) and hydrogen peroxide treatment, both of which are oxidative stress inducers, resulted in AQP1 downregulation which was mediated by MEK/ERK activation (Jiang et al., 2009). In the brain, astrocyte swelling often accompanies vascular edema (Reichenbach et al., 2007). In astrocytes exposed to the oxidative stressor manganese or glial reactive injury, there was an altered expression of AQPs, mediated by the MEK/ERK and p38 MAPKs (McCoy and Sontheimer, 2010; Rao et al., 2010).

In our 5-day rat investigative study, the repression of the Müller cell-specific water channel AQP4 on both day 3 and day 5, coupled with the repression of the inwardly rectifying potassium channels (Kcnj5, Kcnj6), and sodium channels on day 5, signals impaired transcellular fluid transport. Given the evidence for inflammation, intercellular junction disruption and actin cytoskeleton changes in the retina following PD0325901 treatment, this fluid imbalance would contribute to the observed retinal edema in the study and further weaken BRB integrity. The documented involvement of MAPKs in regulating transcellular fluid balance in the retina and the brain raises the possibility that MEK inhibition may play a role in perturbing this equilibrium.

5.2 MAPK, IFN-α and RVO

Given the complex signaling cascades and cross-talk between various MAPK pathways in multiple cell types comprising the BRB, it is conceivable that inhibition of ERK activation by a MEK inhibitor could lead to dysregulated BRB integrity and subsequent development of RVO. Intriguingly, IFN-α treatment, which is also associated with the development of RVO in the clinic (see Section 4.4.1), has been shown to inhibit the activation of ERK and the associated survival effects, and that MEK and ERK inhibitors enhance the anti-proliferative effect of IFN-α in tumor cells or transformed epithelial cells (Battcock et al., 2006; Caraglia et al., 2005; Caraglia et al., 2003; Christian et al., 2009; Li et al., 2004; Romerio et al., 2000; Romerio and Zella, 2002). Cross-talk between the IFN-α and Ras-MAPK pathways converge on the STAT family of transcription factors. STAT proteins are involved in cytokine, hormone, and growth factor signal transduction, mediating biological processes as diverse as cell proliferation, differentiation, apoptosis, transformation, inflammation and immune response (Caraglia et al., 2005). Activated Ras/MEK has been shown to inhibit the antiviral response of IFN-α by reducing STAT2 levels (Christian et al., 2009). It is conceivable that clinical IFN-α usage may perturb the balance of MAPK signaling pathways in the retina, disrupt BRB function, and ultimately contribute to the development of RVO. That both IFN-α therapy and PD0325901 inhibit ERK activation and are linked to clinical development of RVO lends further support to the hypothesis that modulation of the Ras-MAPK pathway and subsequently BRB permeability changes play a role in the pathogenesis of this ocular adverse event.

6. Conclusion

The BRB, consisting of an endothelial and an epithelial barrier, serves to regulate the bidirectional passage of macromolecules through the retina. Oxidative stress can negatively impact the equilibrium across the BRB, leading to cellular disruption and ocular disorders. MAPK pathways involving ERK, p38 and JNK play a central role in the oxidative stress response of the BRB, modulating inflammatory response, actin cytoskeletal dynamics, water transport, as well as inter-epithelial and inter-endothelial adhesion molecule expression and redistribution. Disruption of the ERK signaling pathway by the MEK inhibitor PD0325901 may disrupt the balance and cross-talk between interconnected signaling networks and produce unexpected cellular sequalae. PD0325901-induced RVO could arise as a consequence of disruption of these tightly regulated molecular processes vital for proper functioning of the BRB. The animal models employed in our study serves as an investigative or screening paradigm for pre-clinical compounds suspected of RVO-inducing potential. Finally, while a firm connection between MEK inhibition and the development of RVO has not been established, it would be prudent for clinicians to monitor patients on MEK inhibitor therapy for signs of ocular adverse events.

7. Acknowledgements

The authors would like to thank Patrick Lappin for critical reading of the manuscript, and Constance Benedict for preparation of Fig. 1.

8. References

Abid, M.R., Guo, S., Minami, T., Spokes, K.C., Ueki, K., Skurk, C., Walsh, K., and Aird, W.C. (2004). Vascular endothelial growth factor activates PI3K/Akt/forkhead signaling in endothelial cells. Arterioscler Thromb Vasc Biol 24, 294-300.

Adjei, A.A., Cohen, R.B., Franklin, W., Morris, C., Wilson, D., Molina, J.R., Hanson, L.J., Gore, L., Chow, L., Leong, S., et al. (2008). Phase I pharmacokinetic and pharmacodynamic study of the oral, small-molecule mitogen-activated protein kinase kinase 1/2 inhibitor AZD6244 (ARRY-142886) in patients with advanced cancers. J Clin Oncol 26, 2139-2146.

Albon, J., Karwatowski, W.S., Avery, N., Easty, D.L., and Duance, V.C. (1995). Changes in the collagenous matrix of the aging human lamina cribrosa. Br J Ophthalmol 79, 368-375.

Angayarkanni, N., Barathi, S., Seethalakshmi, T., Punitham, R., Sivaramakrishna, R., Suganeswari, G., and Tarun, S. (2008). Serum PON1 arylesterase activity in relation to hyperhomocysteinaemia and oxidative stress in young adult central retinal venous occlusion patients. Eye (Lond) 22, 969-974.

Arrigo, A.P., Simon, S., Gibert, B., Kretz-Remy, C., Nivon, M., Czekalla, A., Guillet, D., Moulin, M., Diaz-Latoud, C., and Vicart, P. (2007). Hsp27 (HspB1) and alphaB-crystallin (HspB5) as therapeutic targets. FEBS Lett 581, 3665-3674.

Aveleira, C.A., Lin, C.M., Abcouwer, S.F., Ambrosio, A.F., and Antonetti, D.A. (2010). TNF-alpha signals through PKCzeta/NF-kappaB to alter the tight junction complex and increase retinal endothelial cell permeability. Diabetes 59, 2872-2882.

Bajaire, B.J., Paipilla, D.F., Arrieta, C.E., and Oudovitchenko, E. (2011). Mixed vascular occlusion in a patient with interferon-associated retinopathy. Case Report Ophthalmol 2, 23-29.

Basuroy, S., Seth, A., Elias, B., Naren, A.P., and Rao, R. (2006). MAPK interacts with occludin and mediates EGF-induced prevention of tight junction disruption by hydrogen peroxide. Biochem J 393, 69-77.

Battcock, S.M., Collier, T.W., Zu, D., and Hirasawa, K. (2006). Negative regulation of the alpha interferon-induced antiviral response by the Ras/Raf/MEK pathway. J Virol 80, 4422-4430.

Bekaii-Saab, T., Phelps, M.A., Li, X., Saji, M., Goff, L., Kauh, J.S., O'Neil, B.H., Balsom, S., Balint, C., Liersemann, R., et al. (2011). Multi-institutional phase II study of selumetinib in patients with metastatic biliary cancers. J Clin Oncol 29, 2357-2363.

Bodoky, G., Timcheva, C., Spigel, D.R., La Stella, P.J., Ciuleanu, T.E., Pover, G., and Tebbutt, N.C. (2011). A phase II open-label randomized study to assess the efficacy and safety of selumetinib (AZD6244 [ARRY-142886]) versus capecitabine in patients with advanced or metastatic pancreatic cancer who have failed first-line gemcitabine therapy. Invest New Drugs.

Boyd, S.R., Zachary, I., Chakravarthy, U., Allen, G.J., Wisdom, G.B., Cree, I.A., Martin, J.F., and Hykin, P.G. (2002). Correlation of increased vascular endothelial growth factor with neovascularization and permeability in ischemic central vein occlusion. Arch Ophthalmol 120, 1644-1650.

Brafman, A., Mett, I., Shafir, M., Gottlieb, H., Damari, G., Gozlan-Kelner, S., Vishnevskia-Dai, V., Skaliter, R., Einat, P., Faerman, A., et al. (2004). Inhibition of oxygen-induced retinopathy in RTP801-deficient mice. Invest Ophthalmol Vis Sci 45, 3796-3805.

Burke, J.M. (2008). Epithelial phenotype and the RPE: is the answer blowing in the Wnt? Prog Retin Eye Res 27, 579-595.

Busik, J.V., Mohr, S., and Grant, M.B. (2008). Hyperglycemia-induced reactive oxygen species toxicity to endothelial cells is dependent on paracrine mediators. Diabetes 57, 1952-1965.

Caraglia, M., Marra, M., Pelaia, G., Maselli, R., Caputi, M., Marsico, S.A., and Abbruzzese, A. (2005). Alpha-interferon and its effects on signal transduction pathways. J Cell Physiol 202, 323-335.

Caraglia, M., Tagliaferri, P., Marra, M., Giuberti, G., Budillon, A., Gennaro, E.D., Pepe, S., Vitale, G., Improta, S., Tassone, P., et al. (2003). EGF activates an inducible survival response via the RAS-> Erk-1/2 pathway to counteract interferon-alpha-mediated apoptosis in epidermoid cancer cells. Cell Death Differ 10, 218-229.

Chan, C.M., Huang, J.H., Lin, H.H., Chiang, H.S., Chen, B.H., Hong, J.Y., and Hung, C.F. (2008). Protective effects of (-)-epigallocatechin gallate on UVA-induced damage in ARPE19 cells. Mol Vis 14, 2528-2534.

Cheung, N., Klein, R., Wang, J.J., Cotch, M.F., Islam, A.F., Klein, B.E., Cushman, M., and Wong, T.Y. (2008). Traditional and novel cardiovascular risk factors for retinal vein occlusion: the multiethnic study of atherosclerosis. Invest Ophthalmol Vis Sci 49, 4297-4302.

Chibber, R., Ben-Mahmud, B.M., Chibber, S., and Kohner, E.M. (2007). Leukocytes in diabetic retinopathy. Curr Diabetes Rev 3, 3-14.

Christian, S.L., Collier, T.W., Zu, D., Licursi, M., Hough, C.M., and Hirasawa, K. (2009). Activated Ras/MEK inhibits the antiviral response of alpha interferon by reducing STAT2 levels. J Virol 83, 6717-6726.

Chua, C.C., Hamdy, R.C., and Chua, B.H. (1998). Upregulation of vascular endothelial growth factor by H2O2 in rat heart endothelial cells. Free Radic Biol Med 25, 891-897.

Cowan, K.J., and Storey, K.B. (2003). Mitogen-activated protein kinases: new signaling pathways functioning in cellular responses to environmental stress. J Exp Biol 206, 1107-1115.

de Nadal, E., Alepuz, P.M., and Posas, F. (2002). Dealing with osmostress through MAP kinase activation. EMBO Rep 3, 735-740.

Dong, X., Li, Z., Wang, W., Zhang, W., Liu, S., and Zhang, X. (2011). Protective effect of canolol from oxidative stress-induced cell damage in ARPE-19 cells via an ERK mediated antioxidative pathway. Mol Vis 17, 2040-2048.

Du, Y., Tang, J., Li, G., Berti-Mattera, L., Lee, C.A., Bartkowski, D., Gale, D., Monahan, J., Niesman, M.R., Alton, G., et al. (2010). Effects of p38 MAPK inhibition on early stages of diabetic retinopathy and sensory nerve function. Invest Ophthalmol Vis Sci 51, 2158-2164.

El-Remessy, A.B., Behzadian, M.A., Abou-Mohamed, G., Franklin, T., Caldwell, R.W., and Caldwell, R.B. (2003). Experimental diabetes causes breakdown of the blood-retina barrier by a mechanism involving tyrosine nitration and increases in expression of vascular endothelial growth factor and urokinase plasminogen activator receptor. Am J Pathol 162, 1995-2004.

El-Remessy, A.B., Rajesh, M., Mukhopadhyay, P., Horvath, B., Patel, V., Al-Gayyar, M.M., Pillai, B.A., and Pacher, P. (2011). Cannabinoid 1 receptor activation contributes to vascular inflammation and cell death in a mouse model of diabetic retinopathy and a human retinal cell line. Diabetologia 54, 1567-1578.

Esmaeli, B., Koller, C., Papadopoulos, N., and Romaguera, J. (2001). Interferon-induced retinopathy in asymptomatic cancer patients. Ophthalmology 108, 858-860.

Fegan, C.D. (2002). Central retinal vein occlusion and thrombophilia. Eye (Lond) 16, 98-106.

Felinski, E.A., and Antonetti, D.A. (2005). Glucocorticoid regulation of endothelial cell tight junction gene expression: novel treatments for diabetic retinopathy. Curr Eye Res 30, 949-957.

Fraenkl, S.A., Mozaffarieh, M., and Flammer, J. (2010). Retinal vein occlusions: The potential impact of a dysregulation of the retinal veins. EPMA J 1, 253-261.

Frey, T., and Antonetti, D.A. (2011). Alterations to the blood-retinal barrier in diabetes: cytokines and reactive oxygen species. Antioxid Redox Signal 15, 1271-1284.

Funk, M., Kriechbaum, K., Prager, F., Benesch, T., Georgopoulos, M., Zlabinger, G.J., and Schmidt-Erfurth, U. (2009). Intraocular concentrations of growth factors and cytokines in retinal vein occlusion and the effect of therapy with bevacizumab. Invest Ophthalmol Vis Sci 50, 1025-1032.

Furst, R., Zahler, S., and Vollmar, A.M. (2008). Dexamethasone-induced expression of endothelial mitogen-activated protein kinase phosphatase-1 involves activation of the transcription factors activator protein-1 and 3',5'-cyclic adenosine 5'-monophosphate response element-binding protein and the generation of reactive oxygen species. Endocrinology 149, 3635-3642.

Garrido-Urbani, S., Bradfield, P.F., Lee, B.P., and Imhof, B.A. (2008). Vascular and epithelial junctions: a barrier for leucocyte migration. Biochem Soc Trans 36, 203-211.

Glotin, A.L., Calipel, A., Brossas, J.Y., Faussat, A.M., Treton, J., and Mascarelli, F. (2006). Sustained versus transient ERK1/2 signaling underlies the anti- and proapoptotic effects of oxidative stress in human RPE cells. Invest Ophthalmol Vis Sci 47, 4614-4623.

Gonzalez, J.E., DiGeronimo, R.J., Arthur, D.E., and King, J.M. (2009). Remodeling of the tight junction during recovery from exposure to hydrogen peroxide in kidney epithelial cells. Free Radic Biol Med 47, 1561-1569.

Gu, X., Samuel, S., El-Shabrawey, M., Caldwell, R.B., Bartoli, M., Marcus, D.M., and Brooks, S.E. (2002). Effects of sustained hyperoxia on revascularization in experimental retinopathy of prematurity. Invest Ophthalmol Vis Sci 43, 496-502.

Gumus, K., Kadayifcilar, S., Eldem, B., Saracbasi, O., Ozcebe, O., Dundar, S., and Kirazli, S. (2006). Is elevated level of soluble endothelial protein C receptor a new risk factor for retinal vein occlusion? Clin Experiment Ophthalmol 34, 305-311.

Hardy, P., Beauchamp, M., Sennlaub, F., Gobeil, F., Jr., Tremblay, L., Mwaikambo, B., Lachapelle, P., and Chemtob, S. (2005). New insights into the retinal circulation: inflammatory lipid mediators in ischemic retinopathy. Prostaglandins Leukot Essent Fatty Acids 72, 301-325.

Hayreh, S.S., Zimmerman, M.B., and Podhajsky, P. (1994). Incidence of various types of retinal vein occlusion and their recurrence and demographic characteristics. Am J Ophthalmol 117, 429-441.

Houle, F., and Huot, J. (2006). Dysregulation of the endothelial cellular response to oxidative stress in cancer. Mol Carcinog 45, 362-367.

Houle, F., Rousseau, S., Morrice, N., Luc, M., Mongrain, S., Turner, C.E., Tanaka, S., Moreau, P., and Huot, J. (2003). Extracellular signal-regulated kinase mediates phosphorylation of tropomyosin-1 to promote cytoskeleton remodeling in response to oxidative stress: impact on membrane blebbing. Mol Biol Cell 14, 1418-1432.

Huang, W., Yang, A.H., Matsumoto, D., Collette, W., Marroquin, L., Ko, M., Aguirre, S., and Younis, H.S. (2009). PD0325901, a mitogen-activated protein kinase kinase inhibitor, produces ocular toxicity in a rabbit animal model of retinal vein occlusion. J Ocul Pharmacol Ther 25, 519-530.

Huot, J., Houle, F., Rousseau, S., Deschesnes, R.G., Shah, G.M., and Landry, J. (1998). SAPK2/p38-dependent F-actin reorganization regulates early membrane blebbing during stress-induced apoptosis. J Cell Biol 143, 1361-1373.

Janssen, M.C., den Heijer, M., Cruysberg, J.R., Wollersheim, H., and Bredie, S.J. (2005). Retinal vein occlusion: a form of venous thrombosis or a complication of atherosclerosis? A meta-analysis of thrombophilic factors. Thromb Haemost 93, 1021-1026.

Jiang, Q., Cao, C., Lu, S., Kivlin, R., Wallin, B., Chu, W., Bi, Z., Wang, X., and Wan, Y. (2009). MEK/ERK pathway mediates UVB-induced AQP1 downregulation and water permeability impairment in human retinal pigment epithelial cells. Int J Mol Med 23, 771-777.

Kachi, S., Kobayashi, K., Ushida, H., Ito, Y., Kondo, M., and Terasaki, H. (2010). Regression of macular edema secondary to branch retinal vein occlusion during anti-TNF-alpha therapy for rheumatoid arthritis. Clin Ophthalmol 4, 667-670.

Kanehisa, M. and Goto, S. (2000). KEGG: Kyoto Encyclopedia of Genes and Genomes. Nucleic Acids Res 28, 27-30.

Kanehisa, M., Goto, S., Sato, Y., Furumichi, M., and Tanabe, M. (2012). KEGG for integration and interpretation of large-scale molecular datasets. Nucleic Acids Res 40, D109-D114.

Kaur, C., Foulds, W.S., and Ling, E.A. (2008). Blood-retinal barrier in hypoxic ischaemic conditions: basic concepts, clinical features and management. Prog Retin Eye Res 27, 622-647.

Kemeny, N., Childs, B., Larchian, W., Rosado, K., and Kelsen, D. (1990). A phase II trial of recombinant tumor necrosis factor in patients with advanced colorectal carcinoma. Cancer 66, 659-663.

Klein, R., Moss, S.E., Meuer, S.M., and Klein, B.E. (2008). The 15-year cumulative incidence of retinal vein occlusion: the Beaver Dam Eye Study. Arch Ophthalmol 126, 513-518.

Klettner, A., and Roider, J. (2009). Constitutive and oxidative-stress-induced expression of VEGF in the RPE are differently regulated by different Mitogen-activated protein kinases. Graefes Arch Clin Exp Ophthalmol 247, 1487-1492.

Koizumi, H., Ferrara, D.C., Brue, C., and Spaide, R.F. (2007). Central retinal vein occlusion case-control study. Am J Ophthalmol 144, 858-863.

Kotwal, R.S., Butler, F.K., Jr., Murray, C.K., Hill, G.J., Rayfield, J.C., and Miles, E.A. (2009). Central retinal vein occlusion in an Army Ranger with glucose-6-phosphate dehydrogenase deficiency. J Spec Oper Med 9, 59-63.

Kunsch, C., and Medford, R.M. (1999). Oxidative stress as a regulator of gene expression in the vasculature. Circ Res *85*, 753-766.

Larrayoz, I.M., Huang, J.D., Lee, J.W., Pascual, I., and Rodriguez, I.R. (2010). 7-ketocholesterol-induced inflammation: involvement of multiple kinase signaling pathways via NFkappaB but independently of reactive oxygen species formation. Invest Ophthalmol Vis Sci *51*, 4942-4955.

Lee, K.S., Park, S.J., Kim, S.R., Min, K.H., Jin, S.M., Puri, K.D., and Lee, Y.C. (2006). Phosphoinositide 3-kinase-delta inhibitor reduces vascular permeability in a murine model of asthma. J Allergy Clin Immunol *118*, 403-409.

Li, C., Chi, S., He, N., Zhang, X., Guicherit, O., Wagner, R., Tyring, S., and Xie, J. (2004). IFNalpha induces Fas expression and apoptosis in hedgehog pathway activated BCC cells through inhibiting Ras-Erk signaling. Oncogene *23*, 1608-1617.

Lim, L.L., Cheung, N., Wang, J.J., Islam, F.M., Mitchell, P., Saw, S.M., Aung, T., and Wong, T.Y. (2008). Prevalence and risk factors of retinal vein occlusion in an Asian population. Br J Ophthalmol *92*, 1316-1319.

Lip, P.L., Blann, A.D., Jones, A.F., and Lip, G.Y. (1998). Abnormalities in haemorheological factors and lipoprotein (a) in retinal vascular occlusion: implications for increased vascular risk. Eye *12 (Pt 2)*, 245-251.

LoRusso, P.M., Krishnamurthi, S.S., Rinehart, J.J., Nabell, L.M., Malburg, L., Chapman, P.B., DePrimo, S.E., Bentivegna, S., Wilner, K.D., Tan, W., et al. (2010). Phase I pharmacokinetic and pharmacodynamic study of the oral MAPK/ERK kinase inhibitor PD-0325901 in patients with advanced cancers. Clin Cancer Res *16*, 1924-1937.

Lum, H., and Roebuck, K.A. (2001). Oxidant stress and endothelial cell dysfunction. Am J Physiol Cell Physiol *280*, C719-741.

Marcucci, R., Sofi, F., Grifoni, E., Sodi, A., and Prisco, D. (2011). Retinal vein occlusions: a review for the internist. Intern Emerg Med *6*, 307-314.

McAllister, I.L., Vijayasekaran, S., Chen, S.D., and Yu, D.Y. (2009). Effect of triamcinolone acetonide on vascular endothelial growth factor and occludin levels in branch retinal vein occlusion. Am J Ophthalmol *147*, 838-846, 846 e831-832.

McCoy, E., and Sontheimer, H. (2010). MAPK induces AQP1 expression in astrocytes following injury. Glia *58*, 209-217.

Mitchell, P., Smith, W., and Chang, A. (1996). Prevalence and associations of retinal vein occlusion in Australia. The Blue Mountains Eye Study. Arch Ophthalmol *114*, 1243-1247.

Nadir, A., Amin, A., Chalisa, N., and van Thiel, D.H. (2000). Retinal vein thrombosis associated with chronic hepatitis C: a case series and review of the literature. J Viral Hepat *7*, 466-470.

Niessen, C.M. (2007) Tight junctions/adherens junctions: basic structure and function. J Invest Dermatol *127*, 2525-32.

Niwa, K., Inanami, O., Ohta, T., Ito, S., Karino, T., and Kuwabara, M. (2001). p38 MAPK and Ca2+ contribute to hydrogen peroxide-induced increase of permeability in vascular endothelial cells but ERK does not. Free Radic Res *35*, 519-527.

Noma, H., Funatsu, H., Mimura, T., Harino, S., and Hori, S. (2009). Vitreous levels of interleukin-6 and vascular endothelial growth factor in macular edema with central retinal vein occlusion. Ophthalmology *116*, 87-93.

Noma, H., Minamoto, A., Funatsu, H., Tsukamoto, H., Nakano, K., Yamashita, H., and Mishima, H.K. (2006). Intravitreal levels of vascular endothelial growth factor and interleukin-6 are correlated with macular edema in branch retinal vein occlusion. Graefes Arch Clin Exp Ophthalmol 244, 309-315.

Patrassi, G.M., Mares, M., Piermarocchi, S., Santarossa, A., Viero, M., and Girolami, A. (1987). Fibrinolytic behavior in long-standing branch retinal vein occlusion. Ophthalmic Res 19, 221-225.

Peduzzi, M., Debbia, A., Guerrieri, F., and Bolzani, R. (1986). Abnormal blood rheology in retinal vein occlusion. A preliminary report. Graefes Arch Clin Exp Ophthalmol 224, 83-85.

Puli, S.R., and Benage, D.D. (2003). Retinal vein thrombosis after infliximab (Remicade) treatment for Crohn's disease. Am J Gastroenterol 98, 939-940.

Qin, S., McLaughlin, A.P., and De Vries, G.W. (2006). Protection of RPE cells from oxidative injury by 15-deoxy-delta12,14-prostaglandin J2 by augmenting GSH and activating MAPK. Invest Ophthalmol Vis Sci 47, 5098-5105.

Rao, K.V., Jayakumar, A.R., Reddy, P.V., Tong, X., Curtis, K.M., and Norenberg, M.D. (2010). Aquaporin-4 in manganese-treated cultured astrocytes. Glia 58, 1490-1499.

Rehak, J., and Rehak, M. (2008). Branch retinal vein occlusion: pathogenesis, visual prognosis, and treatment modalities. Curr Eye Res 33, 111-131.

Rehak, M., Hollborn, M., Iandiev, I., Pannicke, T., Karl, A., Wurm, A., Kohen, L., Reichenbach, A., Wiedemann, P., and Bringmann, A. (2008). Retinal Gene Expression and Muller Cell Responses after Branch Retinal Vein Occlusion in the Rat. Invest Ophthalmol Vis Sci.

Reichenbach, A., Wurm, A., Pannicke, T., Iandiev, I., Wiedemann, P., and Bringmann, A. (2007). Muller cells as players in retinal degeneration and edema. Graefes Arch Clin Exp Ophthalmol 245, 627-636.

Rinehart, J., Adjei, A.A., Lorusso, P.M., Waterhouse, D., Hecht, J.R., Natale, R.B., Hamid, O., Varterasian, M., Asbury, P., Kaldjian, E.P., et al. (2004). Multicenter phase II study of the oral MEK inhibitor, CI-1040, in patients with advanced non-small-cell lung, breast, colon, and pancreatic cancer. J Clin Oncol 22, 4456-4462.

Rogers, S., McIntosh, R.L., Cheung, N., Lim, L., Wang, J.J., Mitchell, P., Kowalski, J.W., Nguyen, H., and Wong, T.Y. (2010). The prevalence of retinal vein occlusion: pooled data from population studies from the United States, Europe, Asia, and Australia. Ophthalmology 117, 313-319 e311.

Romerio, F., and Zella, D. (2002). MEK and ERK inhibitors enhance the anti-proliferative effect of interferon-alpha2b. Faseb J 16, 1680-1682.

Romerio, F., Riva, A., and Zella, D. (2000). Interferon-alpha2b reduces phosphorylation and activity of MEK and ERK through a Ras/Raf-independent mechanism. Br J Cancer 83, 532-538.

Schweitzer, K.S., Hatoum, H., Brown, M.B., Gupta, M., Justice, M.J., Beteck, B., Van Demark, M.J., Gu, Y., Presson, R.G., Jr., Hubbard, W.C., et al. (2011). Mechanisms of lung endothelial barrier disruption induced by cigarette smoke: role of oxidative stress and ceramides. Am J Physiol Lung Cell Mol Physiol.

Sen, U., Tyagi, N., Patibandla, P.K., Dean, W.L., Tyagi, S.C., Roberts, A.M., and Lominadze, D. (2009). Fibrinogen-induced endothelin-1 production from endothelial cells. Am J Physiol Cell Physiol 296, C840-847.

Serban, D., Leng, J., and Cheresh, D. (2008). H-ras regulates angiogenesis and vascular permeability by activation of distinct downstream effectors. Circ Res *102*, 1350-1358.

Simoncini, S., Sapet, C., Camoin-Jau, L., Bardin, N., Harle, J.R., Sampol, J., Dignat-George, F., and Anfosso, F. (2005). Role of reactive oxygen species and p38 MAPK in the induction of the pro-adhesive endothelial state mediated by IgG from patients with anti-phospholipid syndrome. Int Immunol *17*, 489-500.

Siu, T.L., Morley, J.W., and Coroneo, M.T. (2008). Toxicology of the retina: advances in understanding the defence mechanisms and pathogenesis of drug- and light-induced retinopathy. Clin Experiment Ophthalmol *36*, 176-185.

Sperduto, R.D., Hiller, R., Chew, E., Seigel, D., Blair, N., Burton, T.C., Farber, M.D., Gragoudas, E.S., Haller, J., Seddon, J.M., et al. (1998). Risk factors for hemiretinal vein occlusion: comparison with risk factors for central and branch retinal vein occlusion: the eye disease case-control study. Ophthalmology *105*, 765-771.

Sprague, A.H., and Khalil, R.A. (2009). Inflammatory cytokines in vascular dysfunction and vascular disease. Biochem Pharmacol *78*, 539-552.

Tenenbaum, T., Matalon, D., Adam, R., Seibt, A., Wewer, C., Schwerk, C., Galla, H.J., and Schroten, H. (2008). Dexamethasone prevents alteration of tight junction-associated proteins and barrier function in porcine choroid plexus epithelial cells after infection with Streptococcus suis in vitro. Brain Res *1229*, 1-17.

Tezel, G., Yang, X., Luo, C., Peng, Y., Sun, S.L., and Sun, D. (2007). Mechanisms of immune system activation in glaucoma: oxidative stress-stimulated antigen presentation by the retina and optic nerve head glia. Invest Ophthalmol Vis Sci *48*, 705-714.

The Eye Disease Case-Control Study Group. (1993). Risk factors for branch retinal vein occlusion. Am J Ophthalmol *116*, 286-296

The Eye Disease Case-Control Study Group. (1996). Risk factors for central retinal vein occlusion. Arch Ophthalmol *114*, 545-554.

Trope, G.E., Lowe, G.D., McArdle, B.M., Douglas, J.T., Forbes, C.D., Prentice, C.M., and Foulds, W.S. (1983). Abnormal blood viscosity and haemostasis in long-standing retinal vein occlusion. Br J Ophthalmol *67*, 137-142.

Tsao, Y.P., Ho, T.C., Chen, S.L., and Cheng, H.C. (2006). Pigment epithelium-derived factor inhibits oxidative stress-induced cell death by activation of extracellular signal-regulated kinases in cultured retinal pigment epithelial cells. Life Sci *79*, 545-550.

Tyagi, N., Roberts, A.M., Dean, W.L., Tyagi, S.C., and Lominadze, D. (2008). Fibrinogen induces endothelial cell permeability. Mol Cell Biochem *307*, 13-22.

Uno, K., Prow, T.W., Bhutto, I.A., Yerrapureddy, A., McLeod, D.S., Yamamoto, M., Reddy, S.P., and Lutty, G.A. (2010). Role of Nrf2 in retinal vascular development and the vaso-obliterative phase of oxygen-induced retinopathy. Exp Eye Res *90*, 493-500.

Usatyuk, P.V., and Natarajan, V. (2004). Role of mitogen-activated protein kinases in 4-hydroxy-2-nonenal-induced actin remodeling and barrier function in endothelial cells. J Biol Chem *279*, 11789-11797.

Usatyuk, P.V., Parinandi, N.L., and Natarajan, V. (2006). Redox regulation of 4-hydroxy-2-nonenal-mediated endothelial barrier dysfunction by focal adhesion, adherens, and tight junction proteins. J Biol Chem *281*, 35554-35566.

Ushio-Fukai, M., and Alexander, R.W. (2004). Reactive oxygen species as mediators of angiogenesis signaling: role of NAD(P)H oxidase. Mol Cell Biochem *264*, 85-97.

Vandenbroucke, E., Mehta, D., Minshall, R., and Malik, A.B. (2008). Regulation of endothelial junctional permeability. Ann N Y Acad Sci 1123, 134-145.

Veerappan, S.G., Kennedy, M., O'Morain, C.A., and Ryan, B.M. (2008). Retinal vein thrombosis following infliximab treatment for severe left-sided ulcerative colitis. Eur J Gastroenterol Hepatol 20, 588-589.

Vergou, T., Moustou, A.E., Maniateas, A., Stratigos, A.J., Katsambas, A., and Antoniou, C. (2010). Central retinal vein occlusion following infliximab treatment for plaque-type psoriasis. Int J Dermatol 49, 1215-1217.

Wang, X., Martindale, J.L., Liu, Y., and Holbrook, N.J. (1998). The cellular response to oxidative stress: influences of mitogen-activated protein kinase signalling pathways on cell survival. Biochem J 333 (Pt 2), 291-300.

Wang, Y., Bian, Z.M., Yu, W.Z., Yan, Z., Chen, W.C., and Li, X.X. (2010). Induction of interleukin-8 gene expression and protein secretion by C-reactive protein in ARPE-19 cells. Exp Eye Res 91, 135-142.

Wu, W.C., Hu, D.N., Gao, H.X., Chen, M., Wang, D., Rosen, R., and McCormick, S.A. (2010). Subtoxic levels hydrogen peroxide-induced production of interleukin-6 by retinal pigment epithelial cells. Mol Vis 16, 1864-1873.

Wurm, A., Iandiev, I., Uhlmann, S., Wiedemann, P., Reichenbach, A., Bringmann, A., and Pannicke, T. (2011). Effects of ischemia-reperfusion on physiological properties of Muller glial cells in the porcine retina. Invest Ophthalmol Vis Sci 52, 3360-3367.

Xu, L., Liu, W.W., Wang, Y.X., Yang, H., and Jonas, J.B. (2007). Retinal vein occlusions and mortality: the Beijing Eye Study. Am J Ophthalmol 144, 972-973.

Yafai, Y., Lange, J., Wiedemann, P., Reichenbach, A., and Eichler, W. (2007). Pigment epithelium-derived factor acts as an opponent of growth-stimulatory factors in retinal glial-endothelial cell interactions. Glia 55, 642-651.

Yang, J., Duh, E.J., Caldwell, R.B., and Behzadian, M.A. (2010a). Antipermeability function of PEDF involves blockade of the MAP kinase/GSK/beta-catenin signaling pathway and uPAR expression. Invest Ophthalmol Vis Sci 51, 3273-3280.

Yang, Y., Hayden, M.R., Sowers, S., Bagree, S.V., and Sowers, J.R. (2010b). Retinal redox stress and remodeling in cardiometabolic syndrome and diabetes. Oxid Med Cell Longev 3, 392-403.

Yoshida, Y., Yamagishi, S., Matsui, T., Jinnouchi, Y., Fukami, K., Imaizumi, T., and Yamakawa, R. (2009). Protective role of pigment epithelium-derived factor (PEDF) in early phase of experimental diabetic retinopathy. Diabetes Metab Res Rev 25, 678-686.

Yoshimura, T., Sonoda, K.H., Sugahara, M., Mochizuki, Y., Enaida, H., Oshima, Y., Ueno, A., Hata, Y., Yoshida, H., and Ishibashi, T. (2009). Comprehensive analysis of inflammatory immune mediators in vitreoretinal diseases. PLoS One 4, e8158.

Yuan, S.Y. (2002). Protein kinase signaling in the modulation of microvascular permeability. Vascul Pharmacol 39, 213-223.

Zhang, C., Lam, T.T., and Tso, M.O. (2005). Heterogeneous populations of microglia/macrophages in the retina and their activation after retinal ischemia and reperfusion injury. Exp Eye Res 81, 700-709.

Zhang, J., Wu, G.S., Ishimoto, S., Pararajasegaram, G., and Rao, N.A. (1997). Expression of major histocompatibility complex molecules in rodent retina. Immunohistochemical study. Invest Ophthalmol Vis Sci 38, 1848-1857.

Zheng, Z., Chen, H., Wang, H., Ke, B., Zheng, B., Li, Q., Li, P., Su, L., Gu, Q., and Xu, X. (2010). Improvement of retinal vascular injury in diabetic rats by statins is associated with the inhibition of mitochondrial reactive oxygen species pathway mediated by peroxisome proliferator-activated receptor gamma coactivator 1alpha. Diabetes 59, 2315-2325.

Section 2

Cancer

9

Monensin Induced Oxidative Stress Reduces Prostate Cancer Cell Migration and Cancer Stem Cell Population

Kirsi Ketola[1], Anu Vuoristo[2], Matej Orešič[2],
Olli Kallioniemi[3] and Kristiina Iljin[1]
[1]Medical Biotechnology, VTT Technical Research Centre of
Finland and University of Turku, Turku,
[2]VTT Technical Research Centre of Finland, Espoo,
[3]Institute for Molecular Medicine, Finland (FIMM),
University of Helsinki,
Finland

1. Introduction

Prostate cancer is the most common malignancy and second leading cause of cancer related death in males in developed countries (Jemal et al. 2011). Patients with localized and metastatic prostate cancer are treated with anti-androgens. Although prostate cancer cell proliferation is initially blocked or slowed down with anti-androgen therapy, eventually castration-resistant disease develops (Sharifi, Gulley & Dahut 2010). Therapeutic options for castration-resistant prostate cancer are limited and treatment responses to currently existing therapies are often unsatisfactory (Bracarda et al. 2011). For example, the cytotoxic therapy often causes severe toxicity and eventually leads also to the development of chemo-resistance (Tannock et al. 2004, Berthold et al. 2008, Bracarda et al. 2011). Thus, there is an urgent need for novel agents to block the proliferation and to inhibit the progression of the primary prostate cancer cells to the advanced stage as well as to target advanced and metastatic prostate cancer. Therefore, understanding of disease progression and drug resistance mechanisms may provide valuable insights into the development of novel treatment options to improve the survival of prostate cancer patients.

We have recently performed a high-throughput cell-based screening of 4,910 known drugs and drug-like molecules in four prostate cancer cell models and two non-tumorigenic prostate epithelial cell lines to identify prostate cancer cell growth selective inhibitors (Iljin et al. 2009). Only four compounds, antibiotic ionophore monensin, aldehyde dehydrogenase (ALDH) inhibitor disulfiram, histone deacetylase inhibitor trichostatin A and fungicide thiram inhibited selectively cancer cell growth at nanomolar concentrations. The mechanistic studies indicated that monensin and disulfiram inhibited prostate cancer cell growth by inducing oxidative stress (Iljin et al. 2009, Ketola et al. 2010). In contrast to disulfiram, monensin induced apoptosis, reduced androgen receptor signalling and showed a synergistic anti-proliferative effect with anti-androgens in prostate cancer cells. Moreover, monensin increased the amount of reactive oxygen species (ROS) and induced an oxidative stress signature in prostate cancer cells (VCaP and LNCaP), but not in the non-malignant prostate epithelial cells (RWPE-1, EP156T) (Ketola et al. 2010). Furthermore, antioxidant vitamin C partially rescued the monensin induced growth inhibition, indicating that oxidative stress plays a key role in the antineoplastic effect of monensin in cultured prostate cancer cells (Ketola et al. 2010).

Oxidative stress occurs in the cell when redox regulation is imbalanced. Redox balance depends on the level of intracellular free radicals and reactive oxygen species as well as on the antioxidative capacity of the cell. Figure 1 illustrates the connection between the malignant progression and increase in intracellular ROS (Fig. 1)

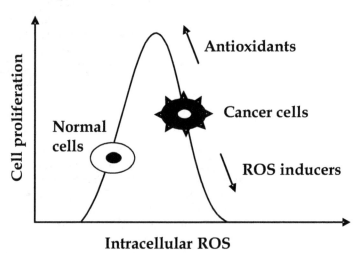

Intracellular ROS

Fig. 1. The balance between cell proliferation and intracellular reactive oxygen species (ROS). The malignant progression and the effects of ROS inducers and antioxidants depend on the intracellular levels of ROS. In cancer cells, intracellular ROS levels are higher than in normal cells making them more vulnerable to agents inducing ROS. Figure idea adapted from Gupte & Mumper 2009.

Cancer cells benefit from the increased mutation rate induced via oxidative stress (Shibutani, Takeshita & Grollman 1991). Therefore, cancer cells need also an active antioxidative mechanism to be able to survive under high oxidative stress. The oxidative stress level is elevated in prostate cancer cells compared to non-malignant prostate epithelial cells (Khandrika et al. 2009, Yossepowitch et al. 2007, Kumar et al. 2008). Many oncogenes are known to protect cells from oxidative stress e.g. androgen receptor (AR), ERG and MYC are known to have antioxidative properties in cancer cells (Pinthus et al. 2007, Tam et al. 2003, Benassi et al. 2006, Swanson et al. 2011, DeNicola et al. 2011). Monensin may sensitize prostate cancer cells to oxidative stress via reducing the expression of these genes (Ketola et al. 2010). In addition, many other anti-neoplastic agents such as vinblastine, cisplatin, mitomycin C, doxorubicin, camptothecin, inostamycin, neocarzinostatin, etoposide, arsenic trioxide and nonsteroidal anti-inflammatory drugs are known to mediate their apoptotic effect by inducing oxidative stress (Fang, Nakamura & Iyer 2007, Rigas, Sun 2008, Sun et al. 2011). The increased sensitivity to oxidative stress combined with dependency on anti-oxidative system may provide a way to selectivity inhibit cancer cell proliferation (Iljin et al. 2009, Ketola et al. 2010).

Recently, redox control including antioxidative defence mechanisms and ROS-scavenging systems has been identified as an important regulator of cancer stem cell potential, metastasis and chemoresistance (Kobayashi, Suda 2011b, Cairns, Harris & Mak 2011, Pani, Galeotti & Chiarugi 2010). Aldehyde dehydrogenase activity is widely used as a marker for cancer stem cells and its activity has been shown to correlate with poor outcome in several cancers such as in prostate cancer (Davydov, Dobaeva & Bozhkov 2004, Burger et al. 2009, Li et al. 2010, Yu et al. 2011, Zhang et al. 2009a). Aldehyde dehydrogenases are detoxifying enzymes that are responsible for the oxidation of intracellular aldehydes (Duester 2000, Magni et al. 1996, Sophos, Vasiliou 2003, Yoshida et al. 1998). Moreover, ALDH oxidize retinol to retinoic acid (vitamin A) known to reduce oxidative stress whereas retinoic acid depletion induces oxidative stress and mitochondrial dysfunction (Ahlemeyer, Krieglstein 1998, Ahlemeyer et al. 2001, Chiu, Fischman & Hammerling 2008, Duester 2000, Chute et al. 2006). Thus, ALDH increases the antioxidative capacity in cells and protects cells from oxidative stress induction. Moreover, the inhibition of ALDH activity has recently been linked to reduced chemotherapy and radiation resistance in cancer stem cells (Croker, Allan 2011). These results suggest that not only cancer cells, but also cancer stem cells could be targeted by oxidative stress induction and/or reduction of antioxidative capacity. Interestingly, the results from mechanistic studies of prostate cancer selective compounds indicated that both disulfiram and monensin reduced the ALDH activity in prostate cancer cells (Iljin et al. 2009, Ketola et al., 2010).

In this study, we explored further the molecular mechanism of monensin induced growth inhibition in cultured prostate cancer cells. Cancer pathway reporter assays and steroid profiling was performed to get insights into altered signalling and metabolite levels in monensin exposed prostate cancer cells. Since monensin reduces ALDH activity, the putative effect on cancer stem cells was evaluated. Furthermore, we studied the effect of monensin on prostate cancer cell differentiation and motility.

2. Materials and methods

2.1 Cells

VCaP prostate carcinoma cells (TMPRSS2-ERG positive, received from Drs. Adrie van Bokhoven, University of Colorado Health Sciences Center, Denver, Colorado and Kenneth Pienta, University of Michigan, Michigan) were grown in Dulbecco's Modified Eagle's Medium (Korenchuk et al. 2001b, Korenchuk et al. 2001a). LNCaP prostate carcinoma cells (received from Dr. Marco Cecchini, University of Bern, Bern, Switzerland) were grown in T-Medium (Invitrogen). PC-3 prostate carcinoma cells were purchased from American Type Culture Collection (LGC Promochem AB) and grown according to provider's instructions. All cells were cultured in appropriate growth media described above including 10% fetal bovine serum (FBS), 2 mM L-glutamine, 100 U/ml penicillin, and 100 μg/ml streptomycin in an incubator with a humidified atmosphere of 95% air and 5% CO_2 at 37°C.

2.2 Compounds

Monensin was purchased from Sigma-Aldrich and diluted in ethanol.

2.3 Cancer 10-pathway reporter array

Cancer 10-pathway Reporter Luciferase Kit (Wnt (TCF/LEF), Notch (RBP-Jκ), p53/DNA damage, TGF-β (SMAD2/3/4), Cell cycle/pRb-E2F (E2F/DP1), NF-κB, Myc/Max, Hypoxia (HIF1A), MAPK/ERK (Elk-1/SRF) and MAPK/JNK (AP-1) was used to study the monensin modulated signalling (SABiosciences, Frederick, MD). The assay was performed according to manufacturer's instructions. In brief, inducible transcription factor responsive firefly luciferase reporters with constitutively expressing Renilla construct transcription factor reporters were plated onto 96-well plates with transfection reagent (siLentFect, Bio-Rad Laboratories), followed by addition of cells and incubation for 24 hours. A mixture of non-inducible firefly luciferase reporter and constitutively expressing Renilla construct was used as the negative control. After 24 hours of transfection, monensin (100 nM) and control treatment were added onto the cells and plates were incubated for 18 hours. The Dual-LuciferaseReporter (DLR™) Assay System (Promega) was utilized in quantitation of reporters and results according to the manufactorer's instructions. The change in the activity of each signalling pathway was determined by comparing the normalized luciferase activity of the reporters in monensin or solvent exposed cells.

2.4 Wound healing assay

The effect of monensin (10 nM, 100 nM and 1 μM) on prostate cancer cell migration was studied using a scratch wound assay. PC-3 cells were plated on 96-well plates (Essen ImageLock, Essen Instruments, UK) and a wound was scratched with wound scratcher (Essen Instruments). Compounds and appropriate controls were added immediately after wound scratching and wound confluence was monitored with Incucyte Live-Cell Imaging System and software (Essen Instruments). Wound closure was calculated for every hour for 24 hours by comparing the mean relative wound density of three biological replicates in each experiment.

2.5 Cell viability assay

Cell viability was determined with CellTiter-Glo (CTG) cell viability assay (Promega, Madison, WI) according to the manufacturer's instructions. Briefly, 2,000 cells per well were plated in 35 μl of their respective growth media and left to attach overnight. Monensin (100 nM) was added to the cells and incubated for 12 or 24 hours. CTG reagent was added and the signals were quantified using Envision Multilabel Plate Reader (Perkin-Elmer, Massachusetts, MA).

2.6 RNA extraction and quantitative real-time PCR

VCaP cells were exposed to monensin for 6 hours, total RNA was extracted and quantitative real-time PCR was done as previously described (Ketola et al. 2010). TaqMan gene expression probes and primers from the Universal Probe Library (Roche Diagnostics, Espoo, Finland) were used to study E-cadherin (5´-cccgggacaacgtttattac-3´ and 5´-gctggctcaagtcaaagtcc-3´) and β-actin (5´-ccaaccgcgagaagatga -3´ and 5´-ccagaggcgtacagggatag -3´) mRNA expression. Three replicate samples were studied.

2.7 Fluorescence-Activated Cell Sorting analysis (FACS)

VCaP and LNCaP cells were exposed to monensin (1 µM) for 6 hours, samples were fixed with 2% paraformaldehyde, and stained with CD44 (FITC-conjugated mouse monoclonal anti-human, BD Pharmingen™ 555478) and CD24 (PE-conjugated rat monoclonal anti-human, Abcam ab25281) antibodies for 45 minutes at 4°C in the dark. Cells were washed and the fluorescence intensity was measured using Accuri C6 Flow Cytometer.

2.8 Immuofluorescence staining

For immunofluorescence stainings, VCaP cells were grown on cover slip slides and exposed to monensin for 6 hours. Cells were fixed with 4 % paraformaldehyde in PBS, permeabilized with 0.2% Triton X-100 in PBS for 15 min, and blocked with 2 % BSA/PBS for 30 min. Cells were stained with E-cadherin antibody (1:100 dilution, polyclonal rabbit anti-human, Cell Signaling Technology, MA, USA) and Alexa-conjugated polyclonal donkey anti-rabbit antibody was used for secondary staining (1:300 dilution, Invitrogen, Molecular Probes, Carlsbad, CA). Cell nuclei were stained with DAPI present in Vectashield mounting medium (Vector Labs) and images were taken with Zeiss Axiovert 200M Microscope with the spinning disc confocal unit Yokogawa CSU22 and a Zeiss Plan-Neofluar 63× oil/1.4 NA objective. Z-stacks with 1 airy unit optical slices were acquired with a step size of 0.5 µm between slices, and the maximum intensity projections were created with SlideBook 4.2.0.7 software (Intelligent Imaging Innovations Inc., CO, USA).

2.9 Statistical analyses

Stars in the figures indicate the significance of the experiments calculated using Student's t-test (*$P<0.05$, **$P<0.01$, *** $P<0.001$).

2.10 Steroid quantification

VCaP cells (1×10^7 cells) were exposed to 1 µM monensin for 6 hours, harvested and counted. An internal standard (labeled C16:0) and chloroform:methanol (2:5) mixture were added, the samples were homogenized with Retsch system (5 min, 20 Hz), centrifuged and the supernatants were collected and evaporated. MOX (25 µl, TS-45950, Thermo Scientific, Helsinki, Finland) was added and the mixture was incubated at 45°C for 60 minutes. Next, 100 µl of MSTFA with 1% trimethylchlorosilane (Fluka, St. Louis, MO) was added and the mixture was incubated at 70°C for 60 minutes. Injection standard was added to the mixture before gas chromatography-mass spectrometry analysis (GC-MS, Agilent 6890 gas chromatograph (GC) combined with Agilent 5973 mass selective detector (MSD)). The injector (injection volume 1 µl with pulsed splitless injection) and MSD temperatures were 230°C (MS Source) and 150°C (MS Quad). The analyses were performed on Supelco 38499-02C capillary column. Selective ion monitoring using specific masses for each target analyte was used in the detection. The following steroids were quantified: 7-ketocholesterol, aldosterone, progesterone, pregnenolone, estrone, 17B-estradiol, 4B-hydroxycholesterol,

25-hydroxycholesterol, 5a,6a-epoxycholesterol (Mono-TMS), dihydrotestosterone and testosterone (the standards were from Steraloids, Newport, RI)).

3. Results

3.1 Monensin reduces NF-κB pathway activity in prostate cancer cells

Here, we studied the effect of monensin exposure on the activities of ten cancer signalling pathways using Cancer pathway Reporter Array in prostate cancer cells. Inducible transcription factor responsive firefly luciferase reporters were transfected to VCaP and LNCaP cells with constitutively active Renilla reporters and incubated for 24 hours. Monensin (100 nM) or solvent control was added onto the transfected cells for 18 hours followed by measure of luciferase activities. The results are presented in Fig. 2. Comparison of the basal pathway activities in VCaP prostate cancer cells indicated that NF-κB was

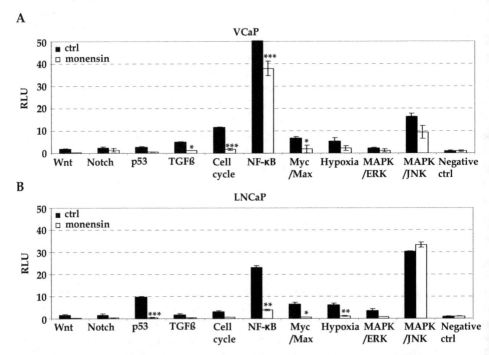

Fig. 2. Cancer pathway activities in VCaP and LNCaP prostate cancer cells. A) VCaP and B) LNCaP cells were exposed to ethanol or 100 nM monensin for 24 hours and pathway activities were measured as described in the text. Results with negative control measuring the background firefly luciferase activity (non-inducible firefly luciferase reporter and constitutively expressing Renilla construct) are indicated and used to determine the basal pathway activities in VCaP and LNCaP cells. The y-axis was set to 50 to allow direct comparison of relative luciferase units (RLU) in VCaP and LNCaP cells, although the basal NK-κB activity in VCaP cells extended RLU 324. Statistical significance of monensin induced changes are shown for the active pathways *P<0.05, **P<0.01, ***P<0.001.

clearly the most active pathway (fold change FC > 300, compared to the negative control). In addition, TGF-ß (FC = 5), cell cycle (FC = 12), Myc/Max (FC = 7) and hypoxia (FC = 5) were active in VCaP cells. In LNCaP cells, MAPK/JNK was the most active pathway (FC = 30 compared to the negative control), followed by NF-κB (FC = 23), p53 (FC = 10), Myc/Max (FC = 7) and hypoxia (FC = 6). Interestingly, NF-κB pathway was 14 times more active in TMPRSS2-ERG fusion positive VCaP cells than in ERG negative LNCaP cells. These results are in agreement with previous data indicating that ERG induces NF-κB activity in prostate cells *in vitro* and *in vivo* (Wang et al. 2010). The results with p53 pathway are also in accordance with the previous literature since VCaP cells are known to have an inactivating mutation in p53 (Trp-248) whereas LNCaP cells express the active form, supporting further the overall functionality of the assay (van Bokhoven et al. 2003).

Monensin exposure inhibited NF-κB activity in both VCaP (by 88%) and LNCaP (by 83%) cells. NF-κB is a transcription factor that regulates various cellular processes such as cellular antioxidant defence capacity (Gloire, Legrand-Poels & Piette 2006). These results suggest that monensin induced oxidative stress may result from reduced NF-κB signalling. In addition, monensin reduced TGF-ß (by 75%), cell cycle (by 85%) and Myc/Max (by 71%) activities in VCaP cells and p53 (by 94%), hypoxia (by 82%) and Myc/Max (by 90%) activities in LNCaP cells. The reduced Myc/Max signalling is supported by reduced MYC mRNA expression seen in monensin exposed VCaP and LNCaP cells (Ketola et al. 2010). Taken together, monensin reduced the activities of multiple signalling pathways such as NF-κB, TGF-ß, hypoxia and Myc/Max, all associated with cancer cell survival, oxidative stress, stem cell potential and metastasis (Jones, Pu & Kyprianou 2009, Blum et al. 2009, Mimeault, M. & Batra, S.K. 2011, Benassi et al. 2006, Koh et al. 2010).

3.2 Monensin reduces the cancer stem cell population in prostate cancer cell cultures

Monensin exposure reduced the activities of multiple pathways maintaining antioxidative capacity and promoting the growth and survival of cancer stem cells such as NF-κB, HIF1A, MYC and ALDH, in cultured prostate cancer cells. Therefore, the effect of monensin on prostate cancer stem cell population was studied. Prostate cancer stem cells can be identified by high expression of CD44 and low expression of CD24 antigens (Klarmann et al. 2009). Interestingly, CD44 cell surface glycoprotein has recently been shown to increase antioxidative capacity in cancer cells, indicating that cancer initiating cells could be targeted by impairing oxidative stress defence mechanisms (Ishimoto et al. 2011). Thus, we analyzed the effect of monensin exposure to the fraction of CD44+/CD24- cells in cultured prostate cancer cells. VCaP and LNCaP cells exposed to monensin (1 μM) for 6 hours were stained with CD44 and CD24 recognizing antibodies and the samples were analyzed using FACS. The results showed that monensin reduced the fraction of CD44+/CD24- cells in both VCaP (from 3 to 1.3%) and LNCaP (3.1 to 2.6%) cells (Fig. 3). In agreement with these results, genome-wide gene expression analysis indicated that monensin decreased CD44 (by 20%) and induced CD24 (by 30%) mRNAs compared to the levels in ethanol control at 6-hour time point (Ketola et al. 2010). Taken together, these results indicate that monensin reduces the fraction of cancer stem cells in prostate cancer cell cultures.

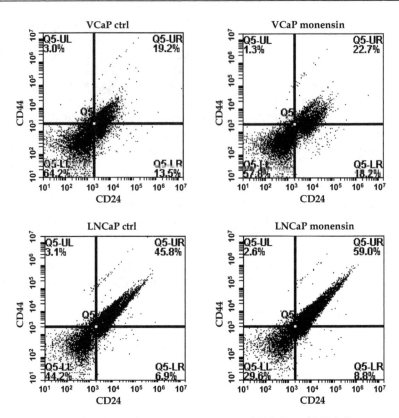

Fig. 3. FACS analysis of CD44 and CD24 immunostained VCaP and LNCaP prostate cancer cells. Monensin reduces the fraction of CD44+/CD24- cells in cultured prostate cancer cells. VCaP and LNCaP cells were stained with CD44 and CD24 antibodies in response to six hour monensin or ethanol exposure. Representative images from one out of four replicates are shown.

3.3 Monensin induces cell differentiation in prostate cancer cells

Cancer stem cells have been hypothesized to arise intrinsically through oncogenic transformation of normal tissue stem or progenitor cells in the early stage of the tumorigenesis or through induction of epithelial-to-mesenchymal transition (EMT) at later stages (Chaffer, Weinberg 2011). The expression of E-cadherin is considered as a marker of epithelial differentiation and it is lost during EMT. Interestingly, ERG knock-down in VCaP cells has been shown to induce E-cadherin expression (Gupta et al. 2010). Moreover, our previous results indicated that monensin reduced ERG expression in VCaP cells (Ketola et al. 2010). Thus, to study whether monensin affects prostate cancer cell differentiation, E-cadherin expression was analysed in monensin exposed VCaP cells. The results from immunochemical staining and quantitative RT-PCR showed that monensin induced E-cadherin expression in VCaP cells (Fig. 4A and B), indicating that monensin promotes cell differentiation in cultured prostate cancer cells.

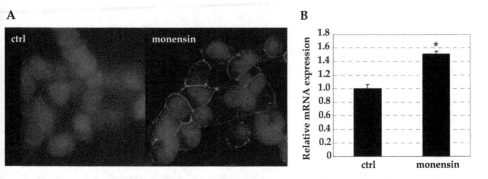

Fig. 4. Monensin induces E-cadherin expression in VCaP prostate cancer cells. A) Immunofluorescence staining of E-cadherin (red) in response to 1 μM monensin or ethanol (ctrl) for 6-hour in VCaP cells. Nuclei are stained with DAPI (blue). B) Relative mRNA expression of E-cadherin in response to 1 μM monensin or ethanol (ctrl) exposure for 6-hour in VCaP cells.

3.4 Monensin reduces migration in cultured prostate cancer cells

Epithelial-to-mesenchymal transition is a perquisite for cancer cell migration (Baum, Settleman & Quinlan 2008). Moreover, in addition to the role in promoting cancer stem cell growth and survival, NF-κB and ALDH activities as well as high CD44 and low CD24 expressions are known to enhance prostate cancer cell migration (van den Hoogen et al. 2010, Klarmann et al. 2009). Since monensin induced cell differentiation, the effect of monensin exposure on prostate cancer cell migration was studied. VCaP and LNCaP cells do not migrate and therefore PC-3 prostate cancer cells were used as a model in migration assay. The results are presented in Fig. 5. Interestingly, already at 10 nM concentration, monensin significantly reduced cell migration. The anti-migratorial effect was stronger at higher concentrations (100 nM and 1 μM) (Fig. 5A). Thus, monensin was able to reduce cell migration at nanomolar concentrations in PC-3 cells although the same concentrations did not significantly decrease cell viability in these cells even at 24 hour time point (Fig. 5B). Pictures showing the decrease in wound closure in monensin exposed PC-3 cells in comparison to control are shown at 12 and 24 hour time points in Fig. 5C.

3.5 Monensin increases oxidative stress inducing steroids as well as reduces the level of androgen precursor and antioxidative steroid

Our previous results indicated that monensin induced oxidative stress and altered the expression of genes involved in cholesterol and steroid biosynthesis (Ketola et al. 2010). Moreover, monensin reduced androgen receptor (AR) signalling and showed synergistic growth inhibitory effects with anti-androgens in prostate cancer cells. To validate the monensin induced changes in cellular steroid levels, steroid profiling was performed in VCaP prostate cancer cells. Cells were exposed to ethanol or monensin (1 μM) for six hours and steroid profiles were studied using gas chromatography - mass spectrometry (GC-MS). The results presented as a heat-map in Fig. 6 show that the most prominent changes in response to monensin exposure were the induction of 7-ketocholesterol and aldosterone

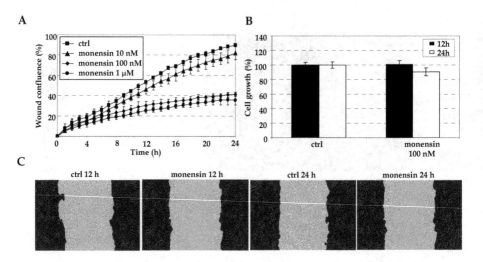

Fig. 5. The effect of monensin exposure on PC-3 prostate cancer cell migration. A) Relative wound confluency in response to monensin (10 nM, 100 nM and 1 µM) or control (ethanol) was monitored for 24 hours. B) Results from cell viability assay in response to monensin exposure (100 nM) for 12 and 24 hours in PC-3 cells. C) Pictures of scratch-wounded wells in response to 100 nM monensin or ethanol exposure at 12- and 24-hour time points. The wound margin in the beginning of the experiment is coloured in dark grey.

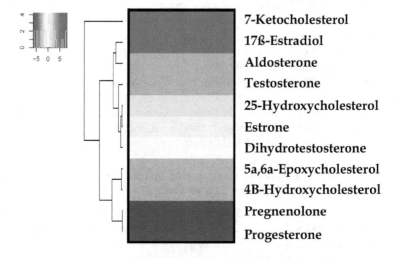

Fig. 6. Steroid profiling heat-map of monensin exposed VCaP prostate cancer cells. The cells were exposed to monensin (1 µM) for 6 hours and the steroid profile was measured with gas chromatography-mass spectrometry (GC-MS). The levels of individual steroids in monensin exposed cells were compared to ethanol exposed samples (presented as fold change, red: induction, blue: reduction).

levels as well as decrease in progesterone and pregnenolone levels. Aldosterone and 7-ketocholesterol are known oxidative stress inducers whereas progesterone has antioxidant properties (Leonarduzzi et al., 2006, Gramajo et al., 2010, Lee et al., 2009, Calo et al., 2010, Queisser et al., 2011, Ozacmak et al., 2009). Moreover, progesterone and pregnenolone are androgen precursors which have been suggested to play a major role in prostate cancer cell survival (Locke et al. 2008). Interestingly, progesterone has also been shown to promote mammary stem cell expansion (Joshi et al. 2010). The steroid profiling validates also the previous Connectivity Map results which indicated that monensin has opposite effects to progesterone and pregnenolone (Ketola et al. 2010). Taken together, monensin induced alterations in cellular steroid profile indicated that monensin induced oxidative stress results from increased aldosterone and 7-ketocholesterol levels as well as decreased progesterone and pregnenolone levels.

4. Conclusion

In this study, we explored the molecular consequences of monensin exposure in cultured prostate cancer cells. Our previous study indicated that monensin inhibited selectively prostate cancer cell viability at nanomolar concentrations by inducing oxidative stress. Cancer cells are constantly under pro-oxidative state (Szatrowski, Nathan 1991, Toyokuni et al. 1995). Long-term oxidative stress stimulates cell growth and proliferation, contributes to metastatic process and promotes cancer cell invasiveness and migration (Mori, Shibanuma & Nose 2004, Sung et al. 2006). Therefore, cancer cells need strong antioxidant mechanisms to survive and profit from these oxidative stress induced changes. Interfering redox balance has been suggested as a potential mean to selectively target cancer cells for example by increasing the cellular ROS level or reducing the expression of antioxidative enzymes (Pelicano, Carney & Huang 2004). Our results with monensin support this hypothesis.

Our previous Connectivity Map results indicated that monensin has agonistic effects to NF-κB inactivator and oxidative stress inducer niclosamide supporting monensin as a potent NF-κB inhibitor. Here, we showed that although monensin reduced the activities of several pathways known to play a role in tumourigenesis, the strongest reduction was seen in NF κB signalling. NF-κB activity promotes cell viability, tumorigenesis and metastasis as well as correlates with poor prognosis in prostate cancer patients (Blum et al. 2009, Sarkar et al. 2008). Importantly, NF-κB regulates the expression of genes responsible for antioxidant defence capacity and its inhibition induces oxidative stress as well as reduces tumourigenesis, metastasis and cancer stem cell potential (Gloire, Legrand-Poels & Piette 2006, Sarkar et al. 2008, Gluschnaider et al. 2010). NF-κB inhibitors are known to decrease AR signalling *in vitro* and reduce the growth of androgen deprivation-resistant prostate cancer xenografts *in vivo* (Jin et al. 2008, Zhang et al. 2009b). However, at present no specific NF-κB inhibitors have reached the stage of clinical trials for prostate cancer treatment (Mahon et al. 2011). Our results support NF-κB as the main mediator of monensin induced oxidative stress, which may also contribute to the reduced androgen signalling and induction of apoptosis in monensin exposed prostate cancer cells.

Recently, cancer stem cell targeting has raised a lot of interest as a prominent way to target cancer drug resistance and metastasic growth (Mimeault, Batra 2011, Clayton, Mousa 2011). In comparison to cancer cells, ROS levels in cancer stem cells are lower due to controlled redox balance system such as high ALDH and CD44 expression protecting cancer stem cells from oxidative stress (Kobayashi, Suda 2011a, Croker, Allan 2011, Ishimoto et al. 2011). Interestingly, NF-κB inhibition induces apoptosis in prostate cancer stem cells and thus NF-κB is considered as an attractive chemotherapeutic target also against cancer stem cells (Jin et al. 2008, Birnie et al. 2008). Since monensin reduced ALDH and NF-κB activities, we studied the fraction of CD44+/CD24- cells in prostate cancer cell cultures in response to monensin exposure. The results confirmed that monensin reduced the amount of prostate cancer stem cells. Moreover, monensin induced epithelial cell differentiation and reduced motility in cultured prostate cancer cells, suggesting that monensin inhibits prostate tumorigenesis by multiple ways. Cancer stem cell inhibitor and cell differentiation inducer salinomycin shares a similar structure as monensin, supporting the functional similarities between these two compounds (Gupta et al. 2009).

Steroidogenic enzymes as well as stem cell markers are induced in castration-resistant prostate cancer both *in vitro* and *in vivo* (Blum et al. 2009, Pfeiffer et al. 2011). Several studies have shown that steroidogenesis is inhibited by ROS (Tsai et al. 2003, Stocco, Wells & Clark 1993, Kodaman, Aten & Behrman 1994, Lee et al. 2009, Abidi et al. 2008). Our previous results indicated that monensin reduced androgen receptor signalling. Here, we showed that monensin increases the levels of oxidative stress inducing steroids, 7-ketocholesterol and aldosterone, and reduces androgen precursor and antioxidative steroid progesterone in cultured prostate cancer cells. Interestingly, 7-ketocholesterol is a ligand for aryl hydrocarbon receptor (AhR) and acts as AhR antagonist (Savouret et al. 2001). AhR expression is elevated in malignant prostate cells and its signalling is activated in prostate cancer stem cells (Blum et al. 2009, Gluschnaider et al. 2010). AhR pathway increases the expression of ALDH proteins and protects cells against oxidative stress and foreign chemicals (Lindros et al. 1998, Vrzal, Ulrichova & Dvorak 2004, Nebert et al. 2000, Kohle, Bock 2007). Interestingly, AhR binds to NF-κB, induces MYC activation and reduces E-cadherin expression in breast cancer cells (Kim et al. 2000, Dietrich, Kaina 2010). Moreover, AhR can form a complex with androgen receptor and protect prostate cancer cells during androgen ablation (Ohtake, Fujii-Kuriyama & Kato 2009, Gluschnaider et al. 2010). Thus, our results indicate that monensin induced oxidative stress is potentially transmitted via reduced AhR signalling. This hypothesis is further supported by our previous gene expression results indicating that although AhR itself was not altered, the expression of AhR target gene mRNAs were decreased in response to monensin exposure (Ketola et al. 2010).

Taken together, we hypothesize that monensin induced anti-neoplastic effects result mainly due to increase in oxidative stress. The overview figure 7 illustrates the various changes occurring in prostate cancer cells in response to monensin exposure. The cancer selectiveness could be explained by increased intracellular ROS due to reduced antioxidative capacity sensitizing prostate cancer cells to oxidative stress. Since normal prostate epithelial cells are not under intensive oxidative stress and therefore, are less dependent on the function of antioxidative genes, they are not as sensitive to monensin exposure as cancer cells. In conclusion, our results support the idea that impairing the redox control, which has a crucial role in cancer cells enabling survival in high intracellular ROS, is a potent way to target prostate cancer cells.

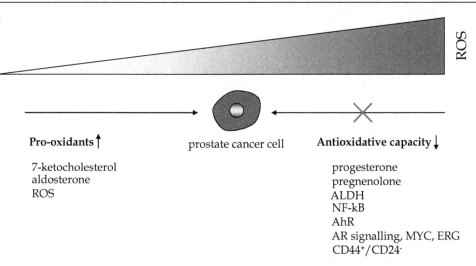

Pro-oxidants ↑

7-ketocholesterol
aldosterone
ROS

prostate cancer cell

Antioxidative capacity ↓

progesterone
pregnenolone
ALDH
NF-kB
AhR
AR signalling, MYC, ERG
CD44+/CD24-

Fig. 7. Overview of monensin induced changes in prostate cancer cells. The figure idea adapted from Cairns, R.A., Harris, I.S. & Mak, T.W. 2011.

5. Acknowledgement

We thank Mika Hilvo, Tuulia Hyötyläinen, Anna-Liisa Ruskeepää (VTT Technical Research Centre of Finland, Espoo, Finland) for their contribution in steroid profiling part of this study.

6. References

Abidi, P., Zhang, H., Zaidi, S.M., Shen, W.J., Leers-Sucheta, S., Cortez, Y., Han, J. & Azhar, S. 2008, "Oxidative stress-induced inhibition of adrenal steroidogenesis requires participation of p38 mitogen-activated protein kinase signaling pathway", *The Journal of endocrinology*, vol. 198, no. 1, pp. 193-207.

Ahlemeyer, B., Bauerbach, E., Plath, M., Steuber, M., Heers, C., Tegtmeier, F. & Krieglstein, J. 2001, "Retinoic acid reduces apoptosis and oxidative stress by preservation of SOD protein level", *Free radical biology & medicine*, vol. 30, no. 10, pp. 1067-1077.

Ahlemeyer, B. & Krieglstein, J. 1998, "Retinoic acid reduces staurosporine-induced apoptotic damage in chick embryonic neurons by suppressing reactive oxygen species production", *Neuroscience letters*, vol. 246, no. 2, pp. 93-96.

Baum, B., Settleman, J. & Quinlan, M.P. 2008, "Transitions between epithelial and mesenchymal states in development and disease", *Seminars in cell & developmental biology*, vol. 19, no. 3, pp. 294-308.

Benassi, B., Fanciulli, M., Fiorentino, F., Porrello, A., Chiorino, G., Loda, M., Zupi, G. & Biroccio, A. 2006, "c-Myc phosphorylation is required for cellular response to oxidative stress", *Molecular cell*, vol. 21, no. 4, pp. 509-519.

Berthold, D.R., Pond, G.R., Soban, F., de Wit, R., Eisenberger, M. & Tannock, I.F. 2008, "Docetaxel plus prednisone or mitoxantrone plus prednisone for advanced prostate cancer: updated survival in the TAX 327 study", *Journal of clinical oncology : official journal of the American Society of Clinical Oncology,* vol. 26, no. 2, pp. 242-245.

Birnie, R., Bryce, S.D., Roome, C., Dussupt, V., Droop, A., Lang, S.H., Berry, P.A., Hyde, C.F., Lewis, J.L., Stower, M.J., Maitland, N.J. & Collins, A.T. 2008, "Gene expression profiling of human prostate cancer stem cells reveals a pro-inflammatory phenotype and the importance of extracellular matrix interactions", *Genome biology,* vol. 9, no. 5, pp. R83.

Blum, R., Gupta, R., Burger, P.E., Ontiveros, C.S., Salm, S.N., Xiong, X., Kamb, A., Wesche, H., Marshall, L., Cutler, G., Wang, X., Zavadil, J., Moscatelli, D. & Wilson, E.L. 2009, "Molecular signatures of prostate stem cells reveal novel signaling pathways and provide insights into prostate cancer", *PloS one,* vol. 4, no. 5, pp. e5722.

Bracarda, S., Logothetis, C., Sternberg, C.N. & Oudard, S. 2011, "Current and emerging treatment modalities for metastatic castration-resistant prostate cancer", *BJU international,* vol. 107 Suppl 2, pp. 13-20.

Burger, P.E., Gupta, R., Xiong, X., Ontiveros, C.S., Salm, S.N., Moscatelli, D. & Wilson, E.L. 2009, "High aldehyde dehydrogenase activity: a novel functional marker of murine prostate stem/progenitor cells", *Stem cells (Dayton, Ohio),* vol. 27, no. 9, pp. 2220-2228.

Cairns, R.A., Harris, I.S. & Mak, T.W. 2011, "Regulation of cancer cell metabolism", *Nature reviews.Cancer,* vol. 11, no. 2, pp. 85-95.

Chaffer, C.L. & Weinberg, R.A. 2011, "A perspective on cancer cell metastasis", *Science (New York, N.Y.),* vol. 331, no. 6024, pp. 1559-1564.

Chiu, H.J., Fischman, D.A. & Hammerling, U. 2008, "Vitamin A depletion causes oxidative stress, mitochondrial dysfunction, and PARP-1-dependent energy deprivation", *The FASEB journal : official publication of the Federation of American Societies for Experimental Biology,* vol. 22, no. 11, pp. 3878-3887.

Chute, J.P., Muramoto, G.G., Whitesides, J., Colvin, M., Safi, R., Chao, N.J. & McDonnell, D.P. 2006, "Inhibition of aldehyde dehydrogenase and retinoid signaling induces the expansion of human hematopoietic stem cells", *Proceedings of the National Academy of Sciences of the United States of America,* vol. 103, no. 31, pp. 11707-11712.

Clayton, S. & Mousa, S.A. 2011, "Therapeutics formulated to target cancer stem cells: Is it in our future?", *Cancer cell international,* vol. 11, pp. 7.

Croker, A.K. & Allan, A.L. 2011, "Inhibition of aldehyde dehydrogenase (ALDH) activity reduces chemotherapy and radiation resistance of stem-like ALDH(hi)CD44 (+) human breast cancer cells", *Breast cancer research and treatment,* .

Dakhova, O., Ozen, M., Creighton, C.J., Li, R., Ayala, G., Rowley, D. & Ittmann, M. 2009, "Global gene expression analysis of reactive stroma in prostate cancer", *Clinical cancer research : an official journal of the American Association for Cancer Research,* vol. 15, no. 12, pp. 3979-3989.

Davydov, V.V., Dobaeva, N.M. & Bozhkov, A.I. 2004, "Possible role of alteration of aldehyde's scavenger enzymes during aging", *Experimental gerontology,* vol. 39, no. 1, pp. 11-16.

DeNicola, G.M., Karreth, F.A., Humpton, T.J., Gopinathan, A., Wei, C., Frese, K., Mangal, D., Yu, K.H., Yeo, C.J., Calhoun, E.S., Scrimieri, F., Winter, J.M., Hruban, R.H., Iacobuzio-Donahue, C., Kern, S.E., Blair, I.A. & Tuveson, D.A. 2011, "Oncogene-induced Nrf2 transcription promotes ROS detoxification and tumorigenesis", *Nature*, vol. 475, no. 7354, pp. 106-109.

Dietrich, C. & Kaina, B. 2010, "The aryl hydrocarbon receptor (AhR) in the regulation of cell-cell contact and tumor growth", *Carcinogenesis*, vol. 31, no. 8, pp. 1319-1328.

Duester, G. 2000, "Families of retinoid dehydrogenases regulating vitamin A function: production of visual pigment and retinoic acid", *European journal of biochemistry / FEBS*, vol. 267, no. 14, pp. 4315-4324.

Fang, J., Nakamura, H. & Iyer, A.K. 2007, "Tumor-targeted induction of oxystress for cancer therapy", *Journal of drug targeting*, vol. 15, no. 7-8, pp. 475-486.

Gloire, G., Legrand-Poels, S. & Piette, J. 2006, "NF-kappaB activation by reactive oxygen species: fifteen years later", *Biochemical pharmacology*, vol. 72, no. 11, pp. 1493-1505.

Gluschnaider, U., Hidas, G., Cojocaru, G., Yutkin, V., Ben-Neriah, Y. & Pikarsky, E. 2010, "beta-TrCP inhibition reduces prostate cancer cell growth via upregulation of the aryl hydrocarbon receptor", *PloS one*, vol. 5, no. 2, pp. e9060.

Gupta, P.B., Onder, T.T., Jiang, G., Tao, K., Kuperwasser, C., Weinberg, R.A. & Lander, E.S. 2009, "Identification of selective inhibitors of cancer stem cells by high-throughput screening", *Cell*, vol. 138, no. 4, pp. 645-659.

Gupta, S., Iljin, K., Sara, H., Mpindi, J.P., Mirtti, T., Vainio, P., Rantala, J., Alanen, K., Nees, M. & Kallioniemi, O. 2010, "FZD4 as a mediator of ERG oncogene-induced WNT signaling and epithelial-to-mesenchymal transition in human prostate cancer cells", *Cancer research*, vol. 70, no. 17, pp. 6735-6745.

Gupte, A. & Mumper, R.J. 2009, "Elevated copper and oxidative stress in cancer cells as a target for cancer treatment", *Cancer treatment reviews*, vol. 35, no. 1, pp. 32-46.

Iljin, K., Ketola, K., Vainio, P., Halonen, P., Kohonen, P., Fey, V., Grafstrom, R.C., Perala, M. & Kallioniemi, O. 2009, "High-throughput cell-based screening of 4910 known drugs and drug-like small molecules identifies disulfiram as an inhibitor of prostate cancer cell growth", *Clinical cancer research : an official journal of the American Association for Cancer Research*, vol. 15, no. 19, pp. 6070-6078.

Ishimoto, T., Nagano, O., Yae, T., Tamada, M., Motohara, T., Oshima, H., Oshima, M., Ikeda, T., Asaba, R., Yagi, H., Masuko, T., Shimizu, T., Ishikawa, T., Kai, K., Takahashi, E., Imamura, Y., Baba, Y., Ohmura, M., Suematsu, M., Baba, H. & Saya, H. 2011, "CD44 variant regulates redox status in cancer cells by stabilizing the xCT subunit of system xc(-) and thereby promotes tumor growth", *Cancer cell*, vol. 19, no. 3, pp. 387-400.

Jemal, A., Bray, F., Center, M.M., Ferlay, J., Ward, E. & Forman, D. 2011, "Global cancer statistics", *CA: a cancer journal for clinicians*, vol. 61, no. 2, pp. 69-90.

Jin, R.J., Lho, Y., Connelly, L., Wang, Y., Yu, X., Saint Jean, L., Case, T.C., Ellwood-Yen, K., Sawyers, C.L., Bhowmick, N.A., Blackwell, T.S., Yull, F.E. & Matusik, R.J. 2008, "The nuclear factor-kappaB pathway controls the progression of prostate cancer to androgen-independent growth", *Cancer research*, vol. 68, no. 16, pp. 6762-6769.

Jones, E., Pu, H. & Kyprianou, N. 2009, "Targeting TGF-beta in prostate cancer: therapeutic possibilities during tumor progression", *Expert opinion on therapeutic targets,* vol. 13, no. 2, pp. 227-234.

Joshi, P.A., Jackson, H.W., Beristain, A.G., Di Grappa, M.A., Mote, P.A., Clarke, C.L., Stingl, J., Waterhouse, P.D. & Khokha, R. 2010, "Progesterone induces adult mammary stem cell expansion", *Nature,* vol. 465, no. 7299, pp. 803-807.

Ketola, K., Vainio, P., Fey, V., Kallioniemi, O. & Iljin, K. 2010, "Monensin is a potent inducer of oxidative stress and inhibitor of androgen signaling leading to apoptosis in prostate cancer cells", *Molecular cancer therapeutics,* vol. 9, no. 12, pp. 3175-3185.

Khandrika, L., Kumar, B., Koul, S., Maroni, P. & Koul, H.K. 2009, "Oxidative stress in prostate cancer", *Cancer letters,* vol. 282, no. 2, pp. 125-136.

Kim, D.W., Gazourian, L., Quadri, S.A., Romieu-Mourez, R., Sherr, D.H. & Sonenshein, G.E. 2000, "The RelA NF-kappaB subunit and the aryl hydrocarbon receptor (AhR) cooperate to transactivate the c-myc promoter in mammary cells", *Oncogene,* vol. 19, no. 48, pp. 5498-5506.

Klarmann, G.J., Hurt, E.M., Mathews, L.A., Zhang, X., Duhagon, M.A., Mistree, T., Thomas, S.B. & Farrar, W.L. 2009, "Invasive prostate cancer cells are tumor initiating cells that have a stem cell-like genomic signature", *Clinical & experimental metastasis,* vol. 26, no. 5, pp. 433-446.

Kobayashi, C.I. & Suda, T. 2011a, "Regulation of reactive oxygen species in stem cells and cancer stem cells", *Journal of cellular physiology,* .

Kobayashi, C.I. & Suda, T. 2011b, "Regulation of reactive oxygen species in stem cells and cancer stem cells", *Journal of cellular physiology,* .

Kodaman, P.H., Aten, R.F. & Behrman, H.R. 1994, "Lipid hydroperoxides evoke antigonadotropic and antisteroidogenic activity in rat luteal cells", *Endocrinology,* vol. 135, no. 6, pp. 2723-2730.

Koh, C.M., Bieberich, C.J., Dang, C.V., Nelson, W.G., Yegnasubramanian, S. & De Marzo, A.M. 2010, "MYC and Prostate Cancer", *Genes & cancer,* vol. 1, no. 6, pp. 617-628.

Kohle, C. & Bock, K.W. 2007, "Coordinate regulation of Phase I and II xenobiotic metabolisms by the Ah receptor and Nrf2", *Biochemical pharmacology,* vol. 73, no. 12, pp. 1853-1862.

Korenchuk, S., Lehr, J.E., MClean, L., Lee, Y.G., Whitney, S., Vessella, R., Lin, D.L. & Pienta, K.J. 2001a, "VCaP, a cell-based model system of human prostate cancer", *In vivo (Athens, Greece),* vol. 15, no. 2, pp. 163-168.

Korenchuk, S., Lehr, J.E., MClean, L., Lee, Y.G., Whitney, S., Vessella, R., Lin, D.L. & Pienta, K.J. 2001b, "VCaP, a cell-based model system of human prostate cancer", *In vivo (Athens, Greece),* vol. 15, no. 2, pp. 163-168.

Kumar, B., Koul, S., Khandrika, L., Meacham, R.B. & Koul, H.K. 2008, "Oxidative stress is inherent in prostate cancer cells and is required for aggressive phenotype", *Cancer research,* vol. 68, no. 6, pp. 1777-1785.

Lee, S.Y., Gong, E.Y., Hong, C.Y., Kim, K.H., Han, J.S., Ryu, J.C., Chae, H.Z., Yun, C.H. & Lee, K. 2009, "ROS inhibit the expression of testicular steroidogenic enzyme genes via the suppression of Nur77 transactivation", *Free radical biology & medicine*, vol. 47, no. 11, pp. 1591-1600.

Leonarduzzi, G., Vizio, B., Sottero, B., Verde, V., Gamba, P., Mascia, C., Chiarpotto, E., Poli, G. & Biasi, F. 2006, "Early involvement of ROS overproduction in apoptosis induced by 7-ketocholesterol", *Antioxidants & redox signaling*, vol. 8, no. 3-4, pp. 375-380.

Li, T., Su, Y., Mei, Y., Leng, Q., Leng, B., Liu, Z., Stass, S.A. & Jiang, F. 2010, "ALDH1A1 is a marker for malignant prostate stem cells and predictor of prostate cancer patients' outcome", *Laboratory investigation; a journal of technical methods and pathology*, vol. 90, no. 2, pp. 234-244.

Lindros, K.O., Oinonen, T., Kettunen, E., Sippel, H., Muro-Lupori, C. & Koivusalo, M. 1998, "Aryl hydrocarbon receptor-associated genes in rat liver: regional coinduction of aldehyde dehydrogenase 3 and glutathione transferase Ya", *Biochemical pharmacology*, vol. 55, no. 4, pp. 413-421.

Locke, J.A., Guns, E.S., Lubik, A.A., Adomat, H.H., Hendy, S.C., Wood, C.A., Ettinger, S.L., Gleave, M.E. & Nelson, C.C. 2008, "Androgen levels increase by intratumoral de novo steroidogenesis during progression of castration-resistant prostate cancer", *Cancer research*, vol. 68, no. 15, pp. 6407-6415.

Magni, M., Shammah, S., Schiro, R., Mellado, W., Dalla-Favera, R. & Gianni, A.M. 1996, "Induction of cyclophosphamide-resistance by aldehyde-dehydrogenase gene transfer", *Blood*, vol. 87, no. 3, pp. 1097-1103.

Mahon, K.L., Henshall, S.M., Sutherland, R.L. & Horvath, L.G. 2011, "Pathways of chemotherapy resistance in castration-resistant prostate cancer", *Endocrine-related cancer*, vol. 18, no. 4, pp. R103-23.

Mimeault, M. & Batra, S.K. 2011, "Frequent Gene Products and Molecular Pathways Altered in Prostate Cancer- and Metastasis-Initiating Cells and their Progenies and Novel Promising Multitargeted Therapies", *Molecular medicine (Cambridge, Mass.)* .

Mori, K., Shibanuma, M. & Nose, K. 2004, "Invasive potential induced under long-term oxidative stress in mammary epithelial cells", *Cancer research*, vol. 64, no. 20, pp. 7464-7472.

Nebert, D.W., Roe, A.L., Dieter, M.Z., Solis, W.A., Yang, Y. & Dalton, T.P. 2000, "Role of the aromatic hydrocarbon receptor and [Ah] gene battery in the oxidative stress response, cell cycle control, and apoptosis", *Biochemical pharmacology*, vol. 59, no. 1, pp. 65-85.

Ohtake, F., Fujii-Kuriyama, Y. & Kato, S. 2009, "AhR acts as an E3 ubiquitin ligase to modulate steroid receptor functions", *Biochemical pharmacology*, vol. 77, no. 4, pp. 474-484.

Pani, G., Galeotti, T. & Chiarugi, P. 2010, "Metastasis: cancer cell's escape from oxidative stress", *Cancer metastasis reviews*, vol. 29, no. 2, pp. 351-378.

Pelicano, H., Carney, D. & Huang, P. 2004, "ROS stress in cancer cells and therapeutic implications", *Drug resistance updates : reviews and commentaries in antimicrobial and anticancer chemotherapy*, vol. 7, no. 2, pp. 97-110.

Pfeiffer, M.J., Smit, F.P., Sedelaar, J.P. & Schalken, J.A. 2011, "Steroidogenic Enzymes and Stem Cell Markers Are Upregulated during Androgen Deprivation in Prostate Cancer", *Molecular medicine (Cambridge, Mass.)*, vol. 17, no. 7-8, pp. 657-664.

Pinthus, J.H., Bryskin, I., Trachtenberg, J., Lu, J.P., Singh, G., Fridman, E. & Wilson, B.C. 2007, "Androgen induces adaptation to oxidative stress in prostate cancer: implications for treatment with radiation therapy", *Neoplasia (New York, N.Y.)*, vol. 9, no. 1, pp. 68-80.

Rigas, B. & Sun, Y. 2008, "Induction of oxidative stress as a mechanism of action of chemopreventive agents against cancer", *British journal of cancer*, vol. 98, no. 7, pp. 1157-1160.

Sarkar, F.H., Li, Y., Wang, Z. & Kong, D. 2008, "NF-kappaB signaling pathway and its therapeutic implications in human diseases", *International reviews of immunology*, vol. 27, no. 5, pp. 293-319.

Savouret, J.F., Antenos, M., Quesne, M., Xu, J., Milgrom, E. & Casper, R.F. 2001, "7-Ketocholesterol is an Endogenous Modulator for the Arylhydrocarbon Receptor", *The Journal of biological chemistry*, vol. 276, no. 5, pp. 3054-3059.

Sharifi, N., Gulley, J.L. & Dahut, W.L. 2010, "An update on androgen deprivation therapy for prostate cancer", *Endocrine-related cancer*, vol. 17, no. 4, pp. R305-15.

Shibutani, S., Takeshita, M. & Grollman, A.P. 1991, "Insertion of specific bases during DNA synthesis past the oxidation-damaged base 8-oxodG", *Nature*, vol. 349, no. 6308, pp. 431-434.

Sophos, N.A. & Vasiliou, V. 2003, "Aldehyde dehydrogenase gene superfamily: the 2002 update", *Chemico-biological interactions*, vol. 143-144, pp. 5-22.

Stocco, D.M., Wells, J. & Clark, B.J. 1993, "The effects of hydrogen peroxide on steroidogenesis in mouse Leydig tumor cells", *Endocrinology*, vol. 133, no. 6, pp. 2827-2832.

Sun, Y., Huang, L., Mackenzie, G.G. & Rigas, B. 2011, "Oxidative stress mediates through apoptosis the anticancer effect of phospho-NSAIDs: Implications for the role of oxidative stress in the action of anticancer agents", *The Journal of pharmacology and experimental therapeutics*, .

Sung, S.Y., Kubo, H., Shigemura, K., Arnold, R.S., Logani, S., Wang, R., Konaka, H., Nakagawa, M., Mousses, S., Amin, M., Anderson, C., Johnstone, P., Petros, J.A., Marshall, F.F., Zhau, H.E. & Chung, L.W. 2006, "Oxidative stress induces ADAM9 protein expression in human prostate cancer cells", *Cancer research*, vol. 66, no. 19, pp. 9519-9526.

Swanson, T.A., Krueger, S.A., Galoforo, S., Thibodeau, B.J., Martinez, A.A., Wilson, G.D. & Marples, B. 2011, "TMPRSS2/ERG fusion gene expression alters chemo- and radio-responsiveness in cell culture models of androgen independent prostate cancer", *The Prostate*, .

Szatrowski, T.P. & Nathan, C.F. 1991, "Production of large amounts of hydrogen peroxide by human tumor cells", *Cancer research*, vol. 51, no. 3, pp. 794-798.

Tam, N.N., Gao, Y., Leung, Y.K. & Ho, S.M. 2003, "Androgenic regulation of oxidative stress in the rat prostate: involvement of NAD(P)H oxidases and antioxidant defense machinery during prostatic involution and regrowth", *The American journal of pathology*, vol. 163, no. 6, pp. 2513-2522.

Tannock, I.F., de Wit, R., Berry, W.R., Horti, J., Pluzanska, A., Chi, K.N., Oudard, S., Theodore, C., James, N.D., Turesson, I., Rosenthal, M.A., Eisenberger, M.A. & TAX 327 Investigators 2004, "Docetaxel plus prednisone or mitoxantrone plus prednisone for advanced prostate cancer", *The New England journal of medicine*, vol. 351, no. 15, pp. 1502-1512.

Toyokuni, S., Okamoto, K., Yodoi, J. & Hiai, H. 1995, "Persistent oxidative stress in cancer", *FEBS letters*, vol. 358, no. 1, pp. 1-3.

Tsai, S.C., Lu, C.C., Lin, C.S. & Wang, P.S. 2003, "Antisteroidogenic actions of hydrogen peroxide on rat Leydig cells", *Journal of cellular biochemistry*, vol. 90, no. 6, pp. 1276-1286.

Vainio, P., Gupta, S., Ketola, K., Mirtti, T., Mpindi, J.P., Kohonen, P., Fey, V., Perala, M., Smit, F., Verhaegh, G., Schalken, J., Alanen, K.A., Kallioniemi, O. & Iljin, K. 2011, "Arachidonic acid pathway members PLA2G7, HPGD, EPHX2, and CYP4F8 identified as putative novel therapeutic targets in prostate cancer", *The American journal of pathology*, vol. 178, no. 2, pp. 525-536.

van Bokhoven, A., Varella-Garcia, M., Korch, C., Johannes, W.U., Smith, E.E., Miller, H.L., Nordeen, S.K., Miller, G.J. & Lucia, M.S. 2003, "Molecular characterization of human prostate carcinoma cell lines", *The Prostate*, vol. 57, no. 3, pp. 205-225.

van den Hoogen, C., van der Horst, G., Cheung, H., Buijs, J.T., Lippitt, J.M., Guzman-Ramirez, N., Hamdy, F.C., Eaton, C.L., Thalmann, G.N., Cecchini, M.G., Pelger, R.C. & van der Pluijm, G. 2010, "High aldehyde dehydrogenase activity identifies tumor-initiating and metastasis-initiating cells in human prostate cancer", *Cancer research*, vol. 70, no. 12, pp. 5163-5173.

Vrzal, R., Ulrichova, J. & Dvorak, Z. 2004, "Aromatic hydrocarbon receptor status in the metabolism of xenobiotics under normal and pathophysiological conditions", *Biomedical papers of the Medical Faculty of the University Palacky, Olomouc, Czechoslovakia*, vol. 148, no. 1, pp. 3-10.

Wang, J., Cai, Y., Shao, L.J., Siddiqui, J., Palanisamy, N., Li, R., Ren, C., Ayala, G. & Ittmann, M.M. 2010, "Activation of NF-kB by TMPRSS2/ERG fusion isoforms through Toll-like receptor-4", *Cancer research*, .

Yoshida, A., Rzhetsky, A., Hsu, L.C. & Chang, C. 1998, "Human aldehyde dehydrogenase gene family", *European journal of biochemistry / FEBS*, vol. 251, no. 3, pp. 549-557.

Yossepowitch, O., Pinchuk, I., Gur, U., Neumann, A., Lichtenberg, D. & Baniel, J. 2007, "Advanced but not localized prostate cancer is associated with increased oxidative stress", *The Journal of urology*, vol. 178, no. 4 Pt 1, pp. 1238-43; discussion 1243-4.

Yu, C., Yao, Z., Dai, J., Zhang, H., Escara-Wilke, J., Zhang, X. & Keller, E.T. 2011, "ALDH Activity Indicates Increased Tumorigenic Cells, But Not Cancer Stem Cells, in Prostate Cancer Cell Lines", *In vivo (Athens, Greece)*, vol. 25, no. 1, pp. 69-76.

Zhang, M., Shoeb, M., Goswamy, J., Liu, P., Xiao, T.L., Hogan, D., Campbell, G.A. & Ansari, N.H. 2009a, "Overexpression of aldehyde dehydrogenase 1A1 reduces oxidation-induced toxicity in SH-SY5Y neuroblastoma cells", *Journal of neuroscience research,* .

Zhang, Q., Helfand, B.T., Jang, T.L., Zhu, L.J., Chen, L., Yang, X.J., Kozlowski, J., Smith, N., Kundu, S.D., Yang, G., Raji, A.A., Javonovic, B., Pins, M., Lindholm, P., Guo, Y., Catalona, W.J. & Lee, C. 2009b, "Nuclear factor-kappaB-mediated transforming growth factor-beta-induced expression of vimentin is an independent predictor of biochemical recurrence after radical prostatectomy", *Clinical cancer research : an official journal of the American Association for Cancer Research,* vol. 15, no. 10, pp. 3557-3567.

10

Oxidative Therapy Against Cancer

Manuel de Miguel and Mario D. Cordero

Departamento de Citología e Histología Normal y Patológica,
Facultad de Medicina, Universidad de Sevilla, Sevilla
Spain

1. Introduction

Although a moderate increase of reactive oxygen species (ROS) may induce cell proliferation, excessive amounts of ROS can cause oxidative damage to lipids, proteins, and DNA, provoking oncogenic transformation, increased metabolic activity, and mitochondrial dysfunction (Dreher and Junod, 1996; Behrend et al., 2003; Pelicano et al, 2003). Many reports suggest that cancer cells are under a continuous oxidative stress (Pervaiz and Clement, 2004; Schumacker et al., 2006; Kryston et al., 2011). Studies with human tumor cell lines clearly show that these cells produce ROS at a much higher rate than healthy cells (Oberley and Buettner, 1979; Lu et al., 2007). ROS have been established as important molecules involved in the multistage process of carcinogenesis (Klaunig and Kamendulis, 2004). Mitochondria are the major consumers of molecular oxygen in cells, representing an important source of ROS. It is well accepted that cancer cells present mitochondrial alterations which result in respiration injury. This mitochondrial dysfunction may induce a low coupling efficiency of the mitochondrial electron chain, increasing electron leakage and leading to enhanced ROS formation. The resulting oxidative stress may cause further damage to both mitochondrial DNA (mtDNA) and the respiratory chain, amplifying the ROS generation (Zorov et al., 2006).

The higher oxidative stress observed in cancer cells can also result from a decrease or inactivation of antioxidants (Huang et al., 2003; Conklin, 2004). The majority of tumor cells usually present very few antioxidative enzymes, such as catalase, superoxide dismutase, and glutathione peroxidase, which are known to play a protective role against ROS in normal cells (Sato et al., 1992; Hasegawa et al., 2002; Pelicano et al., 2004). The lack of proper antioxidant defences makes tumor cells very vulnerable to oxidative stress. Nevertheless, some studies have revealed increased expression of antioxidants, probably as a consequence of selective pressure towards stress adaptation. The sources of ROS in cancer cells and the consequences of oxidative stress in the carcinogenesis process are still under debate.

What seems accepted is that cancer cells have increased ROS steady state level and are likely to be more vulnerable to damage by further ROS insults induced by exogenous agents. Thus, manipulating ROS levels by redox modulation could be a way to selectively kill cancer cells without causing significant toxicity to normal cells. A promising anticancer strategy named "oxidation therapy" has been developed by inducing cytotoxic oxidative stress for cancer treatment. This could be achieved by two different methods: inducing the generation of cytotoxic levels of ROS, and inhibiting the antioxidant system of tumor cells (Fang et al., 2007; Trachootham et al., 2009).

It is well known that ROS, such as hydrogen peroxide (H_2O_2) and superoxide anion, induce apoptosis to a wide range of tumor cells through the activation of the caspase cascade. It has been described that mitochondrial damage induced by the use of certain drugs provokes an increment of oxidative stress and cell death (Chandra et al., 2000; Conklin, 2004). Major ROS-modulating agents are based on the capacity to induce high ROS generation or to reduce the antioxidant defence machinery of cancer cells.

An interesting drug for oxidative cancer therapy is amitriptyline, a tricyclic antidepressant commonly prescribed for depression and therapeutic treatments of several neuropathic and inflammatory illnesses. Chlorimipramine, another tricyclic antidepressant, has been already proposed as a novel anticancer agent targeted to mitochondria as it induces caspase-3-dependent apoptosis (Daley et al., 2005). Several reports showed that the toxicity of this drug is due to an increase in oxidative stress by the generation of high amounts of ROS (Daley et al., 2005; Moreno-Fernández et al., 2008; Cordero et al., 2009). Therefore, amitriptyline has being proposed to be used for anticancer oxidant therapy against tumors that present significant oxidative stress and/or low antioxidant defences (Cordero et al., 2010).

2. Mitochondria and Reactive Oxygen Species (ROS)

Mitochondria are dynamic organelles that play a central role in many cellular functions including the generation of chemical energy (adenosine triphosphate, ATP), heat, and intracellular calcium homeostasis. They are also responsible for the formation of ROS and for triggering the programmed cell death or apoptosis (Turrens, 2003). The primary metabolic function of mitochondria is oxidative phosphorylation, an energy-generating process that couples oxidation of respiratory substrates to the synthesis of ATP (Pieczenik & Neustadt, 2007). The mitochondrial respiratory chain (MRC) is composed of five multisubunit enzyme complexes. Both the mtDNA and the nuclear DNA (nDNA) encode for polypeptide components of these complexes. Electron transport between MRC complexes I–IV is coupled to the extrusion of protons across the inner mitochondrial membrane by proton pump components of the respiratory chain. This movement of protons creates an electrochemical gradient ($\Delta\Psi m$) across the inner mitochondrial membrane. Protons return to the mitochondrial matrix by flowing through ATP synthase (complex V), which utilizes the energy thus produced to synthesize ATP from adenosine diphosphate (ADP) and inorganic phosphate. Both the mtDNA and the nDNA encode for polypeptide components of these complexes. As a consequence, mutations in either genome can cause MRC dysfunction that impairs transport of electrons and/or protons and decreases ATP synthesis. Primary or secondary genetic diseases affecting MRC or secondary mitochondrial dysfunctions usually affect brain and skeletal muscle because of their energy requirements. Besides MRC enzyme complexes, two electron carriers, coenzyme Q_{10} (CoQ) and cytochrome c, are essential for mitochondrial synthesis of ATP. CoQ transports electrons from complexes I and II to complex III and is essential for the stability of complex III. CoQ is a lipid-soluble component of virtually all cell membranes. CoQ also functions as an antioxidant that protects cells both by direct ROS scavenging and by regenerating other antioxidants such as vitamins C and E (Turunen et al., 2004). Given the critical role of CoQ in mitochondria function, it has been suggested that CoQ levels could be a useful biological marker of mitochondrial function (Haas et al., 2008). CoQ deficiency induces decreased mitochondrial respiratory enzymes activity, reduced expression of mitochondrial proteins involved in oxidative phosphorylation, decreased mitochondrial membrane potential, increased production of ROS, mitochondrial permeabilization, mitophagy of dysfunctional mitochondria, reduced growth rates and cell death (Rodriguez-Hernandez et al., 2009, Cotan et al, 2011).

In addition to energy, mitochondrial oxidative phosphorylation also generates ROS. When the MRC becomes highly reduced, the excess electrons from complex I or complex III may increase substantially, passing directly to O_2 to generate high amounts of superoxide anion $(O_2^{\bullet-})$. Superoxide is transformed to hydrogen peroxide (H_2O_2) by the detoxification enzymes manganese superoxide dismutase (MnSOD) or copper/zinc superoxide dismutase (Cu/Zn SOD), and then to water by catalase, glutathione peroxidase (GPx) or peroxiredoxin III (PRX III). However, when these enzymes cannot convert ROS such as the superoxide radical to water fast enough, oxidative damage occurs and accumulates in the mitochondria. If H_2O_2 encounters a reduced transition metal or is mixed with $O_2^{\bullet-}$, the H_2O_2 can be further reduced to hydroxyl radical ($HO^{\bullet}$), the most potent oxidizing agent among ROS. Additionally, nitric oxide ($^{\bullet}NO$) is produced within the mitochondria by mitochondrial nitric oxide synthase (mtNOS) and also freely diffuses into the mitochondria from the cytosol. $^{\bullet}NO$ reacts with $O_2^{\bullet-}$ to produce peroxynitrite ($ONOO^-$). Together, these two radicals as well as others can do great damage to mitochondria and other cellular components (Turrens, 2003).

Under normal physiological conditions, ROS production is highly regulated. However, if the MRC is inhibited, or key mitochondrial components, such as CoQ, are deficient, then, electrons accumulate on the MRC carriers, greatly increasing the rate of a single electron being transferred to O_2 to generate $O_2^{\bullet-}$. An excessive mitochondrial ROS production can exceed the cellular antioxidant defense and the cumulative damage can ultimately destroy the cell by necrosis or apoptosis.

Mitochondria play an important role in cell bioenergetics and life signaling. Mitochondria are necessary for cell survival but they can also trigger cell death, thus exerting decisive control over the biochemistry of several cascades that lead to cell death, specifically the intrinsic pathway of apoptosis. The particular biochemical properties of these organelles are closely related to their segmented structures, which provide an optimal environment for multiple pathways of biosynthesis and bioenergetics. Consequently, it is possible that these organelles are involved in the process of carcinogenesis through alterations in cell metabolism and cell death pathways (Pilkington et al., 2008). Cancer cells show alterations in mtDNA, in oxidative phosphorylation and in energetic metabolism, all triggered by a pro-oxidative change (Indo et al., 2007). The "respiratory damage" in cancer cells predicts low coupling efficiency in electron transfer at the inner membrane of mitochondria and, consequently, greater loss of electrons, leading to the formation of more $O_2^{\bullet-}$ (Pelicano et al., 2003). The superoxide anion will thus generate more free radicals. Generated ROS can be released into cytosol and trigger "ROS-induced ROS-release" (RIRR) in neighbouring mitochondria. This mitochondrion-to-mitochondrion ROS-signaling constitutes a positive feedback mechanism for enhanced ROS production potentially leading to significant mitochondrial injury (Zorov et al., 2006). Recent studies by a number of groups have demonstrated that ROS can directly modify signaling proteins through different modifications, for example by nitrosylation, carbonylation, disulphide bond formation and glutathionylation. Moreover, redox modification of proteins permits further regulation of cell signaling pathways (England and Cotter, 2005).

3. Oxidative stress and cancer

3.1 ROS and carcinogenesis

In general, oxidative/nitrosative stress could be defined as an imbalance between the presence of high levels of ROS and reactive nitrogen species (RNS), and the antioxidative defense mechanisms. These toxic molecules are formed via oxidation-reduction reactions and are highly reactive since they have an odd number of electrons. ROS generated under physiological conditions are essential for life, as they are involved in bactericidal activity of phagocytes, and in signal transduction pathways, regulating cell growth and reduction–oxidation (redox) status (Davies, 1995). ROS includes free radicals, such as hydroxyl and superoxide radicals, and non-radicals, including hydrogen peroxide and singlet oxygen. Oxidative stress and generation of free radicals, as primary or secondary event, have been related in a great number of diseases, including cancer (Floyd, 1990).

At the beginning of the carcinogenic process, tumor cells accumulate mutations that allow them to proliferate in an uncontrolled way. Moreover, these alterations contribute to increase the susceptibility to accumulate additional genetic modifications, facilitating tumor progression and cancer development. Cancer could be defined as a cell-cycle desease, where its misregulation is considered an essential step (Sandhu et al., 2000). Carcinogenesis is a complex process of different sequencies that allow a cell to evolve from a healthy state into a pre-cancerous state and eventually reach a cancerous state. In this sense, there are several theories about the process of carcinogenesis. For instance, an increase of DNA synthesis and mitosis triggered by non-genotoxic agents could induce mutations in new cells. These mutations could spread through new cell divisions, evolving from an initial pre-neoplastic state into a neoplastic state. Another theory explains the existence of an imbalance between proliferation and cell death, where proliferation is favoured. If DNA damage is too high, there are important mechanisms, such as apoptosis, by which the altered cells are selectively eliminated. Protein p53 plays a fundamental role in this process, as it initiates mechanisms that eliminate, for example, those oxidized DNA bases that could cause mutations. If cell damage is too high, p53 initiates the mechanisms of apoptosis, although uncontrolled processes of apoptosis can be harmful for the organism, since healthy cells could also be eliminated. Therefore, there are systems for regulating apoptosis that consist of both pro-apoptotic and anti-apoptotic factors. Alterations that affect the function of gene p53 have been found in more than half of all types of cancer. This fact supports the idea that carcinogenesis would be caused by an imbalance between proliferation and cell death, in favour of proliferation.

Studies of epidemiology and animal experimentation have shown that carcinogenesis could occur in several stages characterized by different mechanisms. Thus, the model of carcinogenesis based on the hypothesis of three stages: initiation, promotion and progression, should be highlighted. Genotoxic agents are mainly chemical substances that damage DNA directly, inducing the generation of a mutation and/or a set of structural changes. On the other hand, there is a second category of non-genotoxic carcinogenic agents, whose role in the carcinogenic process is not related directly to DNA damage. These compounds modulate mechanisms of cell growth and death. Thus, the development process of a cancer would consist of the accumulation of multiple events, and the ROS could act at different levels and in all stages (Klauning and Kamendulis, 2004). Thereby, ROS can induce both genomic instability, caused by DNA damage, and alterations in cell signaling processes related to survival, proliferation, resistance to apoptosis, angiogenesis and metastasis, thus contributing to cancer initiation, promotion and progression.

Initiation involves a DNA mutation that is not lethal, but it produces a cell alteration followed by at least one round of DNA synthesis that allows fixing the damage done. At this point, the cell can stop its cycle temporally to undo DNA damage and then resume cell division. DNA damage may be done by ROS, like hydroxyl radicals formed by the Fenton´s reaction. Several studies have revealed an interesting correlation between tumor size and the amount of 8-OHdG (8-hydroxy-2'-desoxyguanosine; also known as 8-oxo-deoxyguanosine, 8-oxo-dG), a nucleotide modified by the activity of free radicals (Kennedy et al., 1998; Yano et al., 2009). The promotion stage is characterized by the expansion of initiated cells, stimulating cell proliferation and/or apoptosis inhibition. As a result of this process, an identifiable lesion is formed, thus requiring the constant presence of an agent that stimulates promotion. However, it is a reversible process. Many promoter agents have a strong inhibitory capacity against antioxidants like catalases, glutathione, SOD, etc. While a high level of oxidative stress is cytotoxic for cells and stops proliferation inducing apoptosis or even necrosis, moderate levels of oxidative stress may stimulate cell division and, therefore, stimulate tumor growth and promotion (Dreher and Junod, 1996). Progression is the third and last stage of the carcinogenic process. This stage involves cellular and molecular changes that occur from a pre-neoplastic state to a neoplastic state. This stage is irreversible and it is characterized by the accumulation of genetic damage that allows the cell evolving from benign to malignant.

ROS are considered as carcinogenic potentials that facilitate cancer promotion and progression (Pelicano et al., 2004). The DNA molecule is one of the main targets of free radicals activity in the cell, and the modifications performed as a consequence of this activity are relevant for the loss of cell homeostasis. This loss may be extended in time due to the DNA functions of information reservoir. This is why the agents and mechanisms of damage by ROS are studied in depth, because its clarification would lead to elucidate the pathogeny of great morbidity and mortality diseases like cancer. There are different types of oxidative damage to DNA, like modifications and depurinations of DNA bases, DNA chain ruptures, and mutations, which tend to accumulate in an environment with high concentrations of ROS. It is known that DNA damage by ROS occurs spontaneously and there is a normal level of bases modified by ROS in cellular DNA (Okamoto, 2000). The most frequent modification of DNA linked to ROS is the formation of 8-OH-dG, resulting from attack of either a hydroxyl radical or singlet oxygen on deoxyguanosine. 8-OH-dG has a highly mutagenic potential, as 5-hydroxymethyl-2-deoxyuridine (Retel et al., 1993). With relative frequency, the cell will evade DNA damage using specific DNA polymerases and enter DNA replication creating mutations and chromosomal lesions. Alternatively, the presence of unrepaired DNA lesions can induce cell death through the apoptotic pathway. Chronic exposure to DNA lesions can lead to mutations and genomic instability (pre-cancerous state) and eventually to malignant transformations (cancerous state) (Kryston et al., 2011).

3.2 Sources of oxidative stress

Cells from all organisms are exposed to several oxidizing and harmful agents. These attacks can be divided into two groups: exogenous and endogenous. Exogenous sources are related to environmental, medical, diagnostic ionizing and non-ionizing radiations (X- or γ-rays, α-particles from radon decay, UVA radiation) or chemical agents. Endogenous (intracelullar) sources of reactive species are primarily produced by O_2 metabolism, immune responses and inflammation. These processes may result in the production of ROS and RNS that react with DNA and produce several lesions and indirect effects. Ionizing radiations can damage DNA also by direct energy deposition and ionizations (Kryston et al., 2011).

Oxygen metabolism is the major source of ROS in tumor tissues. ROS are continuously formed in mitochondria as respiration byproducts. Cancerous cells are metabolically very active and require a great supply of ATP in order to maintain proliferation and cell growth under control. This high energy demand in the mitocondrial respiratory chain contributes to the generation of ROS. Usually, ATP is produced with high efficiency through oxidative phosphorylation in mitochondria. However, a malfunction of mitochondrial respiration is usually observed in neoplastic cells due to deletions/mutations in mtDNA, to the aberrant expression of some enzymes involved in energy metabolism and to hypoxia (Xu et al., 2005; Verrax et al., 2008). On the other hand, the increase of glycolysis in tumor cells has been shown to be related to the aggressiveness of the tumor (Cuezva et al., 2002). Acquiring a glycolytic phenotype represents a key element for cancer survival and progression, while glycolysis inhibition could be proposed as a goal in antitumor therapy (Pelicano et al., 2004). In this sense, glucose deprivation in laboratory models has been shown to be related to an increase of oxidative stress in tumor cells (Spitz et al., 2000). In cell lines of breast carcinoma, glucose deprivation leads to an intracellular increase of pro-oxidants, a decrease of free radical neutralization, and pyruvate depletion, which leads to an increase of oxidative stress (Spitz et al., 2000; Lee et al., 1999). As a compensation mechanism, glucose deprivation induces the expression of heme oxigenase-1 (HO-1), an enzyme that plays an important role in the antioxidant defense system of the organism and in Fe homeostasis. Chang and collegues (Chang et al., 2003) showed that the generation of ROS in mitochondria induces the overexpression of HO-1, which demonstrates that this is a common mechanism of regulation with the aim of protecting cells against oxidative damage.

As already mentioned, immune response and inflamation are other sources of ROS. It is known that oxidative stress activates inflammatory pathways leading to transformation of a normal cell to tumor cell, tumor cell survival, proliferation, insensitivity to anti-growth signaling, invasion, sustained angiogenesis, and stem cell survival (Reuter et al., 2010). Chronic inflammation is triggered by environmental (extrinsic) factors (eg, infection, tobacco, asbestos) and host mutations (intrinsic) factors (eg, Ras, Myc, p53). Activation of Ras, Myc, and p53 cause mitochondrial dysfunction, resulting in mitochondrial ROS production and downstream signaling (eg, NFkappaB, STAT3, etc.) that promote inflammation-associated cancer (Kamp at al., 2011).

On the other hand, during immune responses of many carcinogenetic processes, several immune-related cells play their roles, like intratumoral lymphocytes killing malignant cells and macrophages and neutrophils degrading cells by using oxidants and enzymes. One of these enzymes is NADPH oxidase, which activity produces ROS. The increased intensity of oxidative stress helps tumor to go on to malignancy. As an example, *in vitro* data suggest that in environments with certain levels of oxidative stress some cytokines, such as interleukin-2 (IL-2) or interferon α (IFNα), induce a decrease of T-lymphocytes or *natural killer* cells, which could supose a tumor evasion to the immune response (Mantovani et al., 2003). Another source of free radical generation, usually underestimated but highlighted by Kryston et al. (2011), is the chronic exposure to viral infections; as in the case of hepatitis viruses, where there is a connection between chronic infection and induction of oxidative stress. There is a variety of viruses associated with increased ROS levels, DNA damage and mutagenic rate. The high intracellular oxidation status in viral infections consists of decreased antioxidant enzymes like catalase, glutathione peroxidase, glutathione reductase as well as high level of hydroxyl radicals (Kryston et al., 2011).

3.3 Influence of ROS in the cell cycle

Since some time already, an increase of ROS and an altered redox state has been observed in cancerous cells (Trachootham et al., 2009). It is known that ROS may serve as cellular messengers in the signal translation pathway and also an increase of ROS may trigger cell growth and proliferation, contributing to cancer development (Filomeno et al., 2005). Uncontrolled tumor cell proliferation requires the up-regulation of multiple intracellular signaling pathways, including cascades involved in survival, proliferation, and cell cycle progression. The most significant effects of oxidants on signaling pathways have been observed in the MAPK/AP-1 and NF-κB pathways (Muller et al., 2010). At the advanced stage of the disease, cancerous cells usually show genetic instability and a sharp increase of ROS, which induces to genetic mutations and failures in metabolic functioning, provoking a greater generation of ROS (Pelicano et al., 2004). Several mechanisms may lead to oxidative stress in cancer patients, such as altered energy metabolism, overproduction of cytokines, which in turn may increase ROS production, and the use of anti-neoplastic drugs. Altered energy metabolism in cancer may explain symptoms such as anorexia/cachexia, nausea and vomiting, which prevent normal nutrition. This deficit in the normal supply of nutrients like glucose, proteins, antioxidants and vitamins, leads eventually to the lack of antioxidant defenses for controlling free radical production.

Protein p53 is known as "the guardian of the genome" because it is essential for maintaining its integrity. This tumor suppressor protein has a fundamental role at detecting and eliminating oxidative damage in nuclear and mitochondrial DNA, at preventing mutations and at genetic instability. On the other hand, p53 acts also as a transcription factor that regulates the expression of many pro-oxidants and antioxidant genes. The functional loss of p53 is related to redox disequilibrium, increase of ROS, greater mutagenesis and tumor growth (Attardi and Donehower, 2005). This loss of p53 function is observed in many human cancers, especially in advanced stages (Bourdon, 2007).

There is evidence that the increase of ROS in tumor cells has a fundamental role in the acquisition of cancer characteristics (Hanahan and Weinberg, 2011): immortalization and transformation, cell proliferation and mitogenic signals, and cell survival and interruption of apoptotic death. Interestingly, in contrast to the effect of tumor promotion, recent studies suggest that the high ROS level has a role in the induction and maintenance of the senescence induced by tumor suppression, through the supported activation of the cell cycle inhibitor p16 (Ramsey and Sharpless, 2006; Takahashi et al., 2006). On the other hand, if the

ROS level increases up to a specific threshold that is incompatible with cell survival, ROS can exert cytotoxic effects that lead to the death of malignant cells and, therefore, limit cancer progression. This double game of ROS effect on cancerous cells may be used for developing new antitumor drugs that provide an increase of lethal oxidative stress in tumor cells, which are more sensitive to this type of attack than normal cells.

3.4 Angiogenesis induced by ROS

In addition to alterations in the cell cycle, an important event in the growth of any tumor is the generation of a new blood supply system that feeds the malignant cells. Migration, proliferation and tubular formation by endothelial cells are essential events in the process of angiogenesis. It has been suggested that ROS play an important role in angiogenesis, although their molecular mechanism remains unknown (Ushio-Fukai, 2004). The vascular endothelial growth factor (VEGF) triggers angiogenesis by stimulating the proliferation of endothelial cells and their migration through the receptor 2 (Flk1/KDR) of VEGF, which has tyrosine kinase activity. ROS derived from NAD(P)H oxidase are important for in vitro VEGF signaling and for in vivo angiogenesis. Arbiser and collegues (Arbiser et al., 2002) showed that ROS increased the expression of the VEGF, triggering the promotion of vascularization mechanisms and the fast expansion of tumors. On the other hand, greater metastatic capacity has been associated with high ROS levels in most tumors (Ishikawa et al., 2008). So, exogenous administration of ROS could increase certain metastatic states, while a treatment with antioxidants could slow down the metastatic progression (Ferraro et al., 2006).

Hypoxia seems to be the most important mechanism for tumor progression through the activation of angiogenesis, which is essential for the tumor growth (Harris, 2002). The neoplastic cells respond to hypoxia with an increase of ROS level; this occurs in the first stages of tumor development as a consequence of the blood supply deficit (Denko et al., 2003). It was shown that despite hypoxia there was a greater production of ROS, a paradox, since the oxygen availability for the formation of $O_2^{\bullet-}$ would be limited. Perhaps the generation of ROS occurs later, throughout tumor development, at the first stages of angiogenesis. When the hypoxia of a tumor is followed by blood supply reperfusion, high levels of ROS are generated; something similar occurs in a myocardial infarction or a brain ischemia.

3.5 ROS and tumor cell invasion

Tumor invasion and metastasis are two important events in cancer in which oxidative stress has shown to have an important role. When mammalian carcinoma cells are treated with hydrogen peroxide before intravenous injection into mice, an increase in lung metastasis formation is observed (Kundu et al., 1995), probably due to a decreased attachment of tumor cells to the basal lamina or it could alternatively be due to the increased activity or expression of proteins that regulate cellular motility.

On the other hand, the matrix metalloproteinases (MMPs) have been involved in the invasion and metastasis of malignant tumors of various histogenetic origins, and are capable of cleaving most components of the basement membrane and extracellular matrix (Westermarck et al., 1999). The activation of MMPs, such as MMP-2, probably occurs by the reaction of ROS with thiol groups in the protease catalytic domain, being also ROS involved in MMP gene expression (Rajagopalan et al., 1996).

Recently, a number of steps in the progression of metastatic disease have been shown to be regulated by redox signaling (Diers et al., 2010). One such redox signaling molecule is the electrophilic cyclopentenone prostaglandin, 15-deoxy-Δ12,14-prostaglandin J2 (15d-PGJ2), which can affect redox signaling through the posttranslational modification of critical cysteine residues in proteins such as actin, vimentin, and tubulin. The fact that 15d-PGJ2 can alter the cytoskeleton coincides with decreased migration and increased focal-adhesion disassembly, which might have important implications in the inhibition of metastatic processes such as invasion, intravasation, and extravasation (Diers et al., 2010).

3.6 ROS adaptation by cancer cells

There is sufficient evidence that cancer cells are under greater oxidative stress (Pervaiz and Clement, 2004; Kryston et al., 2011). High levels of oxidative stress have been found in cancer patients (McEligot et al., 2005; Lu et al., 2007), and it has been reported that different biomarkers of oxidative-stress-mediated events are elevated in cancer-prone tissues (Bartsch and Nair, 2000). *In vitro* studies clearly show that human tumor cell lines produce ROS at a much higher rate than non-transformed cells (Oberley and Buettner, 1979; Lu et al., 2007). In cancerous cells, a high level of oxidative stress is observed, which may result not only from the overproduction of ROS, but also from low levels or inactivation of antioxidants (Huang et al., 2003). Antioxidants are molecules capable of slowing down or even preventing the oxidation of other molecules and play an essential role at protecting cells against ROS aggression. Inhibition of these enzymes seriously endangers the capability of cells to face ROS activity. The first frontier of cellular defense against ROS damage consists of endogenous non-enzymatic radical scavengers like glutathione (GSH) and vitamins like C, E, antioxidant enzymes like SOD, catalase and GPx, as well as specific repair pathways (Kryston et al., 2011).

A decrease of mitochondrial activity and an overexpression of Mn-SOD with a greater production of $O_2^{\bullet-}$ have been detected in some colorectal carcinomas and tumor cells of the pancreas (Van-Driel et al., 1997; Cullen et al., 2003). The accumulation of anion $O_2^{\bullet-}$ stimulates cell growth by altering the redox states of transcription factors and regulators of the protein cell cycle. The increase of ROS in tumor cells may induce an increase of endogenous antioxidants in order to avoid intracellular lesions. On the contrary, a decrease of SOD activity was detected in blood cells of patients with cervical cancer. The decrease of SOD activity observed could be related to the generation of free radicals that cause direct damage to the enzyme by reticulation or mutation induction (Manoharan et al., 2004; Naidu et al., 2007). Besides, this decrease may be caused by an alteration of the oligoelements (essential trace elements) that act as cofactors of this enzyme, which may be considered as a risk factor of tumor growth or carcinogenesis (Naidu et al., 2007). On the other hand, it was observed that the increase in the intensity of oxidative stress in the blood of patients with ovarian cancer is accompained by a decrease of antioxidants like SOD, catalase, vitamin C and vitamin E (Senthil et al., 2004). A high percentage of tumors show low catalase activity, which means an advantageous adaptation for the tumor, which continues to benefit from the high levels of ROS. Although oxidant agents can be toxic to normal cells, the moderate increase of ROS contributes to the growth and survival of many cancers.

The strategy adopted by each type of tumor to increase the intensity of oxidative stress may vary from an increase of ROS to a decrease of antioxidants, going through a combination of both. The concentration of ROS in the tumor cell must be compensated with that of antioxidants in order to obtain an increase of ROS steady state level without reaching lethal levels. However, alterations in the mechanisms of antioxidant production are considered as indicators of oxidative stress in several types of cancer (Schumacker, 2006). Cancerous cells may adapt to survive at certain levels of oxidative stress. For example, greater levels of $O_2^{\bullet -}$ and H_2O_2 were observed in cells transformed with H-ras, showing high levels of antioxidants like peroxiredoxin-3 and thioredoxin peroxidase, compared to normal cells (Young et al., 2004). These adaptive mechanisms keep ROS levels in a distribution space that allows cancerous cells to avoid serious oxidative damage and to survive ROS-mediated stress and mutations. Redox adaptation may be crucial not only for cancer development, but also for drug resistance (Pervaiz and Clement, 2004; Sullivan and Graham, 2008). Redox adaptation, through the increase of endogenous antioxidants and the activation of the cell survival pathway, may confer greater capacity for tolerating the action of exogenous stress, with capacity for increasing DNA repair and decreased apoptosis. Thus, although cancerous cells are under chronic oxidative stress, they possess remodeled antioxidant system which let them to avoid deleterious ROS effects.

4. Antitumor oxidative therapy

4.1 The strategy

The therapeutic selectivity and avoiding resistance to drugs are two important issues in the therapy against cancer. Strategies for improving therapeutic selectivity depend largely on understanding the biological difference between cancerous and normal cells. Tumor cells, compared to normal cells, are under greater oxidative stress related to the oncogenic transformation, to alterations in metabolic activity and to an increase of ROS generation (Toyokuni et al., 1995; Hileman et al., 2003; Behrend et al., 2003). The high concentration of ROS in cancerous cells may have beneficial consequencies, like the stimulation of cell proliferation, mutation multiplication, genetic instability and alterations in the sensitivity to agents against cancer. However, because ROS are chemically active and may inflict serious cell damage, the very fact that cancerous cells usually show greater intrinsic oxidative stress may itself provide a unique opportunity to kill tumor cells according to the vulnerability of ROS action.

In this sense, the high concentration of ROS could function as a double-edged sword. A moderate increase of ROS could trigger both proliferation and differentiation, as well as other tumor characteristics. However, when ROS levels increase up to the lethal threshold it may break through the antioxidant capacity of the cell and trigger its death, by apoptosis or necrosis, depending on the degree of oxidative damage. Under physiological conditions, normal cells maintain redox homeostasis with a low level of basal ROS by controlling the equilibrium between the generation of ROS (pro-oxidant) and their elimination (antioxidant capacity). Exogenous agents that increase the generation of ROS or decrease the antioxidant capacity of cancerous cells will move the redox equilibrium and induce a general increase of ROS levels, which will cause cell death when exceeding the tolerance threshold (Figure 1).

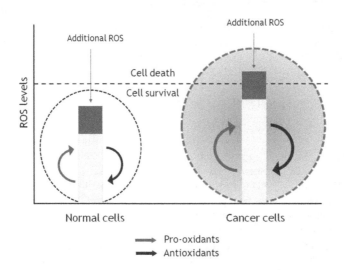

Fig. 1. Biological basis for therapeutic selectivity in oxidative therapy against cancer cells. Adapted from Gupte et al., 2009 and Trachootham et al., 2009.

Anticancer therapies aimed at the mitochondria of tumor cells are being developed in present years, since these organelles are involved in cell death (Daley et al., 2005; Nguyen and Hussain, 2007; Ozben, 2007; Pilkington et al., 2008). Mitochondria are the greatest oxygen consumers in the cell and they are an important source of reactive oxygen mediators. A drug that damages these organelles would cause an increase in ROS production and cell death. Currently, there is an anticancer strategy at its peak known as "oxidative therapy", which consists of inducing high ROS steady state levels in tumor cells. This therapy may be carried out in two different ways: causing the generation of high levels of ROS in solid tumors and inhibiting the antioxidant system of tumor cells (Fang et al., 2007; Trachootham et al., 2009). It is well established that high levels of ROS, like H_2O_2 and $O_2^{\bullet-}$, induce apoptosis in a wide variety of tumor cells activating the caspase cascade (Matsura et al., 1999; Yamakawa et al., 2000).

4.2 Cell death by ROS increment

Neoplastic cells are metabolically more active and require a high supply of ATP in order to keep cell growth and proliferation under control. This high energy demand in the MRC leads to the increase of ROS generation (Behrend et al., 2003). The excessive production of ROS may damage several cellular components like DNA, proteins, lipids and cell membranes. The oxidation of mitochondrial lipids and proteins causes the permeabilization of the mitochondrial membrane, which leads to an alteration of the coupling efficiency of the electron transport chain, resulting in the generation of more free radicals, and the release of cytochrome c, activating the process of programmed cell death (apoptosis) which depends on caspases (Conklin, 2004). There is evidence that the main mechanism by which oxidant agents may kill cells is the activation of apoptosis. In some cases, the high levels of ROS generated may inhibit apoptosis at a caspase level and divert the process toward necrosis (Chandra et al., 2000). The change from apoptosis to necrosis is critical in solid tumors and requires considerable amounts of ROS, a decrease of ATP and alterations in the mitochondrial electron-transport chain (Lee et al., 1999). The harmful consequences of this change lie in the inflammation caused by the rupture of necrotic cells and later release of enzymes that degrade the tissues. Thereby, death by apoptosis is preferred in antineoplastic therapies. Apoptotic cells cause the least damage to nearby tissues, since they do not release their content and are phagocytosed by macrophages.

Cancerous cells evolve with the mediation of endogenous and exogenous oxidative agents, depending on the cell type and the evolution state of the tumor. Tumors adapt to these conditions through the development of powerful antioxidant mechanisms and even by the use of endogenous ROS for proliferation. When ROS levels increase above the tolerance threshold, death of tumor cells is induced. Therefore, the fact that an excess of ROS causes cell damage and even death by apoptosis provides us with a strategy for eliminating cancerous cells, which are more sensitive to exogenous oxidative stress than normal cells, through the generation of free radicals, induced by oxidant pharmacological agents. In some cases, tumor cells attacked with antineoplastic therapies may gain resistance to oxidative stress, which is why combined therapies are promising, and are intended to converge toward improving the oxidative action above the critical threshold, or gathering different cytotoxic mechanisms.

Unfortunately, antitumor therapies may exert harmful effects on normal tissues, partially caused by ROS, which limits the application dose and its antitumor activity. Overcoming these secondary effects, without altering the therapy efficiency, is a priority and a challenge in biomedical research. In this sense, great importance is given to targeted therapies, using vehicles (liposomes, nanoparticles) that recognize specific molecules expressed in tumor cells.

4.3 Antioxidants in oxidative therapy against cancer

Antioxidants play an essential role in cell protection against ROS. The oxidation of antioxidant enzymes reduces the capacity of cells to eliminate free radicals. An important approach in the antitumor therapeutic strategies is to inhibit the antioxidant systems, like catalase, SOD and GPx, which are the main defense lines of the cell. The inhibitors of different antioxidant enzymes have been characterized by their capacity for eliminating neoplastic cells, alone or combined. It has also been described that many pharmacological agents may have more than one mechanism of action and affect multiple biological processes. There are agents that induce apoptosis which are oxidant and others that stimulate cell metabolism. On the other hand, there are apoptosis inhibitors that have antioxidant activities. In the absence of adequate antioxidant defenses, the damage from oxidative stress leads to the activation of the genes responsible for apoptosis.

There are conflicting data in the results obtained by different researchers regarding the levels of antioxidants in tumor tissue and in blood from cancer patients. In some cases, there are differences between different antioxidants in the same patient; some increase and others decrease (Ray et al., 2000). In breast cancer, for example, several studies describe an increase of lipid peroxidation and a decrease of antioxidants (Khanzode et al., 2004; Sener et al., 2007). However, other studies performed in neoplastic tissues have shown a greater presence of ROS and a high expression of antioxidants (Oltra et al., 2001; Gönenç et al.,

2006). So, for instance, an increase in the expression of Mn-SOD has been observed in breast cancer and in blood samples from patients with different types of leukemia (Nishiura et al., 1992; Devi et al., 2000; Ray et al., 2000). This fact may reflect an adaptive mechanism by which cancerous cells respond to an increase of ROS levels produced in mitochondria. ROS can induce the over-regulation of Mn-SOD through the modulation of the redox states of the transcription factors (AP- 1, NF-kappaB). Due to its high expression in certain types of cancer, Mn-SOD has been considered as a tumor marker (Schadendorf et al., 1995). This expression of SOD protects tumor cells against a lethal increase of ROS levels. In fact, it has been demonstrated that the inhibition of SOD with 2-methoxyestradiol would induce apoptosis in leukemia cells through a mechanism mediated by free radicals, without showing significant toxicity in normal lymphocytes (Zhou et al., 2003).

Antitumor therapies mediated by ROS show a promising therapeutic activity in clinical studies (Trachootham et al., 2009). However, some tumor cells, especially in advanced stages of the disease, have adapted to oxidative stress due to their antioxidant capacity. This redox adaptation does not only allow tumor cells surviving under high levels of ROS, it also provides an increase of survival molecules and a greater capacity for drug inactivation. Moreover, it has been suggested that resistance to the agents that induce intracellular ROS production, such as paclitaxel, doxorubicin or other drugs, is correlated to the increase of antioxidants (Glorieux et al., 2011). Thereby, the capability of certain drugs to inhibit or reduce the antioxidant machinery is very useful in oxidative therapy. These drugs could be used in combination with oxidant agents for greater efficiency in antitumor therapies.

5. Amitriptyline as an anti-cancer agent

Amitriptyline is a commonly prescribed tricyclic antidepressant drug that is well known to death investigators, forensic pathologists, and toxicologists. Amitriptyline has sedative effects and is frequently prescribed for patients experiencing symptoms of depression. Amitriptyline, have also been used for therapeutic treatment of neuropathic and inflammatory diseases such as fibromyalgia, chronic fatigue syndrome, migraine, irritable bowel syndrome, and atypical facial pain (Gruber et al., 1996). Besides its anxiolytic properties, amitriptyline has central anticholinergic effects. Amitriptyline inhibits serotonin and noradrenaline uptake in presynaptic nerve ending (Maubach et al., 1999). However, toxicity of amitriptyline has been observed during standard treatments, and frequently during suicidal or accidental overdosage. Tricyclic antidepressant overdosage has toxic effects over cardiovascular, autonomous nervous, and central nervous systems, and may result in cardiotoxicity, cardiac conduction delays, dysrhythmia, hypotension, altered mental status, and seizures (Thanacoody and Thomas, 2005; Kiyan et al., 2006).

In vitro administration of amitriptyline to cell cultures induces several signs of toxicity. Amitriptyline treatment induces alteration of cellular permeability based on its detergent nature (Kitagawa et al., 2006). Furthermore, amitriptyline causes alterations in the glucidic metabolism of neurons resulting in a decrease of both uptake and transport of glucose (Mannerstrom and Tahti, 2004). Additionally, amitriptyline provokes an increase of intracellular lipid peroxidation in mouse 3T3 fibroblasts (Viola et al., 2000) and some mouse tissues (Bautista-Ferrufino et al., 2011), and many of these toxic effects are prevented by antioxidants (Slamon and Pentreath, 2000). Recently, our group has shown that amitriptyline induced toxicity is caused through a mitochondrial dysfunction, and increased ROS level (Moreno-Fernandez et al., 2008; Cordero et al., 2009). Amitriptyline reduced significantly the number of cultured cells; enhanced the production of stimulated lipid peroxidation, inverting the lipid reduced/oxidized ratio; decreased catalase protein levels, cytochrome c, $\Delta\Psi m$, and citrate synthase activity; revealing mitochondrial damage. So, amitriptyline-induced toxicity is caused through mitochondrial dysfunction, and increased mitochondrial ROS production. Moreover, CoQ level was decreased by amitriptyline treatment and CoQ and alpha-tocopherol supplementation ameliorated amitriptyline-induced toxicity in both cultured human primary fibroblasts and zebrafish embryos (Cordero et al., 2009).

Other toxic effects attributed to amitriptyline lie in the alteration of neuron carbohydrate metabolism, which results in a decrease of glucose absorption and transport; causing a total loss of neuron viability in a cell line of neuroblastoma (Mannestrom et al., 2004).

Recent studies have shown that some antidepressants can kill cancerous cells. In fact, tricyclic antidepressants have shown to cause cell death in human normal lymphocytes (Karlson et al., 1998), Hodgkin´s lymphoma cells (Serafeim et al., 2003), neurons (Lirk et al., 2006), glioma cells (Xia et al., 1999; Daley et al., 2005; Levkovitz et al., 2005) and colorectal cancer cells (Arimochi y Morita, 2006). Chlorimipramine exerts its effect via the inhibition of complex III of the MRC (Daley et al., 2005). The same is valid for amitriptyline, as we have already reported (Cordero et al., 2009). We showed in fibroblasts treated with amitriptyline a decrease of expression level of proteins of complex I, complex III, cytochrome c, and reduced CoQ_{10} levels. Deficient mitochondrial protein expression levels and reduced levels of CoQ_{10} may impair normal mitochondrial electron flow and proton pumping, inducing a drop in respiratory complexes activity, and mitochondrial membrane potential. Our data showed that amitriptyline-treated fibroblasts have reduced NADH:cytochrome c reductase (complex I+III) activity, and lower mitochondrial membrane potential, which may contribute to impaired mitochondrial protein import and aggravate mitochondrial dysfunction, ROS production, and oxidative stress. It has been proposed that ROS damage can induce the mitochondrial permeability transition (MPT) by the opening of non-specific high conductance permeability transition (PT) pores in the mitochondrial inner membrane (England and Cotter, 2005). This, in turn, leads to a simultaneous collapse of mitochondrial membrane potential. The activation of MPTcauses mitochondria to become permeable to all solutes up to a molecular mass of about 1500 Da (Forte and Bernardi, 2005). After MPT, mitochondria undergo a dramatic swelling driven by colloid osmotic forces, which culminates in the rupture of the outer membrane and release of proapoptotic mitochondrial intermembrane proteins into the cytosol, such as cytochrome c, apoptosis inducing factor, Smac/Diablo, and others (Cordero et al.,2009).

We have also studied the effect of amitriptyline on tumor cell lines (Cordero et al., 2010). We observed that this drug induced important mitochondrial damage in tumor cell lines (H460: non-small cell lung cancer, HeLa: epithelial cervical cancer, and HepG2: hepatoma), generating high amounts of ROS and provoking apoptotic cell death. Moreover, amitriptyline effects have been compared with three antitumor drugs frequently used in cancer therapy: camptothecin (CPT), doxorubicin (Doxo), and methotrexate (Metho). Interestingly, amitriptyline induced significantly higher ROS generation in comparison with the other drugs, producing a dose-dependent increase of apoptosis in human cancer cells through a mechanism dependent on caspase-3 activation. Apoptosis percentage was significally higher in those cells treated with amitriptyline than in cells treated with CPT, Doxo or Metho (Figure 2A). Moreover, when the cell cycle of synchronized cultures was stopped at the G0/G1 phase by depriving cells from serum, the difference of apoptosis percentage among amitriptyline and the remaining drugs was significantly higher than in normal cultures (Figure 2B). These results suggest that the effect of amitriptyline does not depend on cell cycle stage, whereas CPT, Doxo, and Metho are more harmful in dividing cells, as most chemotherapeutic drugs. These data are of special interest for cancer treatment during the nongrowing phases of certain tumors.

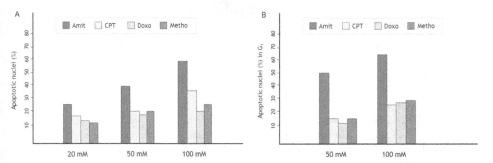

Fig. 2. Comparative study of amitriptyline and different chemotherapeutic drugs for the evaluation of apoptosis. (A) Percentages of apoptotic cells in H460 cell cultures 24 h after administration of drugs at different concentration. (B) Apoptosis assessment in synchronized cultures stopped at the G0/G1 phase. Adapted from Cordero et al., 2010.

After treating cancer cells with amitriptyline, we have found increased ROS level and several signs of mitochondrial damage, as attenuated complex I+III activity, decreased protein levels of complex III, decreased membrane potential, and a significant reduction of the number of this organelle, shown by cytochrome c and citrate synthase determination, and electron microscopy (Figure 3). So, this tricyclic compound provokes oxidative stress in cancer cells, being mitochondria the target of its toxicity. None of the chemotherapeutic drugs tested seemed to damage mitochondria seriously. However, the chemotherapeutic drugs induced apoptosis and increased ROS production in tumor cells, although not with the intensity of amitriptyline.

According to our data, amitriptyline induces a mitochondrial damage characterized by a decrease of the expression levels of complexes I and III of the MRC as well as of cytochrome c and of CoQ levels, which suggests an alteration in the activity, organization and assembly of the mitochondrial complexes, being this reflected in a decrease of the electron flow as well as a decrease of the mitochondrial membrane potential and, therefore, an increase of intramitochondrial ROS prodution. The damage caused by the increase of mitochondrial ROS induces the opening of the mitochondrial permeability transition pore (MPT), thus increasing mitochondrial permeability with the consequent release of proapoptotic proteins to the cytosol such as cytochrome c, Smac/Diablo, etc., initiating the intrinsic pathway of apoptosis dependent on caspase-3 (Figure 4).

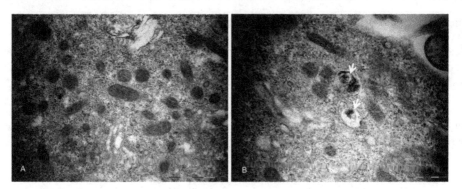

Fig. 3. Transmission electron microscopy showing damage and fewer mitochondria in H460 cells treated with 50 mmol/l of amitriptyline. Degenerating mitochondria (arrows) are observed in treated tumor cells. (A) Nontreated tumor cells. (B) Amitriptyline-treated tumor cells (Cordero et al., 2010).

In general, the increase of ROS production causes, as a response, an increase of the antioxidants activities. However, under the high input rate of ROS, enzyme inactivation prevails, which leads to the reduction of antioxidant enzymes activity and to the process of oxidative damage. Thus, tumor cells frequently possess very little antioxidative enzymes, such as catalase, SOD, and glutathione peroxidase, which are known to play a protective role against ROS in normal cells. We have observed in normal fibroblast treated with amitriptyline a decrease in protein expression of antioxidant enzymes (catalase and MnSOD) 16 h after the treatment, followed by restored levels after 24h, as a mechanism of antioxidant defense (Moreno-Fernández et al., 2008). Interestingly, in cancer cells, the same concentration of amitriptyline provoked an unrestorable decrease of catalase (Cordero et al., 2010). The difference of the antioxidant status observed in cancer cells, in comparison with healthy fibroblasts, may be caused by the lower antioxidant level present in the cancer cell lines used. Besides the decrease of catalase and MnSOD, amitriptyline also produces a significant decrease of CoQ level in tumor cells (Cordero et al., 2010). CoQ plays a critical protective role by either acting as an antioxidant or by the noncompetitive inhibition of the neutral sphingomyelinase of plasma membrane, preventing ceramide production (Mates et al., 1999). Most chemotherapeutic drugs do not provoke any decrement of antioxidants. Instead, they frequently induce an increase of antioxidants as a protecting mechanism against ROS generation, leading to lower cell death (Brea-Calvo et al., 2006). The fact that amitriptyline downregulates both catalase and CoQ activity is very interesting since it destroys the already decreased antioxidant defenses present in cancer cells, making the oxidative stress produced by the amitriptyline-induced ROS generation a more effective weapon.

Thus, amitriptyline promotes enhanced oxidative damage to cancer cells as this drug attacks cells by two different mechanisms: by the production of a high amount of ROS, provoking apoptosis; and by a significant decrease in antioxidant levels, seriously limiting cell reaction to oxidative stress. Therefore, amitriptyline could be used for anticancer oxidant therapy against tumors that present significant oxidative stress and/or low antioxidant defenses. For anticancer therapeutics on those tumors with a similar redox status than normal cells, a drug delivery vehicle should be used.

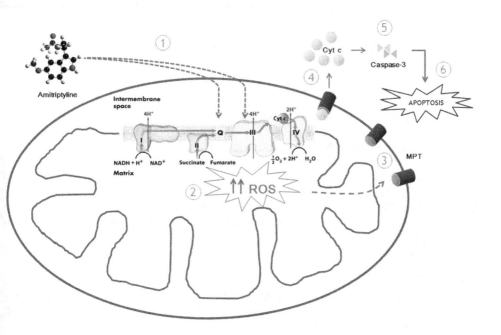

Fig. 4. Mechanism of apoptosis induction by amitriptyline.

6. Conclusion

Evidence exists that cancer cells are under a continuous oxidative stress, facilitating ROS to act as carcinogen, as they were shown to be involved in mutagenesis, cancer promotion and progression. On the other hand, ROS may promote either cell proliferation or cell death, depending on the rate of ROS production and the activity of the antioxidant system. Accordingly, oxidative therapy against cancer arises as a ROS-generating strategy which increase oxidative status above the critical threshold required for cell survival.

In our studies, the tricyclic antidepressant amitriptyline induced high ROS generation as a result of mitochondrial dysfunction, provoking a higher level of apoptosis of tumor cells than common chemotherapeutic drugs. Moreover, it inhibits important antioxidants of the cell-defense machinery, dramatically limiting tumor cell response to ROS production. Thus, amitriptyline, as well as other tricyclic compounds, is being assayed as an anti-cancer drug in oxidative therapy.

Unfortunately, oxidative anti-cancer therapy may exert harmful effects on normal tissues, limiting the applicable dose of drugs and anti-tumor activity. So, differences in ROS-induced cell death and oxidative status between normal and neoplastic cells from different cancer types must be thoroughly investigated, taking into account the complexity of physiological and pathological pathways involved in redox balance. On the other hand, it is not strange that after ROS-generating therapies cancer cells achieve further resistance to oxidative stress. In consequence, combination therapies could be achieved in order to either increase the intensity of oxidative stress above the critical threshold, or to perform different cytotoxic mechanisms.

Nevertheless, although greater efforts must be made, oxidative therapy against cancer is a promising strategy that is worthy of being investigated.

7. References

Arbiser, J.L.; Petros, J.; Klafter R.; Govindajaran, B.; McLaughlin E.R.; Brown L.F.; et al. (2002). Reactive oxygen generated by Nox1 triggers the angiogenic switch. *Proc Null Acad Sci USA*. 99 (Jan) (715-720), ISSN 0027-8424.

Arimochi, H. & Morita, K. (2006). Characterization of cytotoxic actions of tricyclic antidepressants on human HT29 colon carcinoma cells. *Eur J Pharmacol*. 10,541, (Jul) (17- 23), ISSN 0014-2999.

Attardi, L. D. & Donehower, L. A. (2005). Probing p53 biological functions through the use of genetically engineered mouse models. *Mutat. Res.* 576 (Aug) (4-21), ISSN 0027-5107.

Bartsch, H. & Nair, J. (2000). New DNA-based biomarkers for oxidative stress and cancer chemoprevention studies. *Eur J. Cancer* 36,10, (Jun) (1229-1234), ISSN 0959-8049.

Bautista-Ferrufino, M.R.; Cordero M.D.; Sánchez-Alcázar, J.A.; Illanes, M.; Fernández-Rodríguez, A.; Navas, P. & de Miguel, M. (2011). Amitriptyline induces coenzyme Q deficiency and oxidative damage in mouse lung and liver. *Toxicol Lett.* 204 (Jul) (32-37), ISSN 0378-4274.

Behrend, L.; Henderson, G. & Zwacka, R. M. (2003). Reactive oxygen species in oncogenic transformation. *Biochem. Soc. Trans.* 31, (Dec.) (1441–1444), ISSN 0300-512.

Bourdon, J. C. (2007). p53 and its isoforms in cancer. *Cancer* 97 *(Aug)* (277–282), ISSN 0007-0920.

Brea-Calvo, G.; Rodriguez-Hernandez, A.; Fernandez-Ayala, D. J.; Navas, P.; & Sanchez-Alcazar, J.A. (2006). Chemotherapy induces an increase in coenzyme Q10 levels in cancer cell lines. *Free Radic Biol Med* 40 (Apr) (1293–1302), ISSN 0891- 5849.

Chandra, J.; Samali, A.; & Orrenius, S. (2000). Triggering and modulaton of apoptosis by oxidative stress. *Free Radic Biol Med*; 29 (Aug) (323–333), ISSN 0891-5849.

Chang, E.F.; Wong, R.J.; Vreman, H.J.; Igarashi, T.; Galo, E.; Sharp, F.R.; Stevenson, D.K.; &Noble-Haeusslein, L.J. (2003). Heme oxygenase-2 protects against lipid peroxidation-mediated cell loss and impaired motor recovery after traumatic brain injury. *J Neurosci.* 23, 9, (May) (3689-3696), ISSN 0270-6474.

Conklin, K. A. (2004). Free radicals: the pros and cons of antioxidants. Cancer chemotherapy and antioxidants *J. Nutr*; 134 (Nov) (3201-3204), ISSN 0022-3166.

Cordero, M.D.; Moreno-Fernández, A.M.; Gomez-Skarmeta, J.L.; De Miguel, M.; Garrido-Maraver, J.; Oropesa-Ávila, M.; et al. (2009). Coenzyme Q10 and alpha tocopherol protect against amitryptyline toxicity. Toxicol Appl Pharmacol 15 (Mar) (329-337), ISSN 00410-008X.

Cordero, M.D.; Sánchez- Alcázar, J.A.; Bautista-Ferrufino, M.R.; Carmona-Lopez, M.I.; Illanes, M.; Rios, M. J.; et al. (2010). Acute oxidant damage promoted on cancer cells by amitriptyline in comparison with some common chemotherapeutic drugs. *Anticancer drugs* 10 *(Nov)* (932-944), ISSN 0959-4973.

Cotan, D.; Cordero, M.D.; Garrido-Maraver, J.; Oropesa-Ávila, M.; Rodríguez-Hernández, A.; Gómez Izquierdo, L.; et al. (2011). Secondary coenzyme Q10 deficiency triggers mitochondria degradation by mitophagy in MELAS fibroblasts. *FASEB J.* 25 (Aug) (2669-2687), ISSN 0892-6638.

Cuezva, J.M.; Krajewska, M.; de Heredia M.L.; Krajewski, S.; Santamaría, G.; Kim, H.; et al. (2002). The bioenergetic signature of cancer: a marker of tumor progression. *Cancer Res;* 62 (6674-6681), ISSN 0008- 5472.

Cullen, J.J.; Mitros, F.A.; & Oberley, L.W. (2003). Expression of antioxidant enzymes in diseases of the human pancreas: another link between chronic pancreatitis and pancreatic cancer. *Pancreas* 26, 1, (Jan) (23-27), ISSN 0885-3177.

Daley, E.; Wilkie, D.; Loesch, A.; Hargreaves, I.P.; Kendall, D.A.; & Pilkington G. J, et al. (2005); Clomipramine: a novel anticancer agent with a mitochondrial target. *Biochem Biophys Res Commun* 328 (Mar) (623–632), ISSN 0006-291X.

Davies, K. (1995). Breast cancer genes. Further enigmatic variations. *Nature* 378 (Dec) (362-363), ISSN 0028-0836.

Denko, N.C.; Fontana, L.A. Hudson, K.M. *et al.* (2003). Investigating hypoxic tumor physiology through gene expression patterns. *Oncogene* 22 (Sep) (5907-14), ISSN 0950-9232.

Devi, G.S.; Prasad, M.H.; Saraswathi, I.; Raghu, D.; Rao, D.N. & Reddy, P.P. (2000). Free radicals, antioxidant enzymes and lipid peroxidation in different types of leukemias. *Clin Chim Acta* 293 (Marz) (53-62), ISSN 0009-8981.

Diers, A.R.; Higdon, A.N.; Ricart, K.C.; Johnson, M.S.; Agarwal, A.; Kayanaraman, B.; et al. (2010). Mitochondrial targeting of the electrophilic lipid 15-deoxy-Delta prostaglandin J2 increases apoptotic efficacy via redox cell signalling mechanisms. *Biochem J.* 426, (Jan) (31-41), ISSN 0264-6021.

Dreher, D.; & Junod, A.F. (1996). Role of oxygen free radicals in cancer development. *European Journal of Cancer,* 32A *(Jan)* (30-38), ISSN 0959-8049.

England, K. & Cotter, T.G. (2005). Direct oxidative modifications of signalling proteins in mammalian cells and their effects on apoptosis. *Redox Rep* 10 (237-245), ISSN 1351-0002.

Fang, J.; Nakamura, H.; & Iyer, A. K. (2007). Tumor-targeted induction of oxystress for cancer therapy. *J. Drug Target.* 15 (Aug) (475-486), ISSN 1061-186X.

Ferraro, D.; Corso, S.; Fasano, E.; Panieri, E.; Santangelo, R.; Borrello, S.; et al. (2006). Pro-metastatic signaling by c-Met through RAC- 1 and reactive oxygen species (ROS). *Oncogene* 25 (3689-3698), ISSN 0950-9232.

Filomeno, G.; Rotilio, G. & Ciriolo, M.R. (2005) Disulfide relays and phosphorylative cascades: partners in redox-mediated signalling patways. *Cell Death Differ* 12 (Dic) (1555-1563), ISSN 1350-9047.

Floyd, R.A. (1990). Role of oxygen free radicals in carcinogenesis and brain ischemia. *FASEB J.* 4 (Jun) (2587–2597), ISSN 0892-6638.

Forte, M. & Bernardi, P. (2005). Genetic dissection of the permeability transition pore. J. *Bioenerg. Biomembr.* 37, 3 (Jun) (121- 128), ISSN 0145-479X.

Glorieux C, Dejeans N, Sid B, Beck R, Calderon PB, Verrax J. (2011). Catalase overexpression in mammary cancer cells leads to a less aggressive phenotype and an altered response to chemotherapy. *Biochem Pharmacol.* Jun 13. PMID: 21689642, ISSN 0006-2952.

Gönenç, A.; Erten, D.; Aslan, S.; Akinci, M.; Simşek, B.; & Torun, M. (2006). Lipid peroxidation and antioxidant status in blood and tissue of malignant breast tumor and benign breast disease. *Cell Biol Int.* 30, 4, (Apr) (376-380), ISSN 1065-6995.

Gruber, A.J.; Hudson, J.I. & Pope, G. (1996). The management of treatment-resistant depression in disorders on the interface of psychiatry and medicine. Fibromyalgia, chronic fatigue syndrome, migrañe, irritable bowel syndrome, atypical facial pain, and premenstrual dysphoric disorder. *Psychiatr Clin North Am,* 19,2, (351–359), ISSN 0193-953X.

Gupte, A. & Mumper, R.J. (2009). Elevated copper and oxidative stress in cancer cells as a target for cancer treatment. *Cancer Treat rev.* 35(Feb) (32-46), ISSN 0305-7372.

Haas, R.H.; Parikh, S.; Falk, M.J.; Saneto, R.P.; Wolf, N.I.; Darin, N. et al. (2008). The In-depth evaluation of suspected mitochondrial disease. *Mol Genet Metab.* 94, (May) (16-37), ISSN 1096-7193.

Hanahan, D. & Weinberg, R.A. (2011).The hallmarks of cancer. *Cell.* 144, (Mar) (209-219), ISSN 0092-8674.

Harris, A.L. (2002). Hypoxia- A key regulatory factor in tumor growth. *Nat Rev Cancer.* 2, 1, (Jan) (38-47), ISSN 1474-175X.

Hasegawa, Y.; Takano, T.; Miyauchi, A.; Matsuzuka, F.; Yoshida, H. & Kuma K. (2002). Decreased expression of glutathione peroxidase mRNA in thyroid anaplastic carcinoma. Cancer Lett; 182, (Aug) (69–74) ISSN 0304-3835.

Hileman, E.O.; Liu, J.; Albitar, M.; Keating, M.J. & Huang, P. (2003). Intrinsic oxidative stress in cancer cells a biochemical basis for therapeutic selectivity. *Cancer Chemother. Pharmacol.* 53, 3 (Mar) (209-219), ISSN 0344-5704.

Huang, P.; Feng, L.; Oldham, E.A.; Keating, M.J. & Plunkett, W. (2003). Superoxide dismutase as a target for the selective killing of cancer cells. *Nature* 407,5, (Dic) (390-395), ISSN 0028-0836.

Indo, H.P.; Davidson, M.; et al. (2007). Evidence of ROS generation by mitochondria in cells with impaired electron transport chain and mitochondrial DNA damage. *Mitochondrion.* 7, 2, (106-118), ISSN 1567-7249.

Ishikawa, K. *et al.* (2008). ROS-generating mitochondrial DNA mutations can regulate tumor cell metástasis. *Science.* 320, (May) (661–664), ISSN 0036-8075.

Kamp, D.W.; Shacter E. & Weitzman, S.A. (2011). Chronic inflammation and cancer: The role of the mitochondria. *Oncology.* 25 (Apr) (400-410), ISSN 0890-9091.

Karlsson, H.; Gu, Y.; DePierre, J.; Nassberger, L. (1998). Induction of apoptosis in proliferating lymphocytes by tricyclic antidepressants. *Apoptosis.* 3, 4 (Sep) (255–260), ISSN 1360-8185.

Kennedy, C. H.; Cueto, R.; Belinsky, S.A.; Lechner, J.F. & Pryor, W. A. (1998). Overexpression of hMTH1 mRNA: a molecular marker of oxidative stress in lung cancer cells. *FEBS Lett.* 429, 1, (Jun) (17-20), ISSN 0014-5793.

Khanzode, S.S.; Muddeshwar, M.G. & Dakhale, G.N. (2004). Antioxidant enzymes and lipid peroxidation in different stages of breast cancer. *Free Radic. Res.* 38, 1, (Jan) (81-85), ISSN 1071-5762.

Kitagawa, N., Oda, M., Nobutaka, I., Satoh, H., Totoki, T. & Morimoto, M., (2006). A proposed mechanism for amitriptyline neurotoxicity based on its detergent nature. *Toxicol. Appl. Pharmacol.* 15, 217, (Nov) (100–106), ISSN 0041-008X.

Kiyan, S.; Aksay, E.; Yanturali, S.; Atilla, R.; & Ersel, M.; (2006). Acute myocardial infarction associated with amitriptyline overdose. *Basic Clin. Pharmacol. Toxicol.* 98, 45, (May) (462–466), ISSN 1742-7835.

Klauning, J.E. & Kamendulis, L.M. (2004). The role of oxidative stress in carcinogenesis. *Annu Rev Pharmacol Toxicol.* 44, (239-267), ISSN 0362- 1642.

Kryston, T.B.; Georgiev, A.B.; Pissis, P. & Georgakilas, A.G. (2011). Role of oxidative stress and DNA damage in human carcinogenesis. *Mutat Res* 3, 711, (Jun) (193-201), ISSN 0027-5107.

Kundu, N.; Zhang, S. & Fulton, A. M. (1995). Sublethal oxidative stress inhibits tumor cell adhesion and enhances experimental metastasis of murine mammary carcinoma. *Clin. Exp. Metastasis* 13, 1, (Nov) (16–22), ISSN 0262-0898.

Lee, Y.J.; Galaforo, S.S.; Sim, J.E.; et al. (1999). Dominant negative Jun Nterminal protein kinase (JNK-1) inhibits metabolic oxidative stress during glucose deprivation in a human breast carcinoma cell line. *Free Radic Biol Med.* 274, 28, (Feb) (575-584) ISSN 0891-0895.

Levkovitz, Y.; Gil-Adl.; Zeldich, E.; Dayag, M. & Weizman, A. (2005). Differential induction of apoptosis by antidepressants in glioma and neuroblastoma cell lines: evidence for p-c-Jun, cytochrome c, and caspase-3 involvement. *J. Mol Neurosci.* 27, 3, (29-42), ISSN 0895-8696.

Lirk, P.; Haller, I.; Hausott, B.; Ingorokva, S.; Deibl, M.; Gerner, P. & Klimaschewski, L. (2006). The neurotoxic effects of amitriptyline are mediated by apoptosis and are effectively blocked by inhibition of caspase activity. *Anesth Analg.* 102, (Jun) (1728-33), ISSN 0003-2999.

Lu, W.; Ogasawara, M. A. & Huang, P. (2007). Models of reactive oxygen species in cancer. *Drug Discov.* 4, (67–73), ISSN 1740-6757.

Mannerstrom, M. & Tahti, H. (2004). Modulation of glucose uptake in glial and neuronal cell lines by selected neurological drugs. *Toxicol. Lett.* 15, 151, (Jun) (87-97), ISSN 0378-4274.

Manoharan, S.; Klanjiappan, K. & Kayalvizi, M. (2004). Enhanced lipid peroxidation and impaired enizmatic antioxidant activities in the erythrocytes of the patients with cervical carcinoma. *Cell Mol Biol Lett.* 9, 4 A (699-707), ISSN 1425-8153.

Mantovani, G.; Maccio, A.; Madeddu, C. & Massa, E. (2003). Cancer-related cachexia and oxidative stress: beyond current therapeutic options. *Expert Rev Anticancer Ther.* 3,3 (Jun) (381-392), ISSN 1473- 7140.

Mates, J.M.; Perez-Gómez, C. & Nuñez de Castro, I. (1999). Antioxidant enzymes and Human diseases. *Clin Biochem;* 32 (Nov) (595–603), ISSN 0009-9120.

Matsura, T.; Kai, M.; Fujii, Y.; Ito H. & Yamada, K. (1999). Hydrogen peroxide-induced apoptosis in HL-60 cells requires caspase-3 activation. *Free Radic Res.* 30 (Jan) (73-83), ISSN 1071-5762.

Maubach, K.A.; Rupniak, N.M.; Kramer, M.S. & Hill, R.G. (1999). Novel strategies for pharmacotherapy of depression. *Curr. Opin. Chem. Biol.* 3, 4 (Aug) (481–488), ISSN 1367-5931.

McEligot, A. J., Yang, S. & Meyskens, F. L. Jr. (2005). Redox regulation by intrinsic species and extrinsic nutrients in normal and cancer cells. *Annu. Rev. Nutr.* 25 (261–295), ISSN 0199-9885.

Moreno-Fernández, A.M.; Cordero, M.D.; De Miguel, M.; Delgado-Rufino, M.D; Sanchez-Alcazar, J.A.; Navas, P. (2008). Cytotoxic effects of amitriptyline in human fibroblasts. *Toxicology*; 243 (Jan) (51-58), ISSN 0300-483X.

Naidu, M.S.K.; Suryakar, A.N.; Sanjay, C.; Swami, S.C.; Katkam. R.V. & Kumbar, K.M. (2007). Oxidative stress and antioxidant status in cervical cancer patients. *Indian J Clin Biochem*. 22, 5 (140-144), ISSN 0200-423.

Nguyen, D.M.; Hussain, M. (2007). The role of the mitochondria in mediating cytotoxicity of anti-cancer therapies. *J Bioenerg Biomembr*. 39, 1, (Feb) (13–21), ISSN 0145-479X.

Nishiura, T.; Suzuki, K. & Kawaguchi, T. et al. (1992). Elevated serum manganese superoxide dismutase in acute leukemias. *Cancer Lett*. 15, 62 (Mar) (211-215), ISSN 0304-3835.

Oberley, L. W. & Buettner, G. R. (1979). Role of superoxide dismutase in cancer: a review. *Cancer Res*. 39, 4, (Apr) (1141–1149), ISSN 0008-5472.

Okamoto, K.; Kondo-Okamoto, N. & Ohsumi, Y. (2009). Mitochondria-anchored receptor Atg32 mediates degradation of mitochondria via selective autophagy. *Dev. Cell*. 17,1 (Jul) (87-97), ISSN 1534-5807.

Oltra, A. M.; Carbonell, F.; Tormos, C.; Iradi, A. & Saez, G. T. (2001). Antioxidant enzyme activities and the production of MDA and 8-oxo-dG in chronic lymphocytic leukemia. *Free Radic. Biol. Med*. 30 (Jun) (1286–1292), ISSN 0891-5849.

Ozben, T. (2008). Oxidative stress and apoptosis: Impact on cancer therapy. *J Pharm Sci*. 96, 9, (Sep) (2181-2196), ISSN 0022-3549.

Pelicano, H.; Feng, L.; Zhou, Y.; Carew, J.S.; Hileman, E.O.; Plunkett, W.; et al. (2003). Inhibition of mitochondrial respiration: a novel strategy to enhance drug-induced apoptosis in human leukemia cells by a reactive oxygen species-mediated mechanism. *J. Biol. Chem*. 278, (Sep) (37832–37839), ISSN 0021-9258.

Pelicano, H.; Carney, D.; & Huang, P. (2004). Ros stress in cancer cells and therapeutic implications. *Drug Resist Updat*. 7, 2, (Apr) (97-110), ISSN 1368-7646.

Pervaiz, S. & Clement, M. V. (2004). Tumor intracellular redox status and drug resistance serendipity or a causal relationship? *Curr. Pharm. Des*. 10, (1969–1977), ISSN 1381-6128.

Pieczenik, S.R. & Neustadt, J. (2007). Mitochondrial dysfunction and molecular pathways of disease. *Exp Mol Pathol*. 83, 1, (Aug) (84-92), ISSN 0014-4800.

Pilkington, G.J.; Parker, K. & Murray, S.A. (2008). Approaches to mitochondrially mediated cancer therapy *Semin Cancer Biol*. 18, (Jun) (226-235), ISSN 1368-7646.

Rajagopalan, S.; Meng, X.P.; Ramasamy, S.; Harrison D.G.; & Galis, Z. S. (1996). Reactive oxygen species produced by macrophage-derived foam cells regulate the activity of vascular matrix metalloproteinases in vitro. Implications for atherosclerotic plaque stability. J Clin Invest. 98, (Dec) (2572-2579), ISSN 0021-9738.

Ramsey, M. R. & Sharpless, N. E. (2006). ROS as a tumour suppressor? *Nature Cell Biol*. 8,11, (1213–1215), ISSN 1465-7392.

Ray, G.; Batra, S.; Shukla, N.K.; Deo, S.; Raina, V.; Ashok, S.; Husain, S.A. (2000). Lipid peroxidation, free radical production and antioxidant status in breast cancer. *Breast Cancer Res. Treat*. 59, 2, (Jan) (163–170) ISSN 0167-6806.

Retel, J.; Hoebee, B. Braun, J.E.; et al (1993). Mutational specificity of oxidative DNA damage. *Mutat Res*. 299, (May) (165-182), ISSN 0027-5107.

Reuter, S.; Gupta, S.C.; Chaturvedi, M.M.; & Aggarwal, B.B. (2010). Oxidative stress, inflammation, and cancer: how are they linked? *Free Radic Biol Med.* 49, 11, (Dic) (1603-1616), ISSN 0891-5849.

Sandhu, C. & Slingerland, J. (2000). Desregulación of the cell cycle in cancer. *Cancer Detection and Prevention.* 24, (107-118), ISSN 0361-090X.

Sato, K.; Ito, K.; Kohara, H.; Yamaguchi, Y.; Adachi, K. & Endo, H. (1992). Negative regulation of catalase gene expression in hepatoma cells. *Mol Cell Biol.* 12, (Jun) (2525–2533), ISSN 0270-7306.

Schadendorf, D.; Zuberbier, T.; Diehl, S.; Schadendorf, C.; & Czarnetzki, B.M. (1995). Serum manganese superoxide dismutase is a new tumour marker for malignant melanoma. *Melanoma Res.* 5, (Oct) (351-353), ISSN 8960-8981.

Schumacker, P.T. (2006). Reactive oxygen species in cancer cells: live by the sword, die by the sword. *Cancer cell.* 10, (Sep.) (175-176), ISSN 1535-6108.

Sener, D.E.; Gonenc, A.; Akmer, M. & Torun, M. (2007). Lipid peroxidation and total antioxidant status in patients with breast cancer. *Cell Biochem Func.* 25, (Jul) (377–382), ISSN 0263-6484.

Senthil, K.; Aranganathan, S. & Nalini, N. (2004). Evidence of oxidative stress in the circulation of ovarian cancer patients. *Clin Chim Acta.* 339, (Jan) (27-32), ISSN 0009-8981.

Serafeim, A.; Holder, M.J.; Grafton, G.; Chamba, A.; Dravson, M.T. & Luonq, Q.T, et al. (2003). Selective serotonin reuptake inhibitors directly signal for apoptosis in biopsy-like Burkitt lymphoma cells. *Blood.* 101, (Apr) (3212-3219), ISSN 0006-4971.

Slamon, N.D. & Pentreath, V.W. (2000). Antioxidant defense against antidepressants in C6 and 1321N1 cells. *Chem Biol Interact.* 14, 127, (Jul) (181–199), ISSN 0009-2797.

Spitz, D.R.; Sim, J.E.; Ridnour, L.A.; Galoforo, S.S. & Lee, Y.J. (2000). Glucose deprivation-induced oxidative stress in human tumor cells. A fundamental defect in metabolism? Ann N Y *Acad Sci.* 899, (349-362), ISSN 0077-8923.

Sullivan, R. & Graham, C. H. (2008). Chemosensitization of cancer by nitric oxide. *Curr. Pharm. Des.* 14, (1113–1123), ISSN 1381-6128.

Takahashi, A.; Ohtani, N.; Yamakoshi, K.; Iida, S.; Tahara, H.; Nakayama, K.; et al. (2006). Mitogenic signalling and the p16INK4a-Rb pathway cooperate to enforce irreversible cellular senescence. *Nature Cell Biol.* 8, (Nov) (1291–1297), ISSN 1465-7392.

Thanacoody, H.K. & Thomas, S.H. (2005). Tricyclic antidepressant poisoning: cardiovascular toxicity. *Toxicol.* 24, 3, (205–214), ISSN 1176-2551.

Toyokuni, S.; Okamoto, K.; Yodoi, J.; & Hiai, H. (1995). Persistent oxidative stress in cancer. *FEBS Lett.* 358, (Jan) (1–3), ISSN 0014-5793.

Trachootham, D.; Alexandre, J. & Huang, P. (2009). Targeting cancer cells by ROS mediated mechanisms: a radical therapeutic approach? *Nat Rev.* 8, (Jul) (579–591), ISSN 1474-1776.

Turrens, J.F. (2003). Mitochondrial formation of reactive oxygen species. *J Physiol.* 552, 2, (Oct) (335-44), ISSN 0022-3751.

Turunen, M.; Olsson, J. & Dallner, G. (2004). Metabolism and function of coenzyme Q. *Biochim Biophys Acta.* 1660, 1, (Jan) (171-99), ISSN 0006-3002.

Ushio-Fukai, M. & Nakamura, Y. (2008). Reactive oxygen species and angiogenesis: NADPH oxidase as target for cancer therapy. *Cancer Lett.* 266, (Jul) (37–52), ISSN 0304-3835.

Van-Driel, B.E.; Lyon, H.; Hoogenraad, D.C.; Anten, S. & Hansen, U. (1997). Expression of CuZn- and Mn-superoxide dismutase in human colorectal neoplasms. *Free Radic Biol Med.* 23, (435- 344), ISSN 0891-5849.

Verrax, J.; Taper, H. & Calderon, P.B. (2008). Targeting cancer cells by an oxidant based-therapy. *Curr Mol Pharmacol.* 1, (Jan) (80-92), ISSN 1874-4672.

Viola, G.; Miolo, G. Vedaldi, D. Dall'Acqua F. (2000). In vitro studies of the phototoxic potential of the antidepressant drugs amitriptyline and imipramine. *Il Farmaco.* 55, 3, (Mar) (211–218), ISSN 0014-827X.

Vousden, K. H. & Lane, D. P. (2007). p53 in health and disease. *Nature Rev. Mol. Cell Biol.* 8 (Apr) (275–283), ISSN 1471-0072.

Westermarck, J.; Kahari, V.M. (1999). Regulation of matrix metalloproteinase expression in tumor invasion. *FASEB J.* 13, 8, (May) (781-792), ISSN 0892-6638.

Xia, Z.; Bergstrand, A.; DePierre, J.W. & Nassberger, L., (1999). The antidepressants imipramine, clomipramine, and citalopram induce apoptosis in human acute myeloid leukemia HL-60 cells via caspase-3 activation. *J. Biochem biol Toxicol.* 13, 6, (338-347), ISSN 1095-6670.

Xu, R.; Pelicano, H.; Zhou, Y.; Carew, J.S.; Feng, L.; Bhalla, K.N.; et al. (2005). Inhibition of glycolysis in cancer cells: a novel strategy to overcome drug-resistance associated with mitochondrial respiratory defect and hypoxia. *Cancer Res.* 65, 2, (Jan) (613-621), ISSN 0008-5472.

Yamakawa, H.; Ito, Y.; Naganawa, T.; Banno, Y.; Nakashima. S. & Yoshimura, S. et al. (2000) Activation of caspase 9 and 3 during H2O2-induced apoptosis of PC12 cells independent of ceramide formation. *Neurol Res.* 22, 6, (Sept) (556–564), ISSN 0161-6412.

Yano, M.; Ikea, M.; Abe, K.; Kawai, Y.; Kuroki, M.; Mori, K.; et al. (2009).Oxidative stress induces anti-hepatitis C virus status via the activation of extracellular signal-regulated kinase. *Hepathology.* 50, 3, (Sep) (556-564), ISSN 0270-9139.

Young, T. W.; Mei, F.C.; Yang, G.; Thompson-Lanza, J.A.; Liu, J.; & Cheng, X. (2004). Activation of antioxidant pathways in ras-mediated oncogenic transformation of human surface ovarian epithelial cells revealed by Functional proteomics and mass spectrometry. *Cancer Res.* 1, 64, (4577–4584), ISSN 0008-5472.

Zhou, Y.; Hileman, E.O.; Plunkett. W.; Keating, M.J. & Huang, P. (2003). Free radical stress in chronic lymphocytic leukemia cells and its role in cellular sensitivity to ROS generating anticancer agents. *Blood.* 15, 101, (May) (4098-104), ISSN 0006-4971.

Zorov,D.B.; Juhaszova, M.; & Sollott, S.J. (2006). Mitochondrial ROS-induced ROS release: An update and review. *Biochimica et Biophysica Acta (BBA)–Bioenergetics.* 1757, (May) (509-517), ISSN 0006-4971.

Section 3

Antioxidants as Therapeutics

11

Compounds with Antioxidant Capacity as Potential Tools Against Several Oxidative Stress Related Disorders: Fact or Artifact?

P. Pérez-Matute[1*], A.B. Crujeiras[2],
M. Fernández-Galilea[3] and P. Prieto-Hontoria[3]
[1]HIV and Associated Metabolic Alterations Unit,
Infectious Diseases Area,
Center for Biomedical Research of La Rioja (CIBIR)
[2]Laboratory of Molecular and Cellular Endocrinology,
Instituto de Investigación Sanitaria,
Complejo Hospitalario de Santiago and Santiago de Compostela University
[3]Department of Nutrition, Food Science,
Physiology and Toxicology, University of Navarra,
Spain

1. Introduction

Oxidative stress has been generating much recent interest primarily because of its accepted role as a major contributor to the aetiology of both normal senescence and severe pathologies with serious public health implications such as obesity, diabetes, atherosclerosis, metabolic syndrome, cancer etc. However, 'Living with the risk of oxidative stress is a price that aerobic organisms must pay for more efficient bioenergetics' (quoted from V. P. Skulachev).

The term oxidative stress is vaguely defined. In essence, it refers to a serious imbalance between production of reactive species and antioxidant defenses. Thus, oxidative stress can result from diminished levels of antioxidants but can also result from increased production of reactive species (Lushchak, 2011). The consequences of oxidative stress can include: firstly, adaptation of the cell or organism by upregulation of defence systems, which may first, completely protect against damage; second, protect against damage to some extent but not completely; or third, 'overprotect' (e.g. the cell is then resistant to higher levels of oxidative stress imposed subsequently). Secondly, cell injury, which involves damage (oxidative damage) to any or all molecular targets: lipids, DNA, proteins, carbohydrates, etc. Thirdly, cell death as the cell may first, recover from the oxidative damage by repairing it or replacing the damaged molecules, or second, it may survive with persistent oxidative damage or third, oxidative damage, especially to DNA, that may trigger cell death, by apoptosis or necrosis (reviewed by Perez-Matute et al., 2009). There are different types of

* Corresponding Author

reactive species: reactive oxygen species (ROS, thus, oxygen-containing molecules that are highly reactive), reactive chlorine species (RCN) and reactive nitrogen species (RNS). All these reactants contain free radicals as well as nonradicals. Low concentrations of these reactive species are necessary for normal cell redox status, cell function and intracellular signalling (Droge, 2002; Valko et al., 2007; Perez-Matute et al., 2009). However, in some disease states, free radicals are produced in excess and can damage DNA, proteins, carbohydrates and lipid constituents and compromise cell function leading to the development of type 2 diabetes, atherosclerosis, obesity, arthritis etc. Thus, it is clear that excessive production of free radicals causes damage to biological material and is an essential event in the aetiopathogenesis of various diseases. However, the question that has risen in the past years is whether uncontrolled formation of ROS is a primary cause or a downstream consequence of the pathological processes. In other words, it is still not clear what comes first, the chicken or the egg. However, what is clear is that there must be a balance between these reactive species and the antioxidants, whose main function is to counteract the deleterious effects of these reactive species. In fact, antioxidant is defined as any substance that when present at low concentrations compared with those of an oxidizable substrate, significantly delays or prevents oxidation of that substrate (Halliwell & Gutteridge, 1999). These defences include both enzymatic (superoxide dismutases, glutathione peroxidase, catalase, thioredoxin) and non-enzimatic systems (vitamins such as vitamin C, E, A, minerals such as selenium, zinc, cooper, bilirrubine, uric acid, some aminoacids etc).

Several studies have demonstrated an increased oxidative state (either caused by an increased ROS production or diminished levels of antioxidants) in serious pathologies such as obesity, cardiovascular diseases, metabolic syndrome, cancer etc. Thus, oxidative stress actually may be related with the mentioned processes. In this context, it is tempting to suggest that if oxidative damage significantly contributes to disease pathology, then, actions that decrease it (via decreasing ROS production or increasing endogenous levels of antioxidants) might be therapeutically beneficial. In fact, attenuation or complete suppression of oxidative stress as a way to improve several diseases has flourished as one of the main challenges of research in the last years. Thus, several approaches have been carried out in order to either decrease the high levels of ROS generated or boost the endogenous levels of antioxiants. Inhibition of ROS production through the development of inhibitors (natural or chemical) against the main sources of ROS generation offers an interesting approach. Thus, NADPH oxidase and mitochondria have been postulated as the main targets to reduce ROS production (reviewed by Pérez-Matute et al., 2009). Another strategy to decrease the consequences of an increased oxidative state is the investigation that is being carried out in the last years to prove the benefits from usage of antioxidant vitamins, minerals or drinks and foods with bioactive compounds to prevent these oxidative-stress-related diseases. Thus, this chapter will focus on the potential beneficial effects of modulating oxidative stress by several bioactive compounds with antioxidant properties.

2. Counteracting oxidative stress to improve health: Role of antioxidants

As previously mentioned, increasing amount of evidence suggests that oxidative stress is linked to pathophysiological mechanisms concerning multiple acute and chronic human diseases (Dalle-Donne et al., 2006; Halliwell & Gutteridge, 2007). In this context, different strategies have been developed in order to counteract oxidative stress to improve health. We will focus on the modulation antioxidant status through a nutritional approach. Thus, and concerning the conventional antioxidant therapies that have been carried out in the last years, we can underline that there are two main ways to deal with this issue: to promote the ingestion of diets rich in several micro and macronutrients with antioxidant properties that could be beneficial for health (such as the well known Mediterranean diet) or to supplement the diet with specific bioactive compounds with antioxidant properties. In this sense, many diseases have been reported to benefit from antioxidant therapy and covering all of them in one chapter is not possible. However, it is important to note that those pathologies that may benefit the most from this antioxidant therapy are neurodegenerative diseases, Alzheimer disease, Parkinson disease, amyotrophic lateral sclerosis, cancer, stroke, obesity and diabetes (reviewed by Firuzi et al., 2011).

2.1 Diets with recognized benefits on oxidative stress and health: Diets rich in antioxidants

Epidemiological and experimental studies have demonstrated that plant-food intake decreases the risk of chronic diseases and therefore significantly contributes to the maintenance of health. For instance, the lower occurrence of cancer and cardiovascular diseases in the population around the Mediterranean basin has been linked to the dietary habits of this region. This so-called Mediterranean diet is essentially different from the diets consumed in Western and Northern European countries and is rich in nuts, fruits, vegetables, legumes, whole-wheat bread, fish, and olive oil, with moderate amounts of red wine, which is mainly consumed during meals. The components of this diet contain an ample source of molecules with antioxidant and anti-inflammatory actions, among which we can find omega-3 fatty acids, oleic acid, and phenolic compounds (Pauwels, 2011). There are several studies where the health benefits of consuming this diet have been demonstrated. Thus, the study of Dai et al. (2008) has demonstrated that the association between the Mediterranean diet and plasma oxidative stress is robust and is not confounded by genetic or shared environmental factors. Moreover, they demonstrated that a decreased oxidative stress is a plausible mechanism linking the Mediterranean diet ingestion to reduced cardiovascular disease risk (Dai et al., 2008). Moreover, it has been shown that subjects following a Mediterranean diet present low oxidised LDL levels, which seems to be one of the protective effects against cardiovascular events according to a PREDIMED (*Prevención Con Dieta Mediterránea*) cohort trial (Fito et al., 2007). Furthermore, the *French Paradox* is the observation that French people suffer a relatively low incidence of coronary heart diseases, despite having a diet relatively rich in saturated fats along with fruits, vegetables and red wine. In fact, this paradox has been attributed to the consumption of red wine and more specifically to polyphenols (antioxidants such as resveratrol) present in red wine. These effects underline the hypothesis that the Mediterranean diet may also neutralize the deleterious effects caused by the consumption of relatively high amounts of animal fats.

The dietary patterns based on the DASH (Dietary Approaches to Stop Hypertension) emphasizes the consumption of fruits, vegetables, and low-fat dairy products and the reduced ingestion of saturated fat, total fat, and cholesterol (as in the Mediterranean diet) as it has been demonstrated that these patterns substantially lowered blood pressure and low-density lipoprotein cholesterol (Miller et al., 2006). Participants from the SU.VI.MAX (*Supplementation en Vitamines et Minéraux Antioxydants*) cohort who achieved the current daily fruit and vegetable intake recommendations within the DASH diet guidelines presented a lower increase in blood pressure with aging (Dauchet et al., 2007). In addition, a prospective study in the EPIC (European Prospective Investigation into Cancer and Nutrition) cohort evidenced that a high vegetables, legumes, and fruit diet was associated with a reduced risk of all-cause mortality, especially deaths due to cardiovascular disease underling the recommendation for the diabetic population to eat large amounts of vegetables, legumes, and fruit (Nothlings et al., 2008). Furthermore, fruit-enriched hypocaloric diets appear to be more effective against oxidative stress according to the study of Crujeiras et al. (2006). In fact, consumption of antioxidant substances contained in fruit could be a useful strategy in the design of hypocaloric diets that, with the weight reduction, could increase the improvement of cardiovascular risk factors related to obesity. Finally, in a case-control study, an inverse association has been found between the first acute myocardial infarction and the consumption of fruits among the Spanish Mediterranean diet (Martinez-Gonzalez et al., 2002).

Among all the foods included in these healthy diets (such as the Mediterranean diet), legumes have also been suggested to contribute to prevent cardiovascular disease and diabetes mellitus. Indeed, epidemiological studies have shown that Asian people consuming soy in their staple diet present much lower mortality and morbidity from cardiovascular disease than their counterparts in Western counties (Heneman et al., 2007). However, lentils, chickpeas, peas, and beans are the legumes more commonly consumed in Western countries but it has also been demonstrated that a non soybean legumes-based hypocaloric diet induced a higher decrease in blood lipids concentrations as well as lower lipid peroxidation markers related to obesity comorbidities as compared to a conventional and balanced hypocaloric diet (Crujeiras et al., 2007a).

All these studies mentioned above are examples that evidenced the beneficial effects of plant-food intake in promoting health and life-span in part attributed to their high level of antioxidant compounds, which contribute to decrease oxidative stress (Crujeiras et al., 2009). Some studies have also attributed antioxidative properties to fiber-enriched diets, since these compounds enhance the capacity to detoxify free radicals (Diniz et al., 2005). Fiber alters fat absorption from the diet by impairing lipid hydrolysis, resulting in increased fat excretion and as consequence, decreased lipid peroxidation probability. Moreover, fiber secondary metabolites that arise from bacterial fermentation in the colon may have antioxidant properties (Diniz et al., 2005). Reinforcing this idea, a significant correlation between antioxidant power in plasma and dietary fiber plus fructose evidenced the beneficial effects of fruit intake on antioxidant capacity in obese women (Crujeiras et al., 2006). In addition, the fruit (Crujeiras et al., 2006) or legumes (Crujeiras et al., 2007a) hypocholesterolemic effects were in parallel with oxidative stress improvement when evaluated by means of the prooxidant and antioxidant ratio in plasma (Crujeiras et al., 2006) or lipid peroxidation biomarkers (Crujeiras et al., 2007a), suggesting an indirect antioxidant effect of these plant-foods intake mediated by the hypocholesterolemic induction.

The antioxidant effect of plant-food could be also produced by the action of lesser known compounds or by the combination of different compounds occurring in the foods with direct or indirect antioxidant effects (Crujeiras et al., 2007b). In this context, fructose has been proposed to produce specific effects on oxidative stress. Animal models fed with a high content of fructose have shown a significant increase in antioxidant capacity and prevention of lipid peroxidation (Girard et al., 2005). This fruit monosaccharide stimulates uric acid synthesis due to its rapid metabolism by fructokinase (Heuckenkamp & Zollner, 1971). Uric acid has been widely recognized in the scientific literature as a metabolic compound with high antioxidant power participating as an *in vivo* scavenger (Glantzounis et al., 2005). Thus, it has been suggested that urate is responsible for the increase in antioxidant capacity after consuming apples as fruit in healthy subjects (Lotito & Frei, 2004) and after following a fruit-based hypocaloric diet in obese women (Crujeiras et al., 2006). However, the role of uric acid on oxidative stress and health is not clear enough and conflicting results have been provided in different studies, as will be discussed later on in the vitamins section. Taking together these observations, it is conceivable that besides of the direct effect of the antioxidant compounds of plant-foods present in the Mediterranean and other healthy diets, some reported antioxidant health effects can be also associated with the metabolic effect of these foods that indirectly reduces the oxidative damage probability in presence of free radicals. Thus and despite the fact that the Mediterranean diet along with other diets enriched in fruits, fiber or legumes are beneficial for health, it is very difficult to identify which component of the diet is responsible for the positive effects (in fact, in many cases is the association of several compounds). Thus and although the presence of antioxidants has been claimed by many to be responsible for the beneficial effect of vegetables and fruits, it has also been postulated that low content of fat in these foods may be the responsible cause (reviewed by Firuzi et al., 2011). Because of that, several investigations have been carried out to analyze the effects of specific compounds with antioxidant properties more than a food which contains plenty of compounds. In this sense, the most potent antioxidants with beneficial effects on health are presented in the following part of this chapter. It is important to note that we here present a brief review of the most important antioxidants found in foods more than in antioxidants that are currently in clinical use and that have been extensively reviewed elsewhere (Firuzi et al., 2011). Indeed, we have focused the chapter on a nutritional approach of oxidative stress related diseases more than on a pharmacological approach. However, a list with the main antioxidant drugs approved for clinical use is provided in table 1.

Antioxidant	Clinical Use
Edaravone	Ischemic stroke
Idebenone	Alzheimer disease (?)
N-Aceylcysteine	Acetaminophen overdose, mucolytic, dry eye syndrome
α-Lipoic acid*	Diabetic neuropathy
Micronized purified flavonoids fraction (MPFF, Daflon 500®)	Persistent venous ulcers
0-β-hydroxyethyl-rutosides (Venoruton®)	Chronic venous insufficiency
Silibinin (Leaglon®)	Hepatoprotective (?), chemopreventive
Baicalein and catechins (flavocoxid)	Osteoarthritis

*Lipoic acid, due to its dietary source will be deeply discussed in this chapter

Table 1. Antioxidant drugs approved for clinical use in various diseases (Firuzi et al., 2011).

2.2 Supplementation with specific bioactive compounds with antioxidant properties

2.2.1 Lipoic acid

α-Lipoic acid (LA), also known as 1,2-dithiolane-3-pentanoic acid or thioctic acid, is a promising dietary bioactive molecule because of its recognized therapeutic potential on several diseases such as diabetes, vascular disease, hypertension, alzheimer and inflammation (Shay e al., 2009; Firuzi et al., 2011). In fact, LA (dexlipotam) has been clinically approved and used for diabetic neuropathy as pointed out in table 1. In fact, it has been used in Germany for treatment of symptomatic diabetic neuropahty since several years ago.

The two enantiomers of this acid are the R form and the S form. Both R-LA and its reduced form, dihydrolipoic acid or 6,8-dimercaptooctanoic acid (DHLA) exert powerful antioxidant properties although DHLA seems to be more effective (Packer & Suzuki, 1993). Their antioxidant functions involve: quenching ROS (reactive oxygen species), regeneration of endogenous and exogenous antioxidants involving vitamin C, vitamin E and glutathione chelation of redox metal including Cu(II) and Fe (II) and repair of oxidized proteins.

Lipoic Acid can be found in different foods such as spinach and cabbage, liver and meat whole wheat and yeast of beer, but it is also endogenously produced by the liver through the lipoic acid synthase (LASY) machinery. Deficiency of LASY results in an overall disturbance in the antioxidant defence network, leading to increased inflammation, insulin resistance and mitochondrial dysfunction (Padmalayam et al., 2009).

Lipoic Acid is also an essential cofactor for mitochondrial bioenergetic enzymes (Smith et al. 2004). In fact, it is well known the intimate connection of LA with cell metabolism and redox state (Packer et al., 1997) as LA is essential for normal oxidative metabolism and plays a vital role as a cofactor in mitochondrial dehydrogenase reactions (Gilgun-Sherki et al., 2002).

Oxidative stress has been linked to different pathologies such as endothelial dysfunction. In this context, several studies noted that LA plays an important role in the activation of endothelial nitric oxide synthase (eNOS), which is one enzyme responsible for nitric oxide (NO) release/production, which, in turn, is an important regulator and mediator of numerous processes in the nervous, immune and cardiovascular systems. These actions include vascular smooth muscle relaxation resulting in arterial vasodilation and increasing blood flow (Federici et al., 2002; Montagnani et al., 2002). An *in vitro* study in human endothelial cells showed that treatment with LA potentate endothelial NO synthesis and bioactivity by mechanisms that appear to be independent of cellular GSH levels (Visioli et al., 2002). Furthermore, one trial demonstrated that the administration of LA improved vasodilation in patients with metabolic syndrome (Sola et al., 2005), corroborating its positive effects in endothelial dysfunction.

Recent studies also suggest that chronic oxidative stress plays an important role in the aetiology of human obesity (Vincent et al., 2007; Wang et al., 2011). Inadequacy of antioxidant defences probably begins with a low dietary intake of bioactive compounds with antioxidant capacity (Taylor et al., 2006). In fact, it has been demonstrated that obese individuals have a lower intake of bioactive compounds compared with non-obese persons Based on that, different studies suggest a possible nutritional intervention with antioxidants eg. LA for treating obesity which has been associated with an increased oxidative state caused by either an increase in ROS production or a decrease in the antioxidant levels

(Prieto-Hontoria et al., 2009; Carbonelli et al., 2011; Koh et al., 2011). In this context, it has been demonstrated that LA reduces body weight and adiposity in rodents (Kim et al., 2004; Prieto-Hontoria et al., 2009) and humans (Carbonelli et al., 2011). Several mechanisms may contribute to the anti-obesity effects of LA including the suppression of hypothalamic AMPK (adenosine monophosphate-activated protein kinase) activity (Shen et al., 2005), which, in turn, leads to a reduction in food intake. Other mechanism that could also contribute to the anti-obesity effects of LA is the stimulation of energy expenditure by increasing Ucp-1 mRNA levels in brown adipose tissue (Kim et al., 2004). A very recent study has also demonstrated that LA increases energy expenditure by enhancing AMPK in skeletal muscle, a cellular energy sensor that can regulate peroxisome proliferator-activated receptor-gamma coactivator-1alpha (PGC-1alpha), which is a master regulator of mitochondrial biogenesis. Thus, this study demonstrated that LA improves skeletal muscle energy metabolism in aged mice possibly through enhancing AMPK-PGC-1alpha-mediated mitochondrial biogenesis and function (Wang et al., 2010). Furthermore, the inhibitory actions of LA on intestinal sugar transport could also contribute to a lower feed efficiency observed in LA-treated animals (Prieto-Hontoria et al., 2009). Another mechanism that could also contribute in reducing adiposity is the ability of LA to inhibit adipocyte differentiation, as described by Cho et al., (2003). These inhibitory effects of LA on adipocyte differentiation appear to be mediated by reduced levels of PPARγ and C/EBPα, as well as by the activation of MAPK. Another study suggests that the anorexigenic effect of LA are mediated by inhibition the activity of various liver enzymes involved in fatty acid synthesis and desaturation such as glucose 6 - phosphate dehydrogenase, malic enzyme, pyruvate kinase enzyme, ATP-citrate lyase and fatty acid synthase (Huong & Ide, 2008).

In addition, LA has also beneficial actions in both glucose and lipid metabolism and, it has been proposed, as mentioned before, as a potential therapy for insulin resistance and type 2 diabetes. LA positively interacts with the insulin pathway and glucose handling in muscle and adipocytes, by modulating the IR/PI3K/Akt pathway and GLUT4 translocation (Shay et al., 2009). LA also promotes mitochondrial biogenesis in adipocytes and muscle through a stimulation of PGC-1α, contributing to improve the defective mitochondrial function associated to diabetes/obesity (Shen et al., 2008a; Shen et al., 2008b). Furthermore, a very recent study has demonstrated that LA treatment over a period of 2 months improves fasting blood glucose (FBG), insulin resistance (IR), and glutathione peroxidase (GH-Px) activity in type 2 diabetes (T2DM) patients (Ansar et al., 2011).

Furthremore, LA treatment in rats with thioacetamide-induced liver fibrosis, inhibited the development of liver cirrhosis, as indicated by reductions in cirrhosis incidence, hepatic fibrosis, and AST/ALT activities (Foo et al., 2011). Several studies from our group have also demonstrated the beneficial effects of LA supplementation on fatty liver in a diet-induced obesity rat model (Valdecantos et al., 2010b, 2011a,b).

Finally, several trials have also suggested the potential use of LA in cancer therapy (Novotny et al., 2008) due to its ability to induce apoptosis in cancer cells (Shi et al., 2008; Choi et al., 2009). However, the molecular mechanisms underlying the anti-carcinogenic actions of LA are not well understood.

To sum up, LA seems to be a promising candidate against not only diabetes (in fact is one antioxidant approved for clinical use in diabetic neuropathy) but also against obesity and its comorbidities (glucose and lipid impairments) as well as against cardiovascular events, some cancers and liver injuries.

2.2.2 Polyphenolic compunds: Resveratrol

Grapes (*Vitis vinifera L.*) contain high concentrations of polyphenols, especially flavonoids. The amount and composition of biologically active compounds presented in grapes and grape products vary greatly according to the species, variety, maturity, seasonal conditions, production area and yield of the fruit. The main grape polyphenols are anthocyanins in red grapes and flavan-3-ols in the case of white grapes. Red grapes contain more total polyphenols than white grapes. Grape seeds and skins are also an important dietary source of flavonoids, and seeds contain significant amounts of proanthocyanidins or condensed tannins. The most common commercial product derived from grapes is wine, a moderately alcoholic drink made by fermentation of juice extracted from fresh, ripe grapes. Its moderate consumption is suggested in the Mediterranean diet as cited before. The processing of grapes to yield wine transforms the polyphenols present in grapes and as a result the main polyphenols in wine are flavan-3-ols, flavan-3,4-diols, anthocyanins and anthocyanidins, flavonols, flavones, condensed tannins and a characteristic biologically active compound, resveratrol – a stilbene whose concentration can range from 1·5 to 3 mg/1 (reviewed by Perez-Jimenez & Saura-Calixto, 2008). Resveratrol (trans-3,5,4'-trihydroxystilbene) is also found in various plants, including berries and peanuts. Moreover, this compound is now available in tablets on the market as a dietary supplement (not for clinical use).

A remarkable range of biological functions have been ascribed to this molecule. For example, resveratrol has shown cardioprotective actions (Hung et al., 2000), anti-cancer effects (Vanamala et al.) and anti-inflammatory and antioxidant properties (de la Lastra & Villegas, 2007). Its cardiovascular properties, including inhibition of platelet aggregation and promotion of vasodilation by enhancing the production of nitric oxide, have also been described (Cucciolla et al., 2007). It has also been reported to have many biological activities and protect against several neurodegenerative disorders such as Alzheimer's disease (Sun et al., 2010), but also to protect against oxidative stress in liver as well as steatosis in obese rats (Sebai et al., 2010; Gomez-Zorita et al., 2011) and against other diseases including AIDS (James, 2006; Zhang et al., 2009; Touzet & Philips, 2010), age-related illnesses and, more recently, obesity (Macarulla et al., 2009; Alberdi et al., 2011; Lasa et al., 2011). In fact, it seems to mimic the effects of energy restriction, thus leading to reduced body fat and improved insulin sensitivity. The mechanisms underlying these positive effects on obesity include: inhibition of preadipocyte proliferation and adipogenic differentiation, stimulation of basal and insulin-stimulated glucose uptake and inhibition of *de novo* lipogenesis (Fischer-Posovszky et al.). Resveratrol may also influence the secretion and plasma concentrations of some adipokines such as adiponectin and TNF-α and inhibits leptin secretion from rat adipocytes (Baur et al., 2006; Szkudelska et al., 2009). Resveratrol also regulates lipolysis via adipose triglyceride lipase (Lasa et al., 2011).

Several studies have suggested that activation of SIRT1 and AMPK plays a key role in the metabolic effects of resveratrol (Feige et al., 2008; Um et al., 2010). Sirtuins may provide novel targets for treating some diseases associated with oxidative stress. More specifically, SIRT1 has been shown to regulate metabolism and stress response by acting on several transcription factors and cofactors, histones and other chromatin proteins and components of DNA repair machinery. A recent research has also shown that resveratrol modulates tumor cell proliferation and protein translation via SIRT1-dependent AMPK activation (Lin et al.). In this context, resveratrol has been proposed as a potential dietary compound against various cancers including breast and colon tumors. Resveratrol may affect all three discrete stages of carcinogenesis (initiation, promotion, and progression) by modulating signal transduction pathways that control cell division and growth, apoptosis, inflammation, angiogenesis, and metastasis (Bishayee, 2009). Recently, it has been shown that resveratrol suppresses IGF-1 induced cell proliferation and elevates apoptosis in human colon cancer cells, via suppression of IGF-1R/Wnt and activation of p53 signaling pathways (Vanamala et al., 2010).

Tat protein plays a pivotal role in both the human immunodeficiency virus type 1 (HIV-1) replication cycle and the pathogenesis of HIV-1 infection. A very recent study has demonstrated that resveratrol, a SIRT1 activator, attenuates the transactive effects of Tat in HeLA-CD4-long terminal repeat-β-gal cells (MAGI) via NAD(+)-dependent SIRT1 activity suggesting that this antioxidant, through the regulation of different pathways such as SIRT1 activation, could be a novel therapeutic approach in anti-HIV-1 therapy (Zhang et al., 2009).

In addition, resveratrol also induces the activation of genes that encode for proteins involved in oxidative phosphorylation and mitochondrial biogenesis processes (reviewed by Szkudelska & Szkudelski, 2010). In this context, it has been shown that resveratrol improves the functioning of mitochondria in cells. In fact, the capacity of this antioxidant to reduce mitochondrial ROS levels and to induce the biosynthesis of antioxidant molecules, like MnSOD, along with its ability to increase the activity of these antioxidant defences, has been previously demonstrated (Valdecantos et al. 2010a). These actions could also explain the protective role of this antioxidant against situations with an imbalance in the oxidative status such as steatosis, obesity etc.

2.2.3 Vitamins with antioxidant properties: Vitamin E and Vitamin C

Vitamin E is the nature´s most effective lipid-soluble antioxidant, with an important role protecting unsaturated fatty acids residues in cells membranes, which are important for membrane function and structure (Van Gossum et al., 1988). Vitamin E is only produced by photosynthetic organisms. It refers to a group of eight naturally occurring compounds α-, β-, γ-, δ- tocopherols and tocotrienols. α-tocopherol, especially the naturally occurring D-α-tocopherol, is the one with the highest biological activity (Brigelius-Flohe & Traber, 1999). This variant of vitamin E can be found most abundantly in vegetable oils such as wheat germ oil, sunflower, and safflower oils (Reboul et al., 2006). Vitamin E is also found in many foods, mainly of plant origin, especially in leafy green (broccoli, spinach), seeds, including soybeans, wheat germ, some breakfast cereals and yeast beer. It can also be found in animal foods such as egg yolk.

The role of the vitamin E has emerged as a possible therapy for decreasing ROS production or increasing the endogenous levels of antioxidants and for protecting cell membranes at an early stage of free radical attack (Horwitt, 1986). Thus, vitamin E down-regulates NADPH oxidase (Calvisi et al., 2004), which is the major source of ROS in the vascular wall and it also up-regulates eNOS activity which leads to an increase in NO production (Ulker et al., 2003). As vitamin E is a potent antioxidant with anti-inflammatory properties, several lines of evidence suggest that α-tocopherol has also potential beneficial effects with regard to cardiovascular disease (Singh et al., 2005; Rodrigo et al., 2008). A recent study has also demonstrated that natural vitamin E analog alpha-tocopheryl phosphate (alphaTP) modulates atherosclerotic and inflammatory events through the regulation of certain genes (Zingg et al., 2010). However, it is also important to point out that the non-antioxidant activities of tocopherols may also represent the main biological reason for the selective retention of alpha-tocopherol in the body, or vice versa, for the metabolic conversion and consequent elimination of the other tocopherols (Zingg et al., 2004).

Several studies have demonstrated the beneficial effects of vitamin E on obesity and its related disorders such as diabetes. In fact, plasma vitamin E reflects the amount of α-tocopherol in the body. It is interesting to note that lower plasma vitamin E levels have been observed in type 2 diabetic patients (Skrha et al., 1999). In addition, the study from Botella-Carretero et al., (2010) demonstrated that alpha-tocopherol concentrations are inversely associated with body mass index in morbid obesity. Other study has demonstrated that vitamin E intervention increased the plasma activity of several antioxidant enzymes such as superoxide dismutase (SOD), glutathione peroxidase (GPx) and T-AOC (total anti-oxidative capacity) whereas it is able to decrease the levels of Isoprostane 8-epi PGF2alpha, which is a product of oxidative stress that causes potent smooth muscle contraction. The same study demonstrated that vitamin E intervention also decreased plasma glucose, insulin and triglycerides level in obese rats. Therefore, this study demonstrated that vitamin E has positive effects for improvement of oxidative stress status and glucose metabolism in an animal model of diet-induced obesity (Shen et al., 2009). In this context, Manning et al., (2004) showed that vitamin E supplementation decreased plasma peroxide concentration in obese individuals. Other study showed that antioxidant supplementation with vitamin E, C and β-carotene reduced exercise-induced lipid hidroperoxide (ΔPEROX) in overweight young adults. Possible collective mechanisms to explain this finding include a shift in the cytokine profile from a pro-inflammatory to a less inflammatory profile (lowered IL-6, increased adiponectin), an attenuation of cholesterol and triglyceride levels during exercise and a small increase in total antioxidant status (Vincent et al., 2006). On the other hand, vitamin E supplementation decreased concentrations of both 8-isoprostane and lipid peroxides in overweight subjects, indicating a decrease in systemic oxidative stress (Sutherland et al., 2007). Vitamin E supplementation in patients with diabetes decreased the levels of proinflammatory adipokines, such as IL-1, TNF-α, IL-6, and reactive C protein in serum and stimulated monocytes (Devaraj & Jialal, 2000; Upritchard et al., 2000). A recent study demonstrated that supplementing alpha-tocopherol (vitamin E) and vitamin D3 in high fat diet decreases IL-6 production in murine epididymal adipose tissue and 3T3-L1 adipocytes following LPS stimulation (Lira et al., 2011). Thus, this study suggested that vitamin E and D3 supplementation can be used as an adjunctive therapy to reduce the proinflammatory cytokines present in obese patients. A significant role played by oxidative stress and lipid peroxidation in the cascade of events involved in hepatic necroinflammatory damage is supported by an experimental study, which also showed that antioxidant vitamin E reduces fatty liver in obese Zucker rats (Soltys et al., 2001). In this context, the study from Vajro et al., (2004) strengthens the view that antioxidants, and especially vitamin E, may represent a relevant therapeutic tool for the treatment of children with obesity-related dysfunction who are unable to adhere to low-calorie diets (Vajro et al., 2004).

Finally, a very recent study concludes that MitoVES, a mitochondrially targeted analog of α-ocopheryl succinate, is an efficient anti-angiogenic agent of potential clinical relevance, exerting considerably higher activity than its untargeted counterpart. MitoVES may be helpful against cancer but may compromise wound healing (Neuzil et al., 2011). However, it is important to state that there are several controversial effects of vitamin E on cancer and diabetes that will be discuss later.

Vitamin C (Vit C) or ascorbic acid is one of the non-enzymatic antioxidants that can eliminate ROS, thus preventing tissue damage (Fetoui et al., 2008). Moreover, Vitamin C is the most abundant water-soluble antioxidant in the body and acts primarily in cellular fluid having the potential to protect both cytosolic and membrane components of cells from oxidant damage (Talaulikar & Manyonda, 2011). Vit C exerts its antioxidant effects in both direct and indirect ways. In the direct way, Vit C scavenges free radicals formed (Dawson et al., 1990) or interacts with reduced glutathione (Dudek et al., 2005). As an indirect way, it helps recycling vitamin E, thus, supplying active vitamin E (Netke et al., 1997).

Vit C is present in several fruits and vegetables such as citrus fruits, tomato, strawberry, pepper, cabbage, and leafy greens. Vit C can not be stored in the body, and excess Vit C is excreted in urine (Alpsoy & Yalvac, 2011).

Over the years, it has been suggested the usage of Vit C as a remedy against many diseases ranging from common colds to several types of cancers. Moreover, it is known that there is a close relationship between Vit C supply and immune cell activity, especially phagocytosis activity and T-cell function (Strohle et al., 2011) It also contributes to the formation and health of blood vessels, tendons, ligaments, bones, teeth and gums, it helps the body to absorb iron and to recover from wounds and burns, and serious deficiency of this vitamin can lead to scurvy, which is now a rare condition in the Western world (Garriguet, 2010; Strohle et al., 2011).

It has been described that obese patients have lower mean serum concentration of Vit C being even in an inadequate Vit C status, which leads to lower serum antioxidant capacity and greater inflammatory responses (Mah et al., 2011; Aasheim et al., 2008). Thus and regarding its effects on obesity, several studies demonstrated that Vit C dietary supplementation reduced body weight in a cafeteria diet-induced obese rat model, without affecting food intake (Campion et al., 2008; Boque et al., 2009). Moreover, it has been described that Vit C increases lipolysis and decreases triglicerides accumulation by decreasing the activity of glycerophosphate dehydrogenase, a marker of adipose conversion (Hasegawa et al., 2002; Senen et al., 2002). It also has been observed that Vit C supplementation is negatively associated with the occurrence of obesity suggesting that higher waist-to-hip ratios were associated with lower plasma ascorbic acid concentrations and that Vit C depleted individuals may be more resistant to fat mass loss.

Interestingly, these beneficial effects of Vit C seem to be due to a decrease observed in uric acid levels. In fact, it is known that hominoids during the Miocene could not biosynthesize Vit C, as a key gene involved in Vit C production: L-gulono-lactone oxidase had mutated. Hence, this mutation has been proposed to increase uric acid as an antioxidant that could replace the decrease in Vit C availability that may have occurred during this period (Johnson et al., 2009). Moreover, uric acid helps to raise blood pressure, stimulate salt-sensitivity, and induce insulin resistance and mild obesity, and thereby it helps to promote survival during a period of famine or stress which also leads to de development of obesity and its related comorbidities nowadays. Uric acid has been shown to be involved in metabolic pathways that lead to oxidative stress, endothelial dysfunction, and to a vascular and systemic inflammatory response. Moreover, the elevation in uric acid levels observed after fructose ingestion, with a consequent reduction in nitric oxide may lead to a reduced glucose uptake in the skeletal muscle, hyperinsulinemia, and insulin resistance. Thus several clinical studies showed the beneficial effects of lowering uric acid therapies on several markers of cardiovascular and renal disease (Stellato et al. 2011). In this context and supporting this idea, Hunter et al., (2011) concluded that dietary supplementation with Vit C may confer health benefits because of increased antioxidant potential or through mechanisms resulting from increased endogenous Vit C generation or decreased serum uric acid concentrations.

In summary, Vit C is a potent antioxidant that might prevent and improve obesity and several comorbidities by different mechanisms. Besides its antioxidant power, Vit C can also exert its beneficial effects by regenerating other antioxidants such as reduced glutation or vitamin E as well as by lowering uric acid levels.

2.2.4 Selenium

Selenium (Se) is an essential trace element consumed in submilligram amounts. It is primarily found in organically bound forms in the diet. Selenium is naturally found in plants, seafood, meat and meat products. The amount of selenium that is needed to ingest to maximize plasma glutathione peroxidase (GSHPX) activity is established between 55-75 $\mu g/d$ in the EU (Rayman, 2005). The element exists in both organic form of selenium, as part of selenoproteins (selenomethionine and methylated selenocompounds) as well as in inorganic forms such as selenites and selenates (Gromadzinska et al., 2008).

Selenium is required for the function of a number of key selenium-dependent enzymes (selenoproteins). Many of the known selenoproteins, in which selenium is the active site, are necessary for a wide range of metabolic processes, including thyroid hormone regulation, immune function and reproduction and they catalyze redox reactions (Kryukov et al., 2003). Because of the potential of selenoproteins to protect against oxidative stress, selenium functions as a dietary antioxidant and because of that it has been studied for its potential role in chronic diseases such as hypertension, cardiovascular disease, cancer and diabetes mellitus, as well as aging and mortality (Boosalis, 2008). In this context, experimental studies have shown that selenium has carcinostatic effects when added in high levels to the diet of animals treated with carcinogenic chemicals (Gromadzinska et al., 2008). In this context, evaluation of health claims by the FDA in the U.S. concerning the purportedly positive effects of selenium provided certain evidence for permitting a qualified health claim (Trumbo, 2005). Recent results of the SUVIMAX study showed that supplementation with vitamin C, vitamin E, β-carotene, selenium and zinc is able to reduce the rate of prostate cancer in men having normal levels of prostate-specific antigen in their plasma (Meyer et al., 2005).

Observational and interventional studies in humans have demonstrated the beneficial effect of selenium dietary intake. Thus, antioxidant supplementation contained selenium (100 mg) combined with vitamin C (500 mg), vitamin E (200 IU) and co-enzyme Q10 (60 mg) significantly alleviated the atherosclerotic damage caused by excessive production of ROS in patients with multiple cardiovascular risk factors. This beneficial vascular effect was associated with an improvement in glucose and lipid metabolism as well as with a decrease in blood pressure (Shargorodsky et al. 2010). It has also been demonstrated that the selenium supplementation is able to decrease lipid hydroperoxides (LH) post-exercise in overweight subjects, providing preliminary evidence for a potential role of selenium as an effective antioxidant therapy to reduce oxidant stress at rest and following high-intensity exercise in high-risk population groups (Savory et al. 2011).

The mechanisms responsible for the link between selenium and prevention of diseases associated or induced by an excessive production of reactive oxygen species are currently under-known. However, there are experimental evidence of selenium compounds affecting cell growth, cell cycle, DNA repair, gene expression, signal transduction and regulation of the redox status (Gromadzinska et al., 2008). On the other hand and as mentioned before, selenium functions as part of the selenoproteins which are involved in a wide range of metabolic processes. The cellular form of glutathione peroxidase (GPx 1) was the first selenoprotein identified. Several other GPxs containing the amino acid selenocysteine (Sec; analogous to cysteine in which sulfur is replaced by selenium) have been found since then. The glutathione (GSH)-Px system is found in almost all tissues and is believed to play a part in the body's antioxidant defence protecting polyunsaturated fatty acids and proteins from the damaging effects of peroxides and lipid hydroperoxide (LH) (Halliwell B, 2007). The other two major groups of known selenoprotein enzymes are the iodothyronine deiodinases that regulate operation of thyroid hormones, and the thioredoxin reductases (TrxR), involved in catalyzing the reduction of oxidized thioredoxin and other substrates. Additional selenoprotein is the selenoprotein P, the major form of selenium in the plasma and it also acts as an antioxidant in the extracellular space by reducing peroxynitrite and phospolipid hydroperoxides and forming complexes with mercury and cadmium (Gromadzinska et al., 2008).

Therefore and to sum up, there is strong evidence that selenium and the selenoproteins play a regulatory role in the following processes, which underlines its positive effects on health (Gromadzinska et al., 2008):

- ROS-activation of protein kinases in the cytoplasm and nucleus;
- ROS-activated modification of the thiol and hydroxyl groups in the Cys and Tyr;
- Controlling changes in the cell redox potential through inducing activation of the transcriptional factors and initiating *de novo* gene expression;
- Regulating the expression of membrane and nuclear receptors responsible for cell maintenance, intercellular communication, and changes in cell growth;
- Affecting apoptosis, necrosis and cell survival processes.

2.2.5 Green tea

Green tea, a product made from the leaves and buds of the plant *Camellia sinensis*, is, after water, the second most popular beverage worldwide, and a mayor source of dietary polyphenols that are known to render a myriad of health benefits (Rictveld & Wiseman, 2003). Green tea polyphenols are generally known as catechins. These group of compounds includes epicatechin, epigallocatechin, epicatechin gallate and epigallocatechin-3-gallate (EGCG) which is the most active of the major polyphenols and primarily responsible for the effects of green tea (Stewart et al., 2005).

EGCG has a four ring structure with eight hydroxyl groups being, therefore, highly hydrophilic, exhibiting good solubility in aqueous media (Zhong & Shahidi, 2011). EGCG is also a powerful antioxidant, possessing the highest antioxidant potency among all tea catechins, and it plays a protective role against oxidative stress in biological environments. For example, EGCG induced enzymes that play important roles as cellular antioxidant defenses such as SOD and catalase. It also lowers Malonil dialdehide (a product of lipid peroxidation and, therefore, a marker of oxidative stress) and it has also the ability of interacting with singlet molecular-oxygen, superoxide, peroxyl radicals, hydroxyl radicals, and peroxynitrite (Wei & Meng, 2010). Thus, green tea consumption may also show potential preventive effects against several oxidative stress-related disorders such as cardiovascular diseases (Rickman et al., 2010; Plutner et al., 1990; Nantz et al., 2009) and several types of cancer such as breast, prostate, lung, skin, gastric and colon cancer. It also shows neuroprotective effects in Parkinson and Alzeimer's disease (Zhao, 2009), ameliorates several autoimmune diseases such as autoimmune arthritis (Kim, H. R. et al., 2008), and immune-mediated liver injury (Wang et al., 2006) or even it seems to prevent skin cell damage (Jorge et al. 2011). Furthermore, green tea (or its active biomolecule EGCG) could be one potential anti-obesogenic agent (Stefanovic et al., 2008) and might be used in the prevention and treatment of this disease. Moreover, several "*in vivo*" studies demonstrated that green tea extracts or EGCG dietary supplementation decreased both body and adipose tissue weights (Park et al., 2011; Choo, 2003; Hasegawa et al., 2003), improved insulin sensitivity and glucose tolerance (Cao et al., 2007; Serisier et al., 2008) and had beneficial effects on prevention of hypertension (Ihm et al., 2009) and modulation of plasma cholesterol (Bursill et al., 2007), conditions linked to metabolic syndrome. In addition, it lowers the incidence of streptozotocin-induced diabetes (Song et al., 2003) and reduces body weight, body fat, and blood levels of glucose and lipid in leptin receptor-defective obese rats (Kao et al., 2000).

Several mechanisms have been proposed to explain the beneficial effects of EGCG in obesity and diabetes. Thus, EGCG protects pancreatic cells (Song et al., 2003), enhances insulin activity (Dhawan et al., 2002), represses hepatic glucose production (Waltner-Law et al., 2002), reduces food uptake and absorption (Kao et al., 2000), stimulates thermogenesis by increasing the uncoupling protein 2 (UCP2) and lipid excretion (Dulloo et al., 1999; Liao, 2001), and modulates insulin-leptin endocrine systems (Kao et al., 2000). Moreover, EGCG inhibits the sodium-dependent glucose transporter (Kobayashi et al., 2000) and represses various enzymes related to lipid metabolism, such as acetyl-CoA carboxylase, fatty acid synthase, pancreatic lipase, gastric lipase, and lipooxygenase (Liao, 2001; Wang & Tian, 2001) as well as lipolytic genes such as hormone sensitive lipase (HSL) and adipose triglyceride lipase (ATGL) in adipose tissue (Lee et al., 2009). It also reduces serum- or insulin-induced increases in the cell number and the triacylglycerol content of 3T3-L1 adipocytes during a 9-day period of differentiation (Sakurai et al., 2009) and also reviewed by Liu et al. (2006). It also inhibits adipocyte proliferation (Hung et al., 2005). Moreover, EGCG suppressed the differentiation of adipocytes through the inactivation of the forkhead transcription factor class O1 (FoxO1) and sterol regulatory element-binding protein-1 (SREBP1c) which are involved in adipocyte differentiation and lipid synthesis respectively in 3T3-L1 adipocytes (Freise et al. 2010). Regarding adipocytes hyperplasia it has been described that green tea EGCG may act at different concentrations in regulating mitogenesis and apoptosis of 3T3-L1 preadipocytes by inducing a decrease in the phosphorylated

ERK1/2, which are signal elements found to modulate the mitogenic and adipogenic signaling in 3T3-L1 (Wu et al., 2005), as well as decreasing ciclin dependent kinase 2 (Cdk2) activity and protein levels. Moreover, it has also been described that EGCG inhibited the mitogenic effect of insulin on preadipocytes in a dose and time-dependent manner, and that this inhibition might be due to its suppressive effects on the activities of the insulin receptor (Ku et al., 2009). Thus, the traditional knowledge about the anti-obesity effects of green tea can be confirmed and validated by scientific evidence.

It is important to point out that the beneficial effects of EGCG in cancer, but also in obesity and related disorders, are not always due to its antioxidant nature. In fact, it has been demonstrated that EGCG contribute to the beneficial effects of green tea on diabetes, obesity, and cancer by modulating gene expression. In fact, one of the possible mechanisms by which EGCG can inhibit cancer progression is through the modulation of angiogenesis signaling cascade as EGCG treatment leads to the downregulation of genes involved in the stimulation of proliferation, adhesion and motility as well as invasion processes, but also to the upregulation of several genes known to have antagonist effects (Tudoran et al., 2011). Very recent studies have also suggested the ability of EGCG to prevent several types of cancer through epigenetic mechanisms (Berner et al., 2010; Li Y et al., 2010; Nandakumar et al., 2011).

Concerning its effects on obesity, it has been reported that EGCG reduced them RNA levels of several gluconeogenic enzymes, glucose-6-phosphatase (G6Pase) and phosphoenolpyruvate carboxykinase (PEPCK) in the normal mouse liver as well as in the intestine (Yasui et al., 2011a, b). EGCG also improves cholesterol metabolism through the up-regulation of LDL receptor and also reduces extracellular apoB levels (Goto et al., 2011). Finally, it appears that EGCG modulates body weight gain in high fat-fed mice both by increasing the expression of genes related to fat oxidation in skeletal muscle and by modulating fat absorption from the diet (Sae-Tan et al., 2011).

3. However, not everything is positive: Side effects of antioxidants

Despite the initial positive and beneficial effects observed in many studies (some of them mentioned in the first part of this chapter), not all that glitters is gold. Thus, other clinical studies investigating antioxidant effects have been often disappointing given the consistent and promising findings from experimental investigations, clinical observations and epidemiological data. In this context, there are some controversial results, especially in the field of antioxidant supplementation, cancer and cardiovascular events (and mortality associated with these events) as well as when assessing the direct effects of antioxidants on mitochondria which are the main sources of reactive species in the organism apart from NADPH oxidase in the vascular walls. In this context, clinical trials of antioxidant therapeutics in human volunteers have produced negative or inconclusive results or have shown very little benefit. The inability of clinical trials to prove the usefulness of antioxidant therapies shows the failure in translating our knowledge of molecular and cellular mechanisms into efficient clinical remedies (Firuzi et al., 2011). The reason of clinical failure of many antioxidants despite the existence of overwhelming evidence on the involvement of oxidative damage in various pathologies still remains elusive although it is interesting to note that most of these studies generally agree on the notion that antioxidants are much more effective in prevention of disease rather than in the treatment of an already established active pathology (reviewed by Firuzi et al., 2011). In this context, we will review in the following part of the chapter some of these studies where no positive effects where found with the aforementioned antioxidants (lipoic acid, resveratrol, vitamins etc) and we will summarize some potential explanations for these controversial data.

3.1 Neutral or even deleterious effects of antioxidants

3.1.1 Lipoic acid

Some studies concerning the prooxidant potential of LA and DHLA have been performed in recent years. In fact, DHLA exerts prooxidant actions by accelerated iron-dependent hydroxyl radical generation and lipid peroxidation in liposomes, probably by reducing Fe^{3+} to Fe^{2+} (Scott et al., 1994). A study also concluded that LA and DHLA have prooxidant properties on markers of protein oxidation such as protein thiol and carbonyl in heart muscle of aging rat (Cakatay et al., 2005). In addition, DHLA stimulates MPT (mitochondrial permeability transition) by increasing production of ROS in isolated rat liver mitochondria and bovine heart submitochondrial particles (Morkunaite-Haimi et al., 2003). In this sense, Valdecantos et al. (2010a) also found that LA inhibited glutathione peroxidase activity and induced the uncoupling of the electron transport chain suggesting prooxidant actions of this antioxidant under the experimental conditions established in this study (Valdecantos et al., 2010a).

It is also very interesting to know that the beneficial role of LA supplementation in Type 2 diabetes is controversial. In one way, it has been postulated that the beneficial effects could be manifested by a mild prooxidant activity of the compound, leading to cellular adaptation against oxidative stress in addition to the attenuation of reductive stress in diabetes (Roy et al., 1997). In fact, the results derived from the study of Moini et al. (2002) pointed to the fact that the oxidized form of LA activates the insulin signal transduction pathway by acting as a prooxidant (Moini et al., 2002). Lipoic Acid increased tyrosine phosphorylation of immunoprecipitated insulin receptors, presumably by oxidation of critical thiol groups present in the insulin receptor β-subunit. Furthermore, it has been demonstrated that short-term incubation of LA in 3T3-L1 adipocytes induced glucose uptake by facilitating oxidative stress (Krieger-Brauer et al., 2000). However, long-term incubation of 3T3-L1 adipocytes with LA increased intracellular glutathione levels and inhibited the rate of glucose uptake (Mottley & Mason, 2001; Moini et al., 2002), which suggests that the duration of LA treatment is a critical step when analyzing the effects of LA on glucose uptake and insulin sensitivity. In addition, the effects of LA on adiponectin, a key adipokine involved in insulin sensitivity are also controversial, which does not help to postulate if LA beneficial actions on insulin sensitivity are mediated through this adipokine. Thus, Cummings et al. (2010) did not observe any significant change in fasting plasma adiponectin levels in fructose-fed University of California, Davis-Type 2 diabetes mellitus (UCD-T2DM) rats after dietary LA supplementation (Cummings et al., 2010). But not only the effects of LA on adiponectin are controversial, but also its actions on diabetes and obesity are questionable since several studies have not found any positive effects of this antioxidant on these disorders. Thus, supplementation with LA did not exhibit any effect on the lipid profile or insulin sensitivity of patients with diabetes type 2, with no changes in the concentrations of total cholesterol, cholesterol fractions, TG, and HOMA index (de Oliveira et al., 2011). Furthermore, LA administered orally at this dose for 2 weeks did not protect against lipid-induced insulin resistance in overweight and obese humans (Xiao et al., 2011). In a pilot study with adolescents with type 1 diabetes mellitus LA was not an effective treatment for decreasing oxidative damage, total antioxidant status HbA1c or microalbuminuria in type 1 diabetes mellitus (Huang & Gitelman, 2008). In addition, other studies did not even observe the ability of LA to induce weight loss in obese subjects (Koh et al., 2011).

The experiment design is also another point to take into account when describing the actions of LA, as it can influence the sense of the data obtained. Thus, Volchegorskii et al. (2011) studied the correlation between the effect of α-lipoic acid, emoxipin, reamberin, and mexidol on LPO *in vitro* and the action of these drugs on insulin sensitivity and tolerance to glucose load *in vivo*. They found that the preparations producing prooxidant effect *in vitro* (α-lipoic acid and reamberin) are characterized by pronounced insulin-potentiating activity, but only slightly increase (α-lipoic acid) or even decrease (reamberin) tolerance to glucose load suggesting controversial effects depending on experimental procedure: *in vitro vs. in vivo* (Volchegorskii et al., 2011). In this sense, we have also found that LA exerted direct effects on mitochondria oxidative status in a prooxidant manner (Valdecantos et al., 2010a) whereas we also observed that LA increases hepatic mitochondrial defenses through Foxo3a in a diet-induced obesity rat model (Valdecantos et al., 2011a) corroborating the controversial actions found for this fatty acid depending on the experimental procedures.

Finally, it is important to state that the ability of LA and/or DHLA to function as either anti- or prooxidants, at least in part, is also determined by the type of oxidant stress and the physiological circumstances. These prooxidant actions suggest that LA and DHLA act by multiple mechanisms, many of which are only now being explored and it is interesting to declare that prooxidant actions does not necessary mean deleterious effects as previously described for this antioxidant. In fact, α-Lipoic acid was shown to stimulate glucose uptake into 3T3-L1 adipocytes by increasing intracellular oxidant levels and/or facilitating insulin receptor autophosphorylation presumably by oxidation of critical thiol groups present in the insulin receptor β-subunit. Thus, the real meaning of the antioxidant or prooxidant effects of LA as well as the compounds described in this chapter warrants further investigation.

Lipoic Acid has been reported to have a number of potentially beneficial effects in both prevention and treatment of oxygen- related diseases. Selection of appropriate pharmacological doses of LA for use in oxygen-related diseases is also critical apart from experimental design and duration of treatment as previously described. Thus, in further studies, careful evaluation will be necessary for the decision in the biological system whether LA administration is beneficial or harmful (Cakatay, 2006).

3.1.2 Resveratrol

As mentioned before, many beneficial effects on health have been ascribed to this molecule. However, it should be emphasised that a great deal of work has been developed in isolated cells thus limiting the extrapolation of the results to the *in vivo* situation. In this context, Pérez-Jiménez and Saura-Calixto (2008) have reviewed the *in vivo* trials published during the last 13 years (seventy five trials) were the effects of different grape products on different CVD risk factors have been evaluated (Perez-Jimenez & Saura-Calixto, 2008). Most published studies have dealt with some specific aspects of mechanisms of grape flavonoid action or have focused only on one product, such as wine. Thus, it is important to point out that not only resveratrol actions have been evaluated in these trials but also polyphenols alcohol and dietary fibre have been tested. In animal and human studies, grape products have been shown to produce hypotensive, hypolipidaemic and anti-atherosclerotic effects and also to improve antioxidant status as measured in terms of plasma antioxidant capacity oxidation biomarkers, antioxidant compounds or antioxidant enzymes. However, there are several studies where neutral and even negative effects were found regarding its effects on lipid profile and markers of oxidative stress (reviewed by Pérez-Jiménez & Saura-Calixto 2008). It is important to underline that differences in the design of the studies and in the composition of the tested products (not always provided) could explain the different results observed and therefore these results can not been strictly extrapolated to resveratrol actions.

Despite its potential as an anti-obesity compound, data regarding the effects of resveratrol on adipokines are still insufficient to be conclusive. Adipokines are bioactive peptides produced by adipose tissue and involved in the physiological regulation of fat storage energy metabolism, food intake, insulin sensitivity, and immune function among others Several trials have observed that oxidative stress caused dysregulated production of adipokines (Soares et al., 2005; Kamigaki et al., 2006), therefore, it could be very important in the future to analyze the effects of resveratrol on these adipokines in an attempt to restore the optimal concentrations of those which, in turn, could lead to an improvement in obesity and related disorders.

Finally and although long-term effects of using resveratrol are still unknown, it is fair to state that this antioxidant shows a very good profile and could be a potential therapy against a wide range of diseases related to oxidative stress and aging (through SIRT1 actions), although more studies are needed in this field.

3.1.3 Vitamins E and C

Vitamins were selected for antioxidant therapy in several studies in the past decades because they were cheap and available, but they are not the best antioxidant molecules in terms of efficacy. In fact, many studies agree on the lack of evidence on the beneficial effects of antioxidant vitamins and in some cases even point to harmful effects. Thus, observational studies have reported an inverse association between vitamin E and cardiometabolic risk, but also, results from trials studying supplementation with this antioxidant failed to confirm any protective effect of them on cardiovascular disease (Devaraj et al., 2007; Wu et al., 2007).

In the review of Bjelakovic et al. (2007), 68 randomized trials conducted on 232,606 adults who were randomized to receive commonly used antioxidants including β-carotene, selenium, vitamins A, C and E were analyzed for the effect of antioxidant on all cause mortality (Bjelakovic et al., 2007). This review followed the Cochrane Collaboration method and included primary (healthy subjects) and secondary (diseased individuals) prevention studies. When all trials were considered, antioxidants did not seem to significantly affect mortality. However, when 47 "low-bias" trials were separately analyzed, β -carotene, vitamin A and vitamin E administered alone or in combination, significantly enhanced all-cause mortality whereas Vitamin C and selenium did not have any significant effect on mortality. Another meta-analysis performed on 7 large trials of vitamin E involving 81,788 individuals showed that there was no significant difference in cardiovascular mortality when individuals receiving vitamin E were compared to control (Vivekananthan et al., 2003). In another large meta-analysis including 19 trials and 135,967subjects, it was shown that high dose intake of vitamin E (>400IU/day) may increase all-cause mortality (Miller et al., 2005). However, other authors have claimed that the increase in mortality caused by vitamin E is questionable. Large secondary prevention trials of vitamin E including Secondary Prevention with Antioxidants of Cardiovascular Disease in Endstage Renal Disease (SPACE), the Cambridge Heart AntioxidantStudy (CHAOS), the Heart Outcomes Prevention Evaluation (HOPE), Gruppo Italiano per lo Studio de lla Sopravvianzan ell'Infarto Miocardico (GISSI) have evaluated the effects of vitamin E on mortality rates. In a meta-analysis of these trials and other primary and secondary prevention trials, it was concluded that vitamin E supplementation did not significantly affect mortality or risk of cardiovascular diseases (reviewed by Firuzi et al., 2011).

Numerous assays demonstrated that vitamin E decreased atherosclerotic formation (Fruebis et al., 1995; Parker et al., 1995), however, other studies showed no effects on plasma lipids (Nagyova et al., 2002; Cyrus et al., 2003; Hasty et al., 2007) or even an increase in plasma lipids after vitamin E treatment was also observed (Crawford et al., 1998). Mechanistic studies demonstrated that the role of α-tocopherol during the early stages of lipoprotein lipid peroxidation is complex and that the vitamin does not act as a chain-breaking antioxidant (Stocker & Keaney, 2005). It is tempting to suggest that the positive or deleterious effects of vitamin E supplementation or treatment on lipid profile also depend on the population chosen, the study design, types and dosages of antioxidant, and their duration of use. All these factors make the comparison and interpretation of the studies difficult. In addition, in a very recent study, it was demonstrated that vitamin E did not perform any positive effect on heat stress in Japanese quails (Halici et al., 2011). Moreover, there are conflicting results regarding the effects of this vitamin on blood pressure (Plantinga et al., 2007; Ward et al., 2007; Rodrigo et al., 2008).

Apart from the ambiguous effects observed after vitamin E treatment on cardiovascular events and mortality, its effects on cancer are not very clear either. Thus, the Alpha-Tocopherol Beta-Carotene Cancer Prevention Study (ATBC) and the β-Caroteneand Retinol Efficacy Trial, especially on lung cancers did not observe reduction in the incidence of lung cancer among male smokers after five to eight years of dietary supplementation with alpha-tocopherol or β-carotene. In fact, these trials raise the possibility that these supplements may actually have harmful as well as beneficial effects.

Finally, the evidence also suggests no beneficial effect of vitamin E supplementation in improving glycaemic control in unselected patients with type 2 diabetes whereas haemoglobin A$(_{1c})$ (HbA$(_{1c})$) (deeply involved in microvascular complications of diabetes and possibly macrovascular disease) may decrease with vitamin E supplementation in patients with inadequate glycaemic control or low serum levels of vitamin E. This shows the importance of targeting therapy. Due to the limitations of the available evidence, further studies are warranted in the field of vitamin E actions on diabetes and obesity (Suksomboon et al. 2011). On the other way and despite the beneficial effects previously described for vitamin E in obesity, there are also different studies where no significant effects of vitamin E on obesity have been found. Thus, body mass index remained unchanged in patients after 3 months of vitamin E treatment (Skrha et al., 1999; Nagyova et al., 2002; Vincent et al., 2007). Different research groups have examined de effect of vitamin E on F$_2$-isoprostanes (markers of oxidative stress which are increased in obesity), and whereas some of these research groups found statistically significant reductions in F_2-isoprostanes (Kaikkonen et al., 2001; Block et al., 2008), other studies did not find any effect (Meagher & Rader, 2001; Weinberg et al., 2001). Factors that could influence these conflicting results could be the sample size, the degree of obesity and/or presence of elevated F_2-isoprostanes at baseline.

Concerning vitamin C, several studies have also showed controversial results. Thus, the National Health and Nutrition Examination Surveys (NEHENES) reported that low serum levels of Vit C were marginally associated with an increased risk of fatal cardiovascular disease and significantly associated with risk of fatal cardiovascular disease (Schleicher et al., 2009). In contrast, several studies did not find evidence for a protective effect of vitamin C against cardiovascular disease. Thus, Ramos and Martinez-Castelao., (2008) (Ramos & Martinez-Castelao, 2008) failed to demonstrate significant differences on lipoprotein oxidation between vitamin C-treated and not treated patients under haemodialysis. Moreover, other studies found that Vit C further than having benefitial effects it also could have negative effects. Thus, The Physician Health Study (Gaziano et al., 2009) illustrated that vitamin C showed neither health benefits nor safety issues, and Moyad et al., (2008) reported that increased vitamin C intake had adverse effects, such as kidney stones and iron-related disorders. Other reports suggest that it may accelerate atherosclerosis in some people with diabetes, and fail to confer benefit in patients with advanced cancer (Talaulikar & Manyonda, 2011). In fact, vitamin C also seems to have a controversial role in cancer. Thus, many papers have described that millimolar concentrations of ascorbate have a deep inhibitory effect on the growth of several cancer cell lines in vitro. Actually, it seems that such cytotoxic activity of vitamin C relies on its ability to generate reactive oxygen species rather than its popular antioxidant action. This is paradoxical but, the fact is that ascorbic acid may have also prooxidant and even mutagenic effects in the presence of transition metals (reviewed by Verrax & Calderon, 2008). In this sense, Podmore et al., (1998) discovered an increase in a potentially mutagenic lesion, following a typical Vit C supplementation suggesting that prooxidant effects might occur at doses up to 500 mg per day, although at lower doses the antioxidant effect may predominate. In this sense, it is important to mention that the type, dosage and matrix of exogenous antioxidants seem to be determinant in the balance between beneficial or deleterious effects of vit. C. Briefly, the antioxidants in fruit and vegetables may be tightly bound within the tough fibrous material of these foodstuffs and may exert their antioxidant activity not in the blood or tissues but in the gastrointestinal tract where free radicals are constantly generated from food (Kelly et al., 2008) and on the contrary, vitamins ingested as food supplements are probably digested too quickly to replicate these effects. Moreover, in many cases, the equivalent serum levels of vitamin C cannot be achieved if the supplement is given orally since there is an upper limit for absorption of vitamin C of about 500 mg, which is why this is normally the highest dose given (Monsen, 2000). No acute toxic dose has been established but chronic toxicity can occur in those with hereditary glucose-6-phosphate dehydrogenase deficiency given doses of 2 g/day of this vitamin and some of the problems that can occur include kidney stones, diarrhoea, nausea, and red blood hemolysis. There is also the possibility of dental decalcification and rebound scurvy in infants born to women consuming large concentrations of vitamin C and estrogens changes in women (Soni et al., 2010).

Table 2 (modified from Firuzi et al., 2011) shows large meta analysis of randomized controlled clinical trials exploring the efficancy of vitamins E and C in prevention of various diseases (Alkhenizan & Hafez, 2007; Polyzos et al., 2007; Bardia et al., 2008; Evans, 2008; Arain & Qadeer, 2010; Myung et al., 2010). It is important to highlight that only large studies that included at least 4000 subjects were included in this table. Based on the studies summarized in this table and putting all the former findings together, we can led to the conclusion that vitamins cannot be used as effective antioxidant therapeutics for human diseases unless more definitive and comparative studies will be carried out.

Publication	Antioxidants studied	Number of randomized participants	Illness	Results	Conclusions
Arain et al., 2010	Vitamin E	94,069	Prevention of colorectal cancer	No significant effect on prevention of cancer.	No conclusive evidence on the benefit of treatment
Myung et al., 2010	Vitamin E, vitamin C, vitamin A, β-carotene, selenium (alone and in combination)	161,045	Prevention of cancer	No significant effect on prevention of cancer. No significant effect according to the type of antioxidant or type of cancer. Significant increase in the risk of bladder cancer in a subgroup meta-analysis of 4 trials.	No conclusive evidence on the benefit of treatment
Evans et al., 2008	B-carotene and α-tocopherol	23,099	Prevention of age related macular Degeneration (AMD)	No significant effect on prevention or delaying the onset of AMD (all trialsincluded). No significant effect when the analyses were restricted to either β-carotene or α-tocopherol.	No conclusive evidence on the benefit of treatment
Bardia et al., 2008	β-carotene, vitamin E, selenium	104,196	Prevention of cancer and mortality	Significant increase in cancer incidence and cancer mortality among smokers by β-carotene.	Selenium may be beneficial

Publication	Antioxidants studied	Number of randomized participants	Illness	Results	Conclusions
				Vitamin E supplementation had no effect. Selenium supplementation might have anticarcinogenic effects in men and thus requires further research.	
Alkhenizan et al., 2007	Vitamin E	167,025	Prevention of cancer	No significant difference in all-cause mortality, cancer incidence and cancer mortality. Significant reduction in the incidence of prostate cancer.	Beneficial for prostate cancer prevention. Not beneficial for other causes
Polyzos et al., 2007	Combination of vitamin C and vitamin E	4,680	Prevention of preeclampsia	No significant effect on the risk of preeclampsia, fetal or neonatal loss, or small for gestational age infant.	No conclusive evidence on the benefit of treatment

Table 2. Large meta analysis of randomized controlled clinical trials exploring the efficancy of vitamins E and C in prevention of various diseases.

3.1.4 Selenium

Despite the beneficial effects previously mentioned regarding this antioxidant, it is important to note that only selenium-deficient individuals may benefit from selenium supplementation, because such supplementation in selenium-replete individuals may even cause higher risk of diseases such as cancer (Brozmanova et al., 2010). Selenium has a narrow therapeutic window and there is considerable inter-individual variability in terms of metabolic sensitivity and optimal selenium intake. In fact, optimal intake for any individual is likely to depend on polymorphisms in selenoprotein genes that may also affect the risk of disease. Moreover, the baseline levels of each subject could determine the beneficial effect of the selenium intake (Stranges et al., 2010). For instance, no additional benefit of supplementation (even up to 300 µg/d) was found in an elderly population with mild hypothyroidism where selenium status was adequate prior to the start of supplementation (Rayman et al., 2008). High–selenium diets may stimulate the release of glucagon, promoting hyperglycaemia, or may induce over-expression of GPx-1 and other antioxidant selenoproteins resulting in insulin resistance and obesity (Stranges et al., 2010). Moreover,

he increase in developing diabetes or adverse lipid profile among the participants in the
NHANES 2003-2004 study could be associated to their high plasmatic selenium levels (137
ng/L) (Laclaustra et al., 2009).

Although the underlying mechanisms that could explain the detrimental effects of high
selenium are not fully understood yet, they could involve DNA damage and oxidative stress
induction resulting in apoptosis (Brozmanova et al., 2010). Therefore, due to a broad interest to
exploit the positive effects of selenium on human health, studies investigating the negative
effects such as toxicity and DNA damage induction resulting from high selenium intake are
also highly required (Brozmanova et al., 2010). Moreover, urgent need for personalized risk
prediction with regard to cancer and other diseases prevention and treatment activities of
selenium supplementation is highly suggested (Platz & Lippman, 2009).

3.1.5 Green tea

EGCG is regarded as the most active catechin in green tea, but, in spite of its reported
favorable effects, conflicting results have been reported from epidemiological studies
(Boehm et al., 2009) and EGCG appears to act both as an anti-oxidant as well as a prooxidant
agent as previously described for lipoic acid.

In this context, Elbling et al., (2005) concluded that excessive EGCG concentrations induced
toxic levels of ROS *in vivo*, and moreover, they found *in vitro* DNA-damaging effects at
pharmacological concentrations. Thus, hepatic and intestinal toxicities associated with the
consumption of high doses of green tea preparations were reported in animal studies.
Furthermore, EGCG mediated mitochondrial toxicity and ROS formation was implicated as
the possible mechanism for the cytotoxicity to isolated rat hepatocytes and hepatotoxicity in
mice (Galati et al., 2006). Another study found that higher intake of green tea might cause
oxidative DNA damage of hamster pancreas and liver and also found that the major
cytotoxic mechanism found with hepatocytes was mitochondrial membrane potential
collapse and ROS formation (Takabayashi et al., 2004). Moreover, Yun et al. (2006) clarified
that EGCG acts as a prooxidant, rather than an antioxidant, in pancreatic β cells *in vivo*,
suggesting that consumption of green tea and green tea extracts should be monitored in
certain patients. Thus, it should be considered that the effects of green tea and its
constituents may be beneficial up to a certain dose, and higher doses may cause some
unknown adverse effects similarly as what has been observed with selenium.

The harmful effects of tea overconsumption are due to three main factors: its caffeine
content, the presence of aluminum, and the effects of tea polyphenols on iron bioavailability.
Caffeine is the world's most popular drug and can be found in many beverages including
tea. One reason for the popularity of caffeine-containing beverages is the stimulation of the
central nervous system that they provide (MacKenzie et al., 2007). However, caffeine may
have other effects, including metabolic and hormonal ones. With short-term dosing, caffeine
has been shown to impair glucose metabolism in nondiabetic persons (Greer et al., 2001;
Johnston et al., 2003) and in persons with type 2 diabetes mellitus (Lane et al., 2004;
Robinson et al., 2004). The effects on other hormonal systems have not been as well
investigated. However, cortisol levels may increase after short-term administration of
caffeine in healthy subjects or in those with elevated blood pressure (Lovallo et al., 1996).
Regarding green tea aluminium content, several studies described that the negative effect of

green tea decoction, arises from the high absorption of aluminium released in the decoction Some analogies in the competition mechanism between aluminium and iron will be obtained in human nutritional conditions; the regular green tea decoction consumption could constitute an important additional source of dietary aluminium. Then, it could have in a long term, a negative consequence on iron status and erythropoiesis toxicity particularly in patients with high iron requirements or with chronic renal failure like hemodialysis (Marouani et al., 2007). It is also interesting to mention that an iron–catechin complex formation can cause a significant decrease of the iron bioavailability from the diet (Hamdaoui et al., 2003). Moreover, it has been shown that bioactive dietary polyphenols inhibit heme and non-heme iron absorption in human intestinal cells mainly by reducing basolateral release of iron (Kim, E. Y. et al., 2008).

3.2 Why many antioxidants have failed to show efficacy in interventional human studies? Some explanations

The most simple explanation for the controversial studies found is that not all antioxidants behave in the same way or with the same intensity, at least when their direct actions on mitochondria are analyzed as demonstrated in the study from Valdecantos et al. (2010a). Thus, the outcomes found with different antioxidants should be carefully examined since the physical properties of the assayed molecules are different and could affect their ability to enter the mitochondria and, therefore, to affect their functionality, although other mechanisms different from pure physical characteristics of the compounds can not be rule out.

Some of the antioxidants are ineffective and nonspecific and dosage regimen or duration of therapy was inefficient. Thus, several points should be taken into account before making general conclusions. Thus, the fact that the antioxidant molecule could have low bioavailability should also be considered when planning a trial. In this context, some polyphenolics, especially green tea catechins, may have very low bioavailability (Williamson &; Manach, 2005). Thus, optimization of these molecules has been suggested to improve this outcome, but, it is still under investigation. Other point that should also be taken into account is that the antioxidant could have poor target specificity, that the reaction products of the antioxidant could be toxic, that a single antioxidant is not enough to overcome oxidative stress and therefore a combination of several antioxidant compounds is needed or the fact that certain antioxidants are not effective in well-nourished populations (deeply reviewed by Firuzi et al., 2011).

Other possible reasons relate to patient cohort included in trials, that patients do not equally benefit from antioxidant therapy, the trial design itself and the usage of inappropriate or insensitive methodologies to evaluate oxidative state which underlines the urgent need for the development of sensitive and specific biomarkers to correctly assess the oxidant status of patients. Furthermore, oxidative stress is not always the primary cause of the disease and, therefore, it is not the only cause of the disease (reviewed by Firuzi et al., 2011).

4. Conclusion

There has been much enthusiasm in the field of oxidative-stress related disorders and nutritional approaches to improve health. Antioxidants have been advocated for therapy of a vast range of serious diseases in the 1980s and 1990s. Furthermore, the tendency to add bioactive compounds such as antioxidant molecules in foods to improve consumer health, which has been very strong during the past decade, will increase significantly in the future, in parallel with a growing awareness of the impact of food components on human health. However, in the light of recent negative findings, many doubts have now been raised about the usefulness of administration of single antioxidants. What seems to be clear is that although there are many dietary antioxidants and all of them can act as "antioxidant" molecules, not all behave in the same way. Thus and as described in this chapter, some of them seem to have potential as therapy against several diseases (resveratrol) whereas there are other molecules whose results are not very promising (vitamins C and/or E). Thus, once we apply our experience to select the right disease and the right population, design optimized and highly bioavailable antioxidants directed at specific and appropriate targets and choose optimal treatment times, duration and doses, useful therapeutics could emerge for various diseases. On the other hand, as possible negative interactions with antioxidants may rely on the dose consumed by each person, natural antioxidants from natural foods in a balanced diet such as the Mediterranean diet arise as the best way to implement these substances in regular nutrition instead of consuming them as supplements.

Since there are not yet adequately validated markers of the onset, progression and/or regression of any oxidative stress associated chronic diseases there is the urgent need in sorting out which markers or combinations of markers are predictive of human diseases. Ideally one would wish to demonstrate that modulation of a biomarker by a specific antioxidant intervention is predictive of modulation of incidence of some major chronic disease endpoint in humans. To accomplish this, further investigation is also needed.

Furthermore, inhibition of ROS production through the development of inhibitors against the main sources of ROS generation offers an alternative approach to conventional antioxidant therapies due to their controversial results. Thus, NADPH oxidase, as the main source of ROS production in endothelial cells and directly involved in hypertension and cardiovascular disease, has been suggested as a potential target for decreasing ROS generation. A number of clinically important drugs used for the treatment of hypertension, hypercholesterolaemia and coronary artery disease such as the statins, AT1 (angiotensin II receptor type 1) antagonists and ACE inhibitors have been shown to decrease NADPH oxidase-derived superoxide and ROS production. In this context, one area of investigation that has been the focus of much recent interest in the past years is to address mitochondria, and more specifically, to analyze the potential beneficial effects of modulating mitochondrial ROS generation in order to treat or prevent the development of several oxidative-stress associated disorders (reviewed by Pérez-Matute et al., 2009). Again, more studies are needed in this regard.

Finally, and as described in the review from Prieto-Hontoria et al. (2010), the mechanisms by which antioxidant components modulate obesity, cancer and other oxidative stress related disorders are not fully understood, partly because of the lack of appropriate research tools to identify the complex mechanisms involve. With the emergence of Nutrigenomics, it is now possible to exploit genome-wide changes in gene expression profiles related to molecular nutrition. Evolution of `omics´ such as epigenomics, transcriptomics, proteomics and metabolomics will allow a better understanding of how dietary antioxidants may affect both energy metabolism, carcinogenesis etc leading to healthier foods and, in turn, healthier people and lifestyles.

5. References

Aasheim, E.T., et al. (2008). Vitamin status in morbidly obese patients: a cross-sectional study. *Am J Clin Nutr*, 87(2), 362-9.

Alberdi, G., et al. (2011). Changes in white adipose tissue metabolism induced by resveratrol in rats. *Nutr Metab (Lond)*, 8(1), 29.

Alpsoy, L. & Yalvac, M.E. (2011). Key roles of vitamins A, C, and E in aflatoxin B1-induced oxidative stress. *Vitam Horm*, 86287-305.

Ansar, H., et al. (2011). Effect of alpha-lipoic acid on blood glucose, insulin resistance and glutathione peroxidase of type 2 diabetic patients. *Saudi Med J*, 32(6), 584-8.

Baur, J.A., et al. (2006). Resveratrol improves health and survival of mice on a high-calorie diet. *Nature*, 444(7117), 337-42.

Berner, C., et al. (2011). Epigenetic control of estrogen receptor expression and tumor suppressor genes is modulated by bioactive food compounds. *Ann Nutr Metab*, 57(3-4), 183-9.

Bishayee, A. (2009). Cancer prevention and treatment with resveratrol: from rodent studies to clinical trials. *Cancer Prev. Res. (Phila Pa.)* 2 409–418.

Bjelakovic, G., et al. (2007). Mortality in randomized trials of antioxidant supplements for primary and secondary prevention: systematic review and meta-analysis. *Jama*, 297(8), 842-57.

Block, G., et al. (2008). The effect of vitamins C and E on biomarkers of oxidative stress depends on baseline level. *Free Radic Biol Med*, 45(4), 377-84.

Boehm, K., et al. (2009). Green tea (Camellia sinensis) for the prevention of cancer. *Cochrane Database Syst Rev*, (3), CD005004.

Boosalis, M.G. (2008). The role of selenium in chronic disease. *Nutr Clin Pract*, 23(2), 152-60.

Boque, N., et al. (2009). Some cyclin-dependent kinase inhibitors-related genes are regulated by vitamin C in a model of diet-induced obesity. *Biol Pharm Bull*, 32(8), 1462-8.

Botella-Carretero, J.I., et al. (2010). Retinol and alpha-tocopherol in morbid obesity and nonalcoholic fatty liver disease. *Obes Surg*, 20(1), 69-76.

Brigelius-Flohe, R. & Traber, M.G. (1999). Vitamin E: function and metabolism. *Faseb J*, 13(10), 1145-55.

Brozmanova, J., et al. Selenium: a double-edged sword for defense and offence in cancer. *Arch Toxicol*, 84(12), 919-38.

Bursill, C.A., et al. (2007). A green tea extract lowers plasma cholesterol by inhibiting cholesterol synthesis and upregulating the LDL receptor in the cholesterol-fed rabbit. *Atherosclerosis*, 193(1), 86-93.

Cakatay, U. (2006). Pro-oxidant actions of alpha-lipoic acid and dihydrolipoic acid. *Med Hypotheses*, 66(1), 110-7.

Cakatay, U., et al. (2005). Prooxidant activities of alpha-lipoic acid on oxidative protein damage in the aging rat heart muscle. *Arch Gerontol Geriatr*, 40(3), 231-40.

Calvisi, D.F., et al. (2004). Vitamin E down-modulates iNOS and NADPH oxidase in c-Myc/TGF-alpha transgenic mouse model of liver cancer. *J Hepatol*, 41(5), 815-22.

Campion, J., et al. (2008). Vitamin C supplementation influences body fat mass and steroidogenesis-related genes when fed a high-fat diet. *Int J Vitam Nutr Res*, 78(2), 87-95.

Cao, H., et al. (2007). Green tea polyphenol extract regulates the expression of genes involved in glucose uptake and insulin signaling in rats fed a high fructose diet. *J Agric Food Chem*, 55(15), 6372-8.

Carbonelli, M.G., et al. (2011). Alpha-lipoic acid supplementation: a tool for obesity therapy? *Curr Pharm Des*, 16(7), 840-6.

Crawford, R.S., et al. (1998). Dietary antioxidants inhibit development of fatty streak lesions in the LDL receptor-deficient mouse. *Arterioscler Thromb Vasc Biol*, 18(9), 1506-13.

Crujeiras, A.B., et al. (2009). Fruit, vegetables and legumes consumption: Role in preventing and treating obesity. Bioactive foods in promoting health. Fruits and vegetables. R. R. Watson and B. R. Preedy. Oxford, Academic Press: 359-380.

Crujeiras, A.B., et al. (2007a). A hypocaloric diet enriched in legumes specifically mitigates lipid peroxidation in obese subjects. *Free Radic Res*, 41(4), 498-506.

Crujeiras, A.B., et al. (2007b). Functional properties of fruit. *Food*, 30-35.

Crujeiras, A.B., et al. (2006). A role for fruit content in energy-restricted diets in improving antioxidant status in obese women during weight loss. *Nutrition*, 22(6), 593-9.

Cucciolla, V., et al. (2007). Resveratrol: from basic science to the clinic. *Cell Cycle*, 6(20), 2495-510.

Cummings, B.P., et al. (2010). Dietary fructose accelerates the development of diabetes in UCD-T2DM rats: amelioration by the antioxidant, alpha-lipoic acid. *Am J Physiol Regul Integr Comp Physiol*, 298(5), R1343-50.

Cyrus, T., et al. (2003). Vitamin E reduces progression of atherosclerosis in low-density lipoprotein receptor-deficient mice with established vascular lesions. *Circulation*, 107(4), 521-3.

Cho, K.J., et al. (2003). Alpha-lipoic acid inhibits adipocyte differentiation by regulating pro-adipogenic transcription factors via mitogen-activated protein kinase pathways. *J Biol Chem*, 278(37), 34823-33.

Choi, S.Y., et al. (2009). Mechanism of alpha-lipoic acid-induced apoptosis of lung cancer cells. *Ann N Y Acad Sci*, 1171149-55.

Choo, J.J. (2003). Green tea reduces body fat accretion caused by high-fat diet in rats through beta-adrenoceptor activation of thermogenesis in brown adipose tissue. *J Nutr Biochem*, 14(11), 671-6.

Dai, J., et al. (2008). Association between adherence to the Mediterranean diet and oxidative stress. *Am J Clin Nutr*, 88(5), 1364-70.

Dalle-Donne, I., et al. (2006). Biomarkers of oxidative damage in human disease. *Clin Chem*, 52(4), 601-23.

Dauchet, L., et al. (2007). Dietary patterns and blood pressure change over 5-y follow-up in the SU.VI.MAX cohort. *Am J Clin Nutr*, 85(6), 1650-6.

Dawson, E.B., et al. (1990). Relationship between ascorbic acid and male fertility. *World Rev Nutr Diet*, 621-26.

de la Lastra, C.A. & Villegas, I. (2007). Resveratrol as an antioxidant and pro-oxidant agent: mechanisms and clinical implications. *Biochem Soc Trans*, 35(Pt 5), 1156-60.

de Oliveira, A.M., et al. (2011). The effects of lipoic acid and alpha-tocopherol supplementation on the lipid profile and insulin sensitivity of patients with type 2 diabetes mellitus: a randomized, double-blind, placebo-controlled trial. *Diabetes Res Clin Pract*, 92(2), 253-60.

Devaraj, S. & Jialal, I. (2000). Alpha tocopherol supplementation decreases serum C-reactive protein and monocyte interleukin-6 levels in normal volunteers and type 2 diabetic patients. *Free Radic Biol Med*, 29(8), 790-2.

Devaraj, S., et al. (2007). Effect of high-dose alpha-tocopherol supplementation on biomarkers of oxidative stress and inflammation and carotid atherosclerosis in patients with coronary artery disease. *Am J Clin Nutr*, 86(5), 1392-8.

Dhawan, A., et al. (2002). Evaluation of the antigenotoxic potential of monomeric and dimeric flavanols, and black tea polyphenols against heterocyclic amine-induced DNA damage in human lymphocytes using the Comet assay. *Mutat Res*, 515(1-2), 39-56.

Diniz, Y S, et al. (2005). Monosodium glutamate in standard and high-fiber diets: metabolic syndrome and oxidative stress in rats. *Nutrition*, 21(6), 749-55.

Droge, W. (2002). Free radicals in the physiological control of cell function. *Physiol Rev*, 82(1), 47-95.

Dudek, H., et al. (2005). [Concentration of glutathione (GSH), ascorbic acid (vitamin C) and substances reacting with thiobarbituric acid (TBA-rs) in single human brain metastases]. *Wiad Lek*, 58(7-8), 379-81.

Dulloo, A.G., et al. (1999). Efficacy of a green tea extract rich in catechin polyphenols and caffeine in increasing 24-h energy expenditure and fat oxidation in humans. *Am J Clin Nutr*, 70(6), 1040-5.

Elbling, L., et al. (2005). Green tea extract and (-)-epigallocatechin-3-gallate, the major tea catechin, exert oxidant but lack antioxidant activities. *Faseb J*, 19(7), 807-9.

Federici, M., et al. (2002). Insulin-dependent activation of endothelial nitric oxide synthase is impaired by O-linked glycosylation modification of signaling proteins in human coronary endothelial cells. *Circulation*, 106(4), 466-72.

Feige, J.N., et al. (2008). Specific SIRT1 activation mimics low energy levels and protects against diet-induced metabolic disorders by enhancing fat oxidation. *Cell Metab*, 8(5), 347-58.

Fetoui, H., et al. (2008). Oxidative stress induced by lambda-cyhalothrin (LTC) in rat erythrocytes and brain: Attenuation by vitamin C. *Environ Toxicol Pharmacol*, 26(2), 225-31.

Firuzi, O., et al. (2011). Antioxidant Therapy: Current Status and Future Prospects. *Curr Med Chem*,

Fischer-Posovszky, P., et al. (2010). Resveratrol regulates human adipocyte number and function in a Sirt1-dependent manner. *Am J Clin Nutr*, 92(1), 5-15.

Fito, M., et al. (2007). Effect of a traditional Mediterranean diet on lipoprotein oxidation: a randomized controlled trial. *Arch Intern Med*, 167(11), 1195-203.

Foo, N.P., et al. (2011). alpha-Lipoic acid inhibits liver fibrosis through the attenuation of ROS-triggered signaling in hepatic stellate cells activated by PDGF and TGF-beta. *Toxicology*, 282(1-2), 39-46.

Freise, C., et al. (2010). An active extract of Lindera obtusiloba inhibits adipogenesis via sustained Wnt signaling and exerts anti-inflammatory effects in the 3T3-L1 preadipocytes. *J Nutr Biochem*, 21(12), 1170-7.

Fruebis, J., et al. (1995). Effect of vitamin E on atherogenesis in LDL receptor-deficient rabbits. *Atherosclerosis*, 117(2), 217-24.

Galati, G., et al. (2006). Cellular and in vivo hepatotoxicity caused by green tea phenolic acids and catechins. *Free Radic Biol Med*, 40(4), 570-80.

Garriguet, D. (2010). The effect of supplement use on vitamin C intake. *Health Rep*, 21(1), 57-62.

Gaziano, J.M., et al. (2009). Vitamins E and C in the prevention of prostate and total cancer in men: the Physicians' Health Study II randomized controlled trial. *Jama*, 301(1), 52-62.

Gilgun-Sherki, Y., et al. (2002). Antioxidant therapy in acute central nervous system injury: current state. *Pharmacol Rev*, 54(2), 271-84.

Girard, A., et al. (2005). Changes in lipid metabolism and antioxidant defense status in spontaneously hypertensive rats and Wistar rats fed a diet enriched with fructose and saturated fatty acids. *Nutrition*, 21(2), 240-8.

Glantzounis, G.K., et al. (2005). Uric acid and oxidative stress. *Curr Pharm Des*, 11(32), 4145-51.

Gomez-Zorita, S., et al. (2011). Resveratrol attenuates steatosis in obese Zucker rats by decreasing fatty acid availability and reducing oxidative stress. *Br J Nutr*, 1-9.

Goto, T., et al. (2011). Epigallocatechin gallate changes mRNA expression level of genes involved in cholesterol metabolism in hepatocytes. *Br J Nutr*, 1-5.

Greer, F., et al. (2001). Caffeine ingestion decreases glucose disposal during a hyperinsulinemic-euglycemic clamp in sedentary humans. *Diabetes*, 50(10), 2349-54.

Gromadzinska, J., et al. (2008). Selenium and cancer: biomarkers of selenium status and molecular action of selenium supplements. *Eur J Nutr*, 47 Suppl 229-50.

Halici, M., et al. (2011). Effects of alpha-lipoic acid, vitamins E and C upon the heat stress in Japanese quails. *J Anim Physiol Anim Nutr (Berl)*,

Halliwell B, G.J. (2007). Free radicals in biology and medicine. Oxford, UK, Clarendon Press.

Halliwell, B. & Gutteridge, J.M. (1999). Free Radicals in Biology and Medicine. Oxford, Oxford University Press.

Halliwell, B. & Gutteridge, J.M. (2007). Free radicals in biology and medicine. Oxford, UK, Clarendon Press.

Hamdaoui, M.H., et al. (2003). Iron bioavailability and weight gains to iron-deficient rats fed a commonly consumed Tunisian meal 'bean seeds ragout' with or without beef and with green or black tea decoction. *J Trace Elem Med Biol*, 17(3), 159-64.

Hasegawa, N., et al. (2002). Vitamin C is one of the lipolytic substances in green tea. *Phytother Res*, 16 Suppl 1S91-2.

Hasegawa, N., et al. (2003). Powdered green tea has antilipogenic effect on Zucker rats fed a high-fat diet. *Phytother Res*, 17(5), 477-80.

Hasty, A.H., et al. (2007). Effects of vitamin E on oxidative stress and atherosclerosis in an obese hyperlipidemic mouse model. *J Nutr Biochem*, 18(2), 127-33.

Heneman, K.M., et al. (2007). Soy protein with and without isoflavones fails to substantially increase postprandial antioxidant capacity. *J Nutr Biochem*, 18(1), 46-53.

Heuckenkamp, P.U. & Zollner, N. (1971). Fructose-induced hyperuricaemia. *Lancet*, 1(7703), 808-9.

Horwitt, M.K. (1986). Interpretations of requirements for thiamin, riboflavin, niacin-tryptophan, and vitamin E plus comments on balance studies and vitamin B-6. *Am J Clin Nutr*, 44(6), 973-85.

Huang, E.A. & Gitelman, S.E. (2008). The effect of oral alpha-lipoic acid on oxidative stress in adolescents with type 1 diabetes mellitus. *Pediatr Diabetes*, 9(3 Pt 2), 69-73.

Hung, L.M., et al. (2000). Cardioprotective effect of resveratrol, a natural antioxidant derived from grapes. *Cardiovasc Res*, 47(3), 549-55.

Hung, P.F., et al. (2005). Antimitogenic effect of green tea (-)-epigallocatechin gallate on 3T3-L1 preadipocytes depends on the ERK and Cdk2 pathways. *Am J Physiol Cell Physiol*, 288(5), C1094-108.

Hunter, D.C., et al. (2011). Changes in markers of inflammation, antioxidant capacity and oxidative stress in smokers following consumption of milk, and milk supplemented with fruit and vegetable extracts and vitamin C. *Int J Food Sci Nutr*,

Huong, D.T. & Ide, T. (2008). Dietary lipoic acid-dependent changes in the activity and mRNA levels of hepatic lipogenic enzymes in rats. *Br J Nutr*, 100(1), 79-87.

Ihm, S.H., et al. (2009). Catechin prevents endothelial dysfunction in the prediabetic stage of OLETF rats by reducing vascular NADPH oxidase activity and expression *Atherosclerosis*, 206(1), 47-53.

James, J.S. (2006). Resveratrol: why it matters in HIV. *AIDS Treat News*, (420), 3-5.

Johnson, R.J., et al. (2009). Lessons from comparative physiology: could uric acid represent a physiologic alarm signal gone awry in western society? *J Comp Physiol B*, 179(1), 67-76.

Johnston, K.L., et al. (2003). Coffee acutely modifies gastrointestinal hormone secretion and glucose tolerance in humans: glycemic effects of chlorogenic acid and caffeine. *Am J Clin Nutr*, 78(4), 728-33.

Jorge, A.T., et al. A new potent natural antioxidant mixture provides global protection against oxidative skin cell damage. *Int J Cosmet Sci*, 33(2), 113-9.

Kaikkonen, J., et al. (2001). Supplementation with vitamin E but not with vitamin C lowers lipid peroxidation in vivo in mildly hypercholesterolemic men. *Free Radic Res*, 35(6) 967-78.

Kamigaki, M., et al. (2006). Oxidative stress provokes atherogenic changes in adipokine gene expression in 3T3-L1 adipocytes. *Biochem Biophys Res Commun*, 339(2), 624-32.

Kao, Y.H., et al. (2000). Modulation of endocrine systems and food intake by green tea epigallocatechin gallate. *Endocrinology*, 141(3), 980-7.

Kelly, R.P., et al. (2008). Lack of effect of acute oral ingestion of vitamin C on oxidative stress, arterial stiffness or blood pressure in healthy subjects. *Free Radic Res*, 42(5) 514-22.

Kim, E.Y., et al. (2008). Bioactive dietary polyphenolic compounds reduce nonheme iron transport across human intestinal cell monolayers. *J Nutr*, 138(9), 1647-51.

Kim, H.R., et al. (2008). Green tea protects rats against autoimmune arthritis by modulating disease-related immune events. *J Nutr*, 138(11), 2111-6.

Kim, M.S., et al. (2004). Anti-obesity effects of alpha-lipoic acid mediated by suppression of hypothalamic AMP-activated protein kinase. *Nat Med*, 10(7), 727-33.

Kobayashi, Y., et al. (2000). Green tea polyphenols inhibit the sodium-dependent glucose transporter of intestinal epithelial cells by a competitive mechanism. *J Agric Food Chem*, 48(11), 5618-23.

Koh, E.H., et al. (2011). Effects of alpha-lipoic Acid on body weight in obese subjects. *Am J Med*, 124(1), 85 e1-8.

Krieger-Brauer, H.I., et al. (2000). Basic fibroblast growth factor utilizes both types of component subunits of Gs for dual signaling in human adipocytes. Stimulation of adenylyl cyclase via Galph(s) and inhibition of NADPH oxidase by Gbeta gamma(s). *J Biol Chem*, 275(46), 35920-5.

Kryukov, G.V., et al. (2003). Characterization of mammalian selenoproteomes. *Science*, 300(5624), 1439-43.

Ku, H.C., et al. (2009). Green tea (-)-epigallocatechin gallate inhibits insulin stimulation of 3T3-L1 preadipocyte mitogenesis via the 67-kDa laminin receptor pathway. *Am J Physiol Cell Physiol*, 297(1), C121-32.

Laclaustra, M., et al. (2009). Serum selenium concentrations and diabetes in U.S. adults: National Health and Nutrition Examination Survey (NHANES) 2003-2004. *Environ Health Perspect*, 117(9), 1409-13.

Lane, J.D., et al. (2004). Caffeine impairs glucose metabolism in type 2 diabetes. *Diabetes Care*, 27(8), 2047-8.

Lasa, A., et al. (2011). Resveratrol regulates lipolysis via adipose triglyceride lipase. *J Nutr Biochem*,

Lee, M.S., et al. (2009). Green tea (-)-epigallocatechin-3-gallate reduces body weight with regulation of multiple genes expression in adipose tissue of diet-induced obese mice. *Ann Nutr Metab*, 54(2), 151-7.

Li, W., et al. (2011). Synergistic effects of tea polyphenols and ascorbic acid on human lung adenocarcinoma SPC-A-1 cells. *J Zhejiang Univ Sci B*, 11(6), 458-64.

Liao, S. (2001). The medicinal action of androgens and green tea epigallocatechin gallate. *Hong Kong Med J*, 7(4), 369-74.

Lin, J.N., et al. (2010). Resveratrol modulates tumor cell proliferation and protein translation via SIRT1-dependent AMPK activation. *J Agric Food Chem*, 58(3), 1584-92.

Lira, F.S., et al. (2011). Supplementing alpha-tocopherol (vitamin E) and vitamin D3 in high fat diet decrease IL-6 production in murine epididymal adipose tissue and 3T3-L1 adipocytes following LPS stimulation. *Lipids Health Dis*, 1037.

Liu, H.S., et al. (2006). Inhibitory effect of green tea (-)-epigallocatechin gallate on resistin gene expression in 3T3-L1 adipocytes depends on the ERK pathway. *Am J Physiol Endocrinol Metab*, 290(2), E273-81.

Lotito, S.B. & Frei, B. (2004). The increase in human plasma antioxidant capacity after apple consumption is due to the metabolic effect of fructose on urate, not apple-derived antioxidant flavonoids. *Free Radic Biol Med*, 37(2), 251-8.

Lovallo, W.R., et al. (1996). Stress-like adrenocorticotropin responses to caffeine in young healthy men. *Pharmacol Biochem Behav*, 55(3), 365-9.

Lushchak, VI. (2011). Adaptive response to oxidative stress: Bacteria, fungi, plants and animals. *Comp Biochem Physiol C Toxicol Pharmacol*, 153(2), 175-90.

Macarulla, M.T., et al. (2009). Effects of different doses of resveratrol on body fat and serum parameters in rats fed a hypercaloric diet. *J Physiol Biochem*, 65(4), 369-76.

MacKenzie, T., et al. (2007). Metabolic and hormonal effects of caffeine: randomized, double-blind, placebo-controlled crossover trial. *Metabolism*, 56(12), 1694-8.

Mah, E., et al. (2011). Vitamin C status is related to proinflammatory responses and impaired vascular endothelial function in healthy, college-aged lean and obese men. *J Am Diet Assoc*, 111(5), 737-43.

Manning, P.J., et al. (2004). Effect of high-dose vitamin E on insulin resistance and associated parameters in overweight subjects. *Diabetes Care*, 27(9), 2166-71.

Marouani, N., et al. (2007). Both aluminum and polyphenols in green tea decoction (Camellia sinensis) affect iron status and hematological parameters in rats. *Eur J Nutr*, 46(8), 453-9.

Martinez-Gonzalez, M.A., et al. (2002). Mediterranean diet and reduction in the risk of a first acute myocardial infarction: an operational healthy dietary score. *Eur J Nutr*, 41(4), 153-60.

Meagher, E. & Rader, D.J. (2001). Antioxidant therapy and atherosclerosis: animal and human studies. *Trends Cardiovasc Med*, 11(3-4), 162-5.

Meyer, F., et al. (2005). Antioxidant vitamin and mineral supplementation and prostate cancer prevention in the SU.VI.MAX trial. *Int J Cancer*, 116(2), 182-6.

Miller, E.R., 3rd, et al. (2006). The effects of macronutrients on blood pressure and lipids: an overview of the DASH and OmniHeart trials. *Curr Atheroscler Rep*, 8(6), 460-5.

Miller, E.R., 3rd, et al. (2005). Meta-analysis: high-dosage vitamin E supplementation may increase all-cause mortality. *Ann Intern Med*, 142(1), 37-46.

Moini, H., et al. (2002). Antioxidant and prooxidant activities of alpha-lipoic acid and dihydrolipoic acid. *Toxicol Appl Pharmacol*, 182(1), 84-90.

Monsen, E.R. (2000). Dietary reference intakes for the antioxidant nutrients: vitamin C, vitamin E, selenium, and carotenoids. *J Am Diet Assoc*, 100(6), 637-40.

Montagnani, M., et al. (2002). Insulin receptor substrate-1 and phosphoinositide-dependent kinase-1 are required for insulin-stimulated production of nitric oxide in endothelial cells. *Mol Endocrinol*, 16(8), 1931-42.

Morkunaite-Haimi, S., et al. (2003). Reactive oxygen species are involved in the stimulation of the mitochondrial permeability transition by dihydrolipoate. *Biochem Pharmacol*, 65(1), 43-9.

Mottley, C. & Mason, R.P. (2001). Sulfur-centered radical formation from the antioxidant dihydrolipoic acid. *J Biol Chem*, 276(46), 42677-83.

Moyad, M.A., et al. (2008). Vitamin C metabolites, independent of smoking status, significantly enhance leukocyte, but not plasma ascorbate concentrations. *Adv Ther*, 25(10), 995-1009.

Nagyova, A., et al. (2002). Serum ex vivo lipoprotein oxidizability in patients with ischemic heart disease supplemented with vitamin E. *Physiol Res*, 51(5), 457-64.

Nandakumar, V., et al. (2011). (-)-Epigallocatechin-3-gallate reactivates silenced tumor suppressor genes, Cip1/p21 and p16INK4a, by reducing DNA methylation and increasing histones acetylation in human skin cancer cells. *Carcinogenesis*, 32(4), 537-44.

Nantz, M.P., et al. (2009). Standardized capsule of Camellia sinensis lowers cardiovascular risk factors in a randomized, double-blind, placebo-controlled study. *Nutrition*, 25(2), 147-54.

Netke, S.P., et al. (1997). Ascorbic acid protects guinea pigs from acute aflatoxin toxicity. *Toxicol Appl Pharmacol*, 143(2), 429-35.

Neuzil, J., et al. (2011). Mitochondrially targeted -tocopheryl succinate is antiangiogenic: Potential benefit against tumor angiogenesis but caution against wound healing. *Antioxid Redox Signal*,

Nothlings, U., et al. (2008). Intake of vegetables, legumes, and fruit, and risk for all-cause, cardiovascular, and cancer mortality in a European diabetic population. *J Nutr*, 138(4), 775-81.

Novotny, L., et al. (2008). alpha-Lipoic acid: the potential for use in cancer therapy. *Neoplasma*, 55(2), 81-6.

Packer, L., et al. (1997). Alpha-lipoic acid: a metabolic antioxidant and potential redox modulator of transcription. *Adv Pharmacol*, 3879-101.

Packer, L. & Suzuki, Y.J. (1993). Vitamin E and alpha-lipoate: role in antioxidant recycling and activation of the NF-kappa B transcription factor. *Mol Aspects Med*, 14(3), 229-39.

Padmalayam, I., et al. (2009). Lipoic acid synthase (LASY): a novel role in inflammation, mitochondrial function, and insulin resistance. *Diabetes*, 58(3), 600-8.

Park, H.J., et al. (2011). Green tea extract attenuates hepatic steatosis by decreasing adipose lipogenesis and enhancing hepatic antioxidant defenses in ob/ob mice. *J Nutr Biochem*, 22(4), 393-400.

Parker, R.A., et al. (1995). Relation of vascular oxidative stress, alpha-tocopherol, and hypercholesterolemia to early atherosclerosis in hamsters. *Arterioscler Thromb Vasc Biol*, 15(3), 349-58.

Pauwels, E.K. (2011). The protective effect of the Mediterranean diet: focus on cancer and cardiovascular risk. *Med Princ Pract*, 20(2), 103-11.

Perez-Jimenez, J. & Saura-Calixto, F. (2008). Grape products and cardiovascular disease risk factors. *Nutr Res Rev*, 21(2), 158-73.

Perez-Matute, P., et al. (2009). Reactive species and diabetes: counteracting oxidative stress to improve health. *Curr Opin Pharmacol*, 9(6), 771-9.

Plantinga, Y., et al. (2007). Supplementation with vitamins C and E improves arterial stiffness and endothelial function in essential hypertensive patients. *Am J Hypertens*, 20(4), 392-7.

Platz, E.A. & Lippman, S.M. (2009). Selenium, genetic variation, and prostate cancer risk: epidemiology reflects back on selenium and vitamin E cancer prevention trial. *J Clin Oncol*, 27(22), 3569-72.

Plutner, H., et al. (1990). Synthetic peptides of the Rab effector domain inhibit vesicular transport through the secretory pathway. *Embo J*, 9(8), 2375-83.

Podmore, I.D., et al. (1998). Vitamin C exhibits pro-oxidant properties. *Nature*, 392(6676), 559.

Prieto-Hontoria, P.L., et al. (2009). Lipoic acid prevents body weight gain induced by a high fat diet in rats: effects on intestinal sugar transport. *J Physiol Biochem*, 65(1), 43-50.

Prieto-Hontoria, P.L., et al. (2010). Role of obesity-associated dysfunctional adipose tissue in cancer: a molecular nutrition approach. *Biochim Biophys Acta*, 1807(6), 664-78.

Ramos, R. & Martinez-Castelao, A. (2008). Lipoperoxidation and hemodialysis. *Metabolism*, 57(10), 1369-74.

Rayman, M.P. (2005). Selenium in cancer prevention: a review of the evidence and mechanism of action. *Proc Nutr Soc*, 64(4), 527-42.

Rayman, M.P., et al. (2008). Randomized controlled trial of the effect of selenium supplementation on thyroid function in the elderly in the United Kingdom. *Am J Clin Nutr*, 87(2), 370-8.

Reboul, E., et al. (2006). Bioaccessibility of carotenoids and vitamin E from their main dietary sources. *J Agric Food Chem*, 54(23), 8749-55.

Rickman, C., et al. (2010). Green tea attenuates cardiovascular remodelling and metabolic symptoms in high carbohydrate-fed rats. *Curr Pharm Biotechnol*, 11(8), 881-6.

Rietveld, A. & Wiseman, S. (2003). Antioxidant effects of tea: evidence from human clinical trials. *J Nutr*, 133(10), 3285S-3292S.

Robinson, L.E., et al. (2004). Caffeine ingestion before an oral glucose tolerance test impairs blood glucose management in men with type 2 diabetes. *J Nutr*, 134(10), 2528-33.

Rodrigo, R., et al. (2008). Decrease in oxidative stress through supplementation of vitamins C and E is associated with a reduction in blood pressure in patients with essential hypertension. *Clin Sci (Lond)*, 114(10), 625-34.

Roy, S., et al. (1997). Modulation of cellular reducing equivalent homeostasis by alpha-lipoic acid. Mechanisms and implications for diabetes and ischemic injury. *Biochem Pharmacol*, 53(3), 393-9.

Sae-Tan, S., et al. (2011). (-)-Epigallocatechin-3-gallate increases the expression of genes related to fat oxidation in the skeletal muscle of high fat-fed mice. *Food Funct*, 2(2), 111-6.

Sakurai, N., et al. (2009). (-)-Epigallocatechin gallate enhances the expression of genes related to insulin sensitivity and adipocyte differentiation in 3T3-L1 adipocytes at an early stage of differentiation. *Nutrition*, 25(10), 1047-56.

Savory, L.A., et al. (2011). Selenium Supplementation and Exercise: Effect on Oxidant Stress in Overweight Adults. *Obesity (Silver Spring)*,

Scott, B.C., et al. (1994). Lipoic and dihydrolipoic acids as antioxidants. A critical evaluation. *Free Radic Res*, 20(2), 119-33.

Schleicher, R.L., et al. (2009). Serum vitamin C and the prevalence of vitamin C deficiency in the United States: 2003-2004 National Health and Nutrition Examination Survey (NHANES). *Am J Clin Nutr*, 90(5), 1252-63.

Sebai, H., et al. (2010). Resveratrol, a red wine polyphenol, attenuates lipopolysaccharide-induced oxidative stress in rat liver. *Ecotoxicol Environ Saf*, 73(5), 1078-83.

Senen, D., et al. (2002). Contribution of vitamin C administration for increasing lipolysis. *Aesthetic Plast Surg*, 26(2), 123-5.

Serisier, S., et al. (2008). Effects of green tea on insulin sensitivity, lipid profile and expression of PPARalpha and PPARgamma and their target genes in obese dogs. *Br J Nutr*, 99(6), 1208-16.

Shargorodsky, M., et al. (2010). Effect of long-term treatment with antioxidants (vitamin C, vitamin E, coenzyme Q10 and selenium) on arterial compliance, humoral factors and inflammatory markers in patients with multiple cardiovascular risk factors. *Nutr Metab (Lond)*, 755.

Shay, K.P., et al. (2009). Alpha-lipoic acid as a dietary supplement: molecular mechanisms and therapeutic potential. *Biochim Biophys Acta*, 1790(10), 1149-60.

Shen, Q.W., et al. (2005). Effect of dietary alpha-lipoic acid on growth, body composition, muscle pH, and AMP-activated protein kinase phosphorylation in mice. *J Anim Sci*, 83(11), 2611-7.

Shen, W., et al. (2008a). A combination of nutriments improves mitochondrial biogenesis and function in skeletal muscle of type 2 diabetic Goto-Kakizaki rats. *PLoS One*, 3(6), e2328.

Shen, W., et al. (2008b). R-alpha-lipoic acid and acetyl-L-carnitine complementarily promote mitochondrial biogenesis in murine 3T3-L1 adipocytes. *Diabetologia*, 51(1), 165-74.

Shen, X., et al. (2009). Effect of vitamin E supplementation on oxidative stress in a rat model of diet-induced obesity. *Int J Vitam Nutr Res*, 79(4), 255-63.

Shi, D.Y., et al. (2008). Alpha-lipoic acid induces apoptosis in hepatoma cells via the PTEN/Akt pathway. *FEBS Lett*, 582(12), 1667-71.

Singh, U., et al. (2005). Vitamin E, oxidative stress, and inflammation. *Annu Rev Nutr*, 25151-74.

Skrha, J., et al. (1999). Insulin action and fibrinolysis influenced by vitamin E in obese Type 2 diabetes mellitus. *Diabetes Res Clin Pract*, 44(1), 27-33.

Smith, A.R., et al. (2004). Lipoic acid as a potential therapy for chronic diseases associated with oxidative stress. *Curr Med Chem*, 11(9), 1135-46.

Soares, A.F., et al. (2005). Effects of oxidative stress on adiponectin secretion and lactate production in 3T3-L1 adipocytes. *Free Radic Biol Med*, 38(7), 882-9.

Sola, S., et al. (2005). Irbesartan and lipoic acid improve endothelial function and reduce markers of inflammation in the metabolic syndrome: results of the Irbesartan and Lipoic Acid in Endothelial Dysfunction (ISLAND) study. *Circulation*, 111(3), 343-8.

Soltys, K., et al. (2001). Oxidative stress in fatty livers of obese Zucker rats: rapid amelioration and improved tolerance to warm ischemia with tocopherol. *Hepatology*, 34(1), 13-8.

Song, E.K., et al. (2003). Epigallocatechin gallate prevents autoimmune diabetes induced by multiple low doses of streptozotocin in mice. *Arch Pharm Res*, 26(7), 559-63.

Soni, M.G., et al. 2010. Safety of vitamins and minerals: controversies and perspective. *Toxicol Sci*, 118(2), 348-55.

Stefanovic, A., et al. (2008). The influence of obesity on the oxidative stress status and the concentration of leptin in type 2 diabetes mellitus patients. *Diabetes Res Clin Pract*, 79(1), 156-63.

Stellato, D., et al. (2011). Uric acid: a starring role in the intricate scenario of metabolic syndrome with cardio-renal damage? *Intern Emerg Med*,

Stewart, A.J., et al. (2005). On-line high-performance liquid chromatography analysis of the antioxidant activity of phenolic compounds in green and black tea. *Mol Nutr Food Res*, 49(1), 52-60.

Stocker, R. & Keaney, J.F., Jr. (2005). New insights on oxidative stress in the artery wall. *J Thromb Haemost*, 3(8), 1825-34.

Stranges, S., et al. (2010). Selenium status and cardiometabolic health: state of the evidence. *Nutr Metab Cardiovasc Dis*, 20(10), 754-60.

Strohle, A., et al. (2011). Micronutrients at the interface between inflammation and infection-ascorbic acid and calciferol: part 1, general overview with a focus on ascorbic acid. *Inflamm Allergy Drug Targets*, 10(1), 54-63.

Suksomboon, N., et al. (2011). Effects of vitamin E supplementation on glycaemic control in type 2 diabetes: systematic review of randomized controlled trials. *J Clin Pharm Ther*, 36(1), 53-63.

Sun, A.Y., et al. (2010). Resveratrol as a therapeutic agent for neurodegenerative diseases. *Mol Neurobiol*, 41(2-3), 375-83.

Sutherland, W.H., et al. (2007). Vitamin E supplementation and plasma 8-isoprostane and adiponectin in overweight subjects. *Obesity (Silver Spring)*, 15(2), 386-91.

Szkudelska, K., et al. (2009). The inhibitory effect of resveratrol on leptin secretion from rat adipocytes. *Eur J Clin Invest*, 39(10), 899-905.

Takabayashi, F., et al. (2004). Effect of green tea catechins on oxidative DNA damage of hamster pancreas and liver induced by N-Nitrosobis(2-oxopropyl)amine and/or oxidized soybean oil. *Biofactors*, 21(1-4), 335-7.

Talaulikar, V.S. & Manyonda, I.T. (2011). Vitamin C as an antioxidant supplement in women's health: a myth in need of urgent burial. *Eur J Obstet Gynecol Reprod Biol* 157(1), 10-3.

Taylor, A.G., et al. (2006). Inflammation and oxidative stess are associated with a nove dietary "phytochemical Index" in obese young adults. *North American Research Conference on Complementary and Alternative Medicine*, 24-27.

Touzet, O. & Philips, A. (2010). Resveratrol protects against protease inhibitor-induced reactive oxygen species production, reticulum stress and lipid raft perturbation *Aids*, 24(10), 1437-47.

Trumbo, P.R. (2005). The level of evidence for permitting a qualified health claim: FDA's review of the evidence for selenium and cancer and vitamin E and heart disease. *Nutr*, 135(2), 354-6.

Tudoran, O., et al. (2011). Early transcriptional pattern of angiogenesis induced by EGCG treatment in cervical tumor cells. *J Cell Mol Med*, (In press).

Ulker, S., et al. (2003). Vitamins reverse endothelial dysfunction through regulation of eNOS and NAD(P)H oxidase activities. *Hypertension*, 41(3), 534-9.

Um, J.H., et al. (2010). AMP-activated protein kinase-deficient mice are resistant to the metabolic effects of resveratrol. *Diabetes*, 59(3), 554-63.

Upritchard, J.E., et al. (2000). Effect of supplementation with tomato juice, vitamin E, and vitamin C on LDL oxidation and products of inflammatory activity in type 2 diabetes. *Diabetes Care*, 23(6), 733-8.

Vajro, P., et al. (2004). Vitamin E treatment in pediatric obesity-related liver disease: a randomized study. *J Pediatr Gastroenterol Nutr*, 38(1), 48-55.

Valdecantos, M.P., et al. (2010a). Vitamin C, resveratrol and lipoic acid actions on isolated rat liver mitochondria: all antioxidants but different. *Redox Rep*, 15(5), 207-16.

Valdecantos, M.P., et al. (2010b). Lipoic acid improves hepatic mitochondrial function in a model of obesity. *4th International Congress of Nutrigenetics and Nutrigenomics* Pamplona, Spain.

Valdecantos, M., P., et al. (2011a). Lipoic acid increases hepatic mitochondrial defenses through Foxo3a deacetylation by SIRT3 in a diet-induced obesity rat model. *8th MiPconference on Mitochondrial Physiology and Pathology*, Bordeaux, France.

Valdecantos, M.P., et al. (2011b). Lipoic acid prevents fatty liver in a diet-induced obesity rat model. *19th European Congress of Obesity*, Istambul (Turkey).

Valko, M., et al. (2007). Free radicals and antioxidants in normal physiological functions and human disease. *Int J Biochem Cell Biol*, 39(1), 44-84.

Van Gossum, A., et al. (1988). Increased lipid peroxidation after lipid infusion as measured by breath pentane output. *Am J Clin Nutr*, 48(6), 1394-9.

Vanamala, J., et al. (2010). Resveratrol suppresses IGF-1 induced human colon cancer cell proliferation and elevates apoptosis via suppression of IGF-1R/Wnt and activation of p53 signaling pathways. *BMC Cancer*, 10238.

Verrax, J. & Calderon, PB. (2008). The controversial place of vitamin C in cancer treatment *Biochem Pharmacol*, 76(12),1644-52.

Vincent, H.K., et al. (2006). Antioxidant supplementation lowers exercise-induced oxidative stress in young overweight adults. *Obesity (Silver Spring)*, 14(12), 2224-35.

Vincent, H.K., et al. (2007). Oxidative stress and potential interventions to reduce oxidative stress in overweight and obesity. *Diabetes Obes Metab*, 9(6), 813-39.

Visioli, F., et al. (2002). Lipoic acid and vitamin C potentiate nitric oxide synthesis in human aortic endothelial cells independently of cellular glutathione status. *Redox Rep*, 7(4), 223-7.

Vivekananthan, D.P., et al. (2003). Use of antioxidant vitamins for the prevention of cardiovascular disease: meta-analysis of randomised trials. *Lancet*, 361(9374), 2017-23.

Volchegorskii, I.A., et al. (2011). Effect of pro- and antioxidants on insulin sensitivity and glucose tolerance. *Bull Exp Biol Med*, 150(3), 327-32.

Waltner-Law, M.E., et al. (2002). Epigallocatechin gallate, a constituent of green tea, represses hepatic glucose production. *J Biol Chem*, 277(38), 34933-40.

Wang, H., et al. (2011). Obesity Modifies the Relations Between Serum Markers of Dairy Fats and Inflammation and Oxidative Stress Among Adolescents. *Obesity (Silver Spring)*, doi: 10.1038/oby.2011.234.

Wang, X. & Tian, W. (2001). Green tea epigallocatechin gallate: a natural inhibitor of fatty-acid synthase. *Biochem Biophys Res Commun*, 288(5), 1200-6.

Wang, Y., et al. (2010). alpha-Lipoic acid increases energy expenditure by enhancing adenosine monophosphate-activated protein kinase-peroxisome proliferator-activated receptor-gamma coactivator-1alpha signaling in the skeletal muscle of aged mice. *Metabolism*, 59(7), 967-76.

Wang, Y., et al. (2006). (-)-Epigallocatechin-3-gallate protects mice from concanavalin A-induced hepatitis through suppressing immune-mediated liver injury. *Clin Exp Immunol*, 145(3), 485-92.

Ward, N.C., et al. (2007). The effect of vitamin E on blood pressure in individuals with type 2 diabetes: a randomized, double-blind, placebo-controlled trial. *J Hypertens*, 25(1), 227-34.

Wei, H. & Meng, Z. (2010). Protective effects of epigallocatechin-3-gallate against lead-induced oxidative damage. *Hum Exp Toxicol*,

Weinberg, R.B., et al. (2001). Pro-oxidant effect of vitamin E in cigarette smokers consuming a high polyunsaturated fat diet. *Arterioscler Thromb Vasc Biol*, 21(6), 1029-33.

Williamson, G. & Manach, C. (2005). Bioavailability and bioefficacy of polyphenols in humans. II. Review of 93 intervention studies. *Am J Clin Nutr*, 81(1 Suppl), 243S-255S.

Wu, B.T., et al. (2005). The apoptotic effect of green tea (-)-epigallocatechin gallate on 3T3-L1 preadipocytes depends on the Cdk2 pathway. *J Agric Food Chem*, 53(14), 5695-701.

Wu, J.H., et al. (2007). Effects of alpha-tocopherol and mixed tocopherol supplementation on markers of oxidative stress and inflammation in type 2 diabetes. *Clin Chem*, 53(3), 511-9.

Xiao, C., et al. (2011). Short-term oral {alpha}-lipoic acid does not prevent lipid-induced dysregulation of glucose homeostasis in obese and overweight non-diabetic men. *Am J Physiol Endocrinol Metab*,

Yasui, K., et al. (2011a). Effects of a catechin-free fraction derived from green tea on gene expression of gluconeogenic enzymes in rat hepatoma H4IIE cells and in the mouse liver. *Biomed Res*, 32(2), 119-25.

Yasui, K., et al. (2011b). Effects of (-)-epigallocatechin-3-O-gallate on expression of gluconeogenesisrelated genes in the mouse duodenum. *Biomed Res*, 32(5), 313-20.

Yun, S.Y., et al. (2006). Effects of (-)-epigallocatechin-3-gallate on pancreatic beta-cell damage in streptozotocin-induced diabetic rats. *Eur J Pharmacol*, 541(1-2), 115-21.

Zhang, H.S., et al. (2009). Resveratrol inhibited Tat-induced HIV-1 LTR transactivation via NAD(+)-dependent SIRT1 activity. *Life Sci*, 85(13-14), 484-9.

Zhao, B. (2009). Natural antioxidants protect neurons in Alzheimer's disease and Parkinson's disease. *Neurochem Res*, 34(4), 630-8.

Zhong, Y. & Shahidi, F. (2011). Lipophilized epigallocatechin gallate (EGCG) derivatives as novel antioxidants. *J Agric Food Chem*, 59(12), 6526-33.

Zingg, JM., et al. (2010). Modulation of gene expression by alpha-tocopherol and alpha-tocopheryl phosphate in THP-1 monocytes. *Free Radic Biol Med*, 49(12), 1989-2000.

Zingg, JM., & Azzi A. (2004). Non-antioxidant activities of vitamin E. *Curr Med Chem*, 11(9), 1113-33.

References from table 2

Alkhenizan, A. & Hafez, K. (2007). The role of vitamin E in the prevention of cancer: a meta-analysis of randomized controlled trials. *Ann Saudi Med*, 27(6), 409-14.

Arain, M.A. & Abdul Qadeer, A. (2010). Systematic review on "vitamin E and prevention of colorectal cancer". *Pak J Pharm Sci*, 23(2), 125-30.

Bardia, A., et al. (2008). Efficacy of antioxidant supplementation in reducing primary cancer incidence and mortality: systematic review and meta-analysis. *Mayo Clin Proc*, 83(1), 23-34.

Evans, J. (2008). Antioxidant supplements to prevent or slow down the progression of AMD: a systematic review and meta-analysis. *Eye (Lond)*, 22(6), 751-60.

Myung, S.K., et al. (2010). Effects of antioxidant supplements on cancer prevention: meta-analysis of randomized controlled trials. *Ann Oncol*, 21(1), 166-79.

Polyzos, N.P., et al. (2007). Combined vitamin C and E supplementation during pregnancy for preeclampsia prevention: a systematic review. *Obstet Gynecol Surv*, 62(3), 202-6.

12

Microalgae of the Chlorophyceae Class: Potential Nutraceuticals Reducing Oxidative Stress Intensity and Cellular Damage

Blas-Valdivia Vanessa, Ortiz-Butron Rocio,
Rodriguez-Sanchez Ruth, Torres-Manzo Paola,
Hernandez-Garcia Adelaida and Cano-Europa Edgar
Escuela Nacional de Ciencias Biológicas,
Instituto Politécnico Nacional Departamento de Fisiología,
México

1. Introduction

Nutraceutical is a term combining the words nutrition and pharmaceutical. It is a food or food product that provides health and medical benefits, including the prevention and treatment of disease. A nutraceutical has beneficial effects because it possesses many compounds with antioxidant and intracellular signalling-pathway modulator effects. In recent years, it has been demonstrated that microalgae of the Chlorophyceae class could be excellent nutraceuticals because they contain polyphenols, chlorophyll, β-carotene, ascorbic acid, lycopene, α-tocopherol, xanthophylls, and PUFAs. For this reason, some research groups, including ours, have studied the nutraceutical properties of the genera *Dunalliela*, *Haematococcus*, and *Chlorella*. However, our research group has put special emphasis on the genera *Chlorella* and *Chlamydomonas*. For these genera, we present new results that reveal antioxidant effects in different models of oxidative stress and cell damage

2. Nutraceuticals

For a long time, natural products obtained from plants have been used as prominent sources of prophylactic agents for the prevention and treatment of disease in humans, animals, and in plants. Hippocrates (460-370 BC) started "let food be your medicine and medicine be your food". Now, the relationship between food and drugs is getting closer.

As we enter the third millennium, with increased life expectancy and greater media coverage of the health care issue, consumers are understandably more interested in the potential benefits of nutritional support for disease control or prevention. A recent survey in Europe concluded that diet is rated more highly by consumers than exercise or the hereditary factor for achieving good health (Hardy, 2000). For that reason, many entrepreneurs seek to introduce different products into the health and nutritional market. Marketing strategies have exploited the words "functional food" and "nutraceuticals" in their advertisements. Nutraceuticals and functional foods are the fastest growing segment of today´s food industry, although nutraceuticals should be treated as pharmaceutical products as we will detail. Nutraceuticals and functional foods are a market estimated at between $6 billion US and $60 billion US and it is growing at 5% per annum. Unfortunately, entrepreneurs in an effort to make money attract, as irresponsible market entrants, products that do not comply with biosafety tests. This is because there are few laws that regulate the production and sale of such products. Because the products are not submitted for standardized toxicology testing, sometimes they may be toxic for human consumption. There are no specific regulation in any country to control nutraceuticals, and they need to be established and should be considered under the same laws that regulate pharmaceuticals and food (Bernal et al., 2011). For our purposes, we will first define "nutraceuticals" and "functional foods" and how the microalgae could be excellent nutraceuticals.

The term nutraceutical was first mentioned in 1989 to describe the union between nutrition and pharmaceuticals, both key contributors to human wellness. Stephen DeFelice MD is the founder and chairman of the Foundation for Innovation in Medicine (FIM) and he defined a nutraceutical as a food (or part of the food) that provides medicinal health benefits, including the prevention or treatment of a disease. It was proposed that a nutraceutical is not a drug, which is a pharmacologically active substance that potentiates, antagonizes, or otherwise modifies any physiological function. A nutraceutical may be a single natural nutrient in powder, tablet, capsule, or liquid form. It is not necessarily a complete food but equally not a drug (Hardy, 2000). Also, it was proposed that a nutraceutical is a product that delivers a concentrated form of a presumed bioactive agent from a food, presented in a nonfood matrix, and it is used with the purpose of enhancing health in a dosage that exceeds those that could be obtained from normal food (Zeisel, 1999).

Functional food and nutraceutical are terms used incorrectly and indiscriminately for nutrients or nutrient-enriched food that can prevent or treat disease. Functional food is a product that resembles traditional food but it possesses demonstrated physiological benefits (Shahidi, 2009). For example a functional food could be a lutein-rich food as chicken, spinach, tomatoes, or oranges, or the omega-3 fatty acids found in fish oil. All functional foods are processed and consumed as food. A nutraceutical is not a nutritional supplement because the latter are nutrients that are added to the diet to correct or prevent deficiencies of vitamins, minerals, and proteins, and often used in the recovery of a patient suffering an illness or has undergone surgery, and also taken to improve overall health (Mandel et al., 2005). The beneficial effects of nutraceuticals and functional foods have been attributed to their components, such as polyphenols, polyunsatured fatty acids (PUFAs), terpenes, chlorophyll, and accessory pigments of the photosynthetic apparatus in cyanobacteria such as *Spirullina*. In general these compounds are antioxidants that reduce intensity of oxidative stress or modulate intracellular communication

3. Nutraceutical effects of polyphenols, particularly flavonoids

The polyphenols are compounds characterized by a benzene ring bearing one or more hydroxyl groups attached to the ring. They are ubiquitous in the plants, vegetables, fruit, vines, tea, coffee and microalgae. The polyphenols in food originate from one of the main classes of secondary metabolites in plants. They are involved in the growth and reproduction and are produced as a response to defend injured plants against pathogens, and to participate in the defense mechanism against ultraviolet radiation (Biesalski, 2007). Polyphenols have different nutraceutical properties, such as an antioxidant, antiinflammatory (Biesalski, 2007), anticancer (Oz & Ebersole, 2010), antibacterial (Du et al.,

2011), antiatherogenic, and antiangiogenic (Rimbach et al., 2009). There are now polyphenols with therapeutic properties for which the mechanism of action at the molecular level has been discovered and they are used in clinical trials, e.g. flavonoids.

Flavonoids comprise the most common group of polyphenols and provide much of the flavor and color to fruit and vegetables. More than 6000 different flavonoids have been described and it is estimated that humans consume about 1 g/day.

The structure of flavonoids is C6-C3-C6 and they consist of two aromatic rings linked through three carbons usually forming an oxygenated heterocycle nucleus, named the flavan nucleus, and shown in figure 1. In general, the flavonoids are classified into six groups (Grassi et al., 2009).

1. **Flavones**: These kinds of flavonoids are used by angiosperms to color their flowers. Natural flavones include apigenin (4',5,7-trihydroxyflavone), (3',4',5,7-tetrahydroxyflavone), (4',5,6,7,8-pentamethoxyflavone), chrysin (5,7-dihydroxyflavone), baicalein (5,6,7-trihydroxyflavone), scutellarein (5,6,7,4'-tetrahydroxyflavone), wogonin (5,7-Dihydroxy-8-methoxyflavone). There are synthetic flavones such as diosmin and flavoxate.

2. **Flavonols**: These compounds are used by organisms to protect them from UV radiation. Their diversity stems from the different positions of the hydroxyl groups on the benzene rings (show figure 1). There are flavonols as kaempferol (3,4',5,7-tetrahydroxy-2-phenylchromen-4-one), quercetin (3,3',4',5,7-pentahydroxy-2-phenylchromen-4-one), myricetin (3,3',4',5',5,7-hexahydroxy-2-phenylchromen-4-one), galangin (3,5,7-trihydroxy-2-phenylchromen-4-one), and morin (2-(2,4-dihydroxyphenyl)-3,5,7-trihydroxychromen-4-one). **Flavanones**: These flavonoids are the direct precursors of the vast majority of flavonoids. Some examples of flavanones are: naringenin (4',5,7-trihydroxyflavanone) and butin (7,3',4'-trihydroxyflavanone).

3. **Catechin or flavanols**: These flavonoids have two chiral centers on the molecule on carbons 2 and 3, yielding four diastereoisomers. Two of the isomers are in the *trans* configuration and are called catechins and the other two are in the *cis* configuration and are called epicatechins. These flavonoids are present in food as a complexs or oligomerics and polymerics as procyanidins or proantocyanidins. The catechins are found in different fruits, i.e. apples, apricots, blackberries, and grapes. Catechins are also in red wine, but black tea and cocoa are the richest sources. The flavanols in finished food products depend on the cultivar type, geographical origin, agriculture practice, postharvesting handling, and food processing (Scalbert et al., 2005).

4. **Antocyanidins**: Antocyanidins are a large group of natural colorants. The color of most fruits, flowers, and berries are made from a combination of anthocyanins and anthocyanidins. Anthocyanins always contain a carbohydrate molecule, whereas anthocyanidins do not. Examples of antocyanidins are cyanidin (3,3',4',5,7-pentahydroxyflavylium chloride), pelargonidin (3,5,7-trihydroxy-2-(4-hydroxyphenyl) benzopyrylium chloride), and malvidin (3,5,7,4'-tetrahydroxy-3',5'-dimethoxyflavylium)

5. **Isoflavones**: This group is a class of organic compounds that sometimes act as phytoestrogens in mammals and are called antioxidants because of their ability to trap a singlet oxygen. Genistein (4',5,7-trihydroxyisoflavone) and daidzein (4',7-dihydroxyisoflavone) are two examples of isoflavones.

Some authors have proposed that aurones are another flavonoid group, however we consider that aurones are derived from chalcones (Fowler & Koffas, 2009).

The flavonoid synthesis is shown in figure 1. It begins when a cell transforms phenylananine or tyrosine into phenylpropanoic acid or cinnamic acid by phenylalanine-tyrosine ammonia lyase (PAL; EC 4.3.1.25/TAL; EC 4.3.1.25). Then cytochrome-P450 cinnamate 4-hydroxylase (C4H; EC 1.14.13.11) adds a 4′-hydroxyl group to form p-coumaric acid. The CoA esters are subsequently synthetized from cinnamic acid, caffeic acid, or *p*-coumaric acid by 4 coumaryl:CoA ligase (4CL; EC 6.2.1.12). The type III polyketide chalcone synthase (CHS; EC 2.3.1.74) catalyzes the sequential condensation of three malonyl-CoA moieties with one CoA-ester molecule to form chalcones. The flavanones are formed when chalcones are isomerized into (2S)-flavanones by chalcone isomerase (CHI; EC 5.5.1.6). Many enzymes can modify the flavanones. For example the flavanones could be reduced to form isoflavones by isoflavone synthase (IFS; EC 1.14.13.86). After that, isoflavones are modified by different enzymatic systems to produce hydroxylation, reduction, alkylation, oxidation, and glucosylation alone or in combination in the three-ring phenylpropanoid core. Enzymes such as O-methyltransferse (IOMT, EC 2.1.1.150), isoflavone 2′-reductase (I2′R; EC 1.3.1.45) andisoflavone reductase (IFR; EC 1.3.1.45) can yield over 8000 different chemical structures from isoflavone (Winkel-Shirley, 2001; Fowler & Koffas, 2009). Another branch of the biosynthetic pathway of flavonoids is the flavones that are synthesized from flavanones through the action of the flavone synthase type I and II (FSI; EC 1.14.11.22). Flavonones are hydroxylated and then with flavonol synthase (FLS; EC 1.14.11.23) form flavonols. These compounds are the precursors of anthocyanins.

The beneficial effects can be divided into

1. **Antioxidants**: Flavonoids suppress the formation of reactive oxygen species (ROS) either by inhibiting enzymes or chelating trace elements involved in free radical production. Thus flavonoids help maintain an ROS steady state in the case of physical and chemical injury of the cell (Corradini et al., 2011). Not all flavonoids are ROS scavengers because some flavonoids, as nucleophiles, trap electrons from the ROS and become a free radical themselves, which then propagate a chain reaction causing a deleterious effect in the cell (Grassi et al., 2009).

2. **Modulators of intracellular communication**: The flavonoids and their metabolites act in the phosphoinositide 3-kinase (PI3K), Akt-protein kinase B (Akt-PKB), tyrosine kinase, and protein kinase C (PKC) signalling cascade. The inhibition or activation of these cascades modifies cellular function by altering the phosphorylation state of target molecules that modulate the expression of genes. This can explain the anticancer and neuroprotector flavonoid activities (Williams et al., 2004).

3. **Enzyme activity modulator**: Flavonoids offer cardiovascular protection because of their indirect inhibition of the angiotensin-converting enzyme (ACE; EC 3.4.15.1) (Actis-Goretta et al., 2006). Other enzymes inhibited by flavonoids are aromatase (EC 1.14.14.1) and α−amylase (EC 3.2.2.1) (Hargrove et al., 2011). The inhibition of enzymes that have a Fe-S cluster has been demonstrated (Mena et al., 2011).

In general, flavonoids are molecules responsible of some of the beneficial effect of nutraceuticals and functional foods. The different effects of flavonoids are described in table 1.

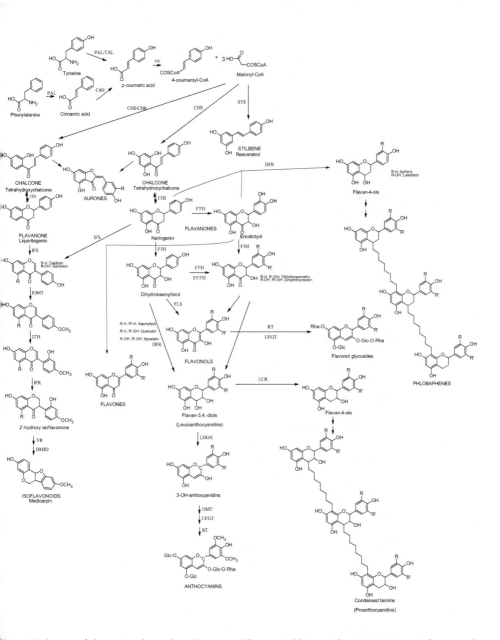

Fig. 1. Scheme of the major branch pathways of flavonoid biosynthesis, starting with general phenylpropanoid metabolism and leading to the nine major subgroups; chalcones, aurones, isoflavonoids, flavones, flavonols, flavandiols, anthocyanins, condensed tannins, and phlobaphene pigments

Flavonoid	Nutraceutic application	Reference
Flavanones	LDL oxidation (atherosclerosis) Cyclooxygenase inhibitory ability (cancer) Malaria chemotherapy (malaria) Inflammation response trigger (reduce inflammation) Antiangiogenic effect Reduce lung metastases Hepatoprotective action Antibacterial action Genoprotective action Inhibitors of NOS and COX in microglia Promote apoptosis in C6 glioma cells	(Miranda et al., 2000) (Kinghorn et al., 2004) (Kumar et al., 2003) (Kontogiorgis et al., 2008) (Mojzis et al., 2008) (Qin et al., 2011) (Prabu et al., 2011) (Celiz et al., 2011) (Orsolic et al., 2011) (Chao et al., 2010) (Sabarinathan et al., 2010)
Flavones	GLUT inhibitors (diabetes) Cyclooxygenase inhibitory ability (cancer) Antitumoral activity Pancreatic cholesterol esterase inhibitor Reduce neurodegeneration Produce apoptosis in melanoma cells Colitis treatment Antiinflammatory effect	(Kwon et al., 2007) (Kinghorn et al., 2004) (Balk, 2011; Polier et al., 2011) (Peng et al., 2011) (Gasiorowski et al., 2011) (Mohan et al., 2011) (Ganjare et al., 2011) (Funakoshi-Tago et al., 2011)
Flavonols	GLUT inhibitors (diabetes) Pancreatic lipase inhibitors (diabetes) Inhibitors of cell cycle control kinases (cancer) Regulate lipid profile in diabetic rats Regulate serum glucose Reduce apoptosis in cell culture Hepatoprotective action Promote new bone formation Anti-inflammatory effect Reduce neuronal damage	(Kwon et al., 2007; Park & Levine, 2000) (Nakai et al., 2005) (Hsu & Yen, 2006) (Liu et al., 2012) (Fontana Pereira et al., 2011) (Jang et al., 2011) (Singab et al., 2010) (Yang et al., 2010) (Mahat et al., 2010) (Lagoa et al., 2009) (Hirose et al., 2009)
Isoflavonoids	Alpha-glucosidase inhibitor (diabetes) GLUT inhibitor (diabetes) Improves cholesterol regulation (diabetes) Inhibitor of tyrosine kinase and antiinflammatory effect in kidney Induce apoptosis in leukemia Neuroprotective action Antiinflammatory action	(Kim et al., 2000) (Kwon et al., 2007; Song et al., 2002) (Lee, 2006) (Elmarakby et al., 2011) (Li et al., 2011a) (Xi et al., 2011) (Neelakandan et al., 2011)
Anthocyanins	Pancreatic lipase and glucosidase	(Kim et al., 2000)

Flavonoid	Nutraceutic application	Reference
	inhibitor (diabetes)	(Tsuda, 2008)
	Regulate adipocyte function (Obesity)	(Wolfram et al., 2006)
	Improves glucose and lipid metabolism	(Sternberg et al., 2008)
	(diabetes)	(Tokimitsu, 2004)
	Modulate blood hormone levels	(Mirshekar et al., 2010)
	(multiple sclerosis)	(Wang et al., 2010)
	Suppress body fat accumulation(obesity)	(Roghani et al., 2010)
	Reduce neuropathic hyperalgesia in	(Cvorovic et al., 2010)
	diabetic rats	
	Antioxidant effect	
	Neuroprotective action	
	produce cytotoxicity in colon cancer cells	

Table 1. Nutraceutical applications of flavonoids.

The mechanism of bioavailability and metabolism of particular flavonoids has been demonstrated in mammals. In general it has been shown that flavonoid absorption and metabolism occurs in a common pathway and it begins in the stomach and intestinal tract. In the small intestine flavonoids pass into the bloodstream in the form of glycosides, though esters or polymers cannot be absorbed. Some intestine cell enzymes or microorganisms of microflora hydrolyze them to be absorbed. In the bloodstream there are different thermodynamic pathways. They could interact with cells to modify intracellular communication. The polyphenols can be conjugated in the intestine or liver to form methylated, glucuronidated, or sulphated metabolites that reach the body via urinary and biliary excretion. The microflora also metabolizes some metabolized flavonoids that are secreted in the bile into the small intestine. Thus, there is a recycling of polyphenols that allow them more time in the plasma (Erdman et al., 2007; Manach et al., 2004). In general, the microalgae produce low quantities of polyphenols. For this reason, in the following parts of this chapter we give special attention to pigments and PUFAs.

4. Nutraceutical effects of terpenes

The terpenes are other secondary metabolites that have nutraceutical properties. The terpenes are not only the largest group of plant natural products, comprising at least 30,000 compounds, but also contain the widest assortment of structural types. Hundreds of different monoterpene (C10), sesquiterpene (C15), diterpene (C20), and triterpene (C30) carbon skeletons are known. The wealth of terpene carbon skeletons can be attributed to an enzyme class known as the terpene synthases (EC 4.2.3.20). These catalysts convert the acyclic prenyl diphosphates and squalene into a multitude of cyclic and acyclic forms. The chief causes of terpene diversity are the large number of different terpene synthases and that some terpene synthases produce multiple products. An excellent review of terpene synthase and the diversity of products were published by Degenhard and coworkers (Degenhardt et al., 2009). Microalgae produce terpenes in the form of carotenoids. These compounds offer therapeutic effects. Carotenoids are tetraterpenoid organic pigments that are naturally occurring in the chloroplasts and chromoplasts of photosynthetic organisms. The use of carotenoids by animals is because they cannot synthetize them. Animals obtain carotenoids in their diets, and they may employ them in various ways in their metabolism.

There are over 600 known carotenoids and they are divided into two classes, xanthophylls (that contain oxygen) and carotenes (that are purely hydrocarbons and contain no oxygen). Carotenoids in general absorb blue light. They serve two key roles in plants and algae; they absorb light energy for use in photosynthesis and they protect chlorophyll from photodamage (Armstrong & Hearst, 1996).

The biosynthesis of carotenes is explained in figure 2. The carotenogenesis differ somewhat among organisms and the current knowledge on the biosynthesis of carotenoids has been gained mainly from studies of bacteria and vascular plants (Armstrong & Hearst, 1996). In Figure 2, we proposed the model of Lohr for the carotenogenesis in *Chlamydomonas*. This is probably related to other microalgae of Chlorophyceae class (Lohr et al., 2005; Lohr, 2008). There are other major divisions in different organisms, such as diatoms (Bertrand, 2010) or plants (Cazzonelli & Pogson, 2010; Zhu et al., 2010), which references the readers can check to deepen their knowledge in this area.

There has been much interest in carotenoids, especially their effect on human health, because they have a market value of several hundred million Euros. Their chemical synthesis is still a demanding challenge for chemists. The major dietary source of vitamin A for mammals, including humans, is derived from carotenoids. Vitamin A is an essential micronutrient for cell growth, embryonic development, vision, and the function of the immune system (Jackson et al., 2008).

In general carotenoids exert their mechanism on health via an antioxidant pathway or by modulating intracellular communication.

1. **Antioxidant properties:** This property of carotenoids was characterized by the ability to quench singlet oxygen, the inhibition of peroxide formation, and the correlation of antioxidant dependency with oxygen partial pressures. The ketocarotenoids, such as astaxanthin and canthaxanthin, were the best radical scavengers that did not contain conjugated terminal carbonyl functions (see figure 2). These findings suggest that the keto function in conjugation with the polyene backbone is able to stabilize carbon-centered radicals more effectively than the polyene backbone alone (Jackson et al., 2008).

2. **Modulation of intracellular communication:** Carotenes modulate the intracellular communication because they or their metabolites interact with nuclear receptors like the pregnant-X-receptor (PXR) or retinoic acid receptor (RAR). For PXR it has been postulated that β-carotene activated the PXR more than its metabolites. Following this pathway, the β-carotene-PXR enhanced the metabolism of xenobiotics, bile acids, and retinoids (Ruhl, 2005). The carotenoids can be converted into two molecules of 9-*cis*-retinal, which is oxidized to 9-*cis*-retinoic acid. The RXR binds the 9-*cis*-retinoic acid with high affinity to modulate cell functions (Heyman et al., 1992). Carotenoids like lycopene modulate mevalonate and Ras pathways to modify cell growth inhibition of cancerous cells (Palozza et al., 2010), and it changes Wnt and hedgehog proteins in those cells (Sarkar et al., 2010). The PI3K-Akt and MAPK pathways are stimulated in kidney by lycopene (Chan et al., 2009).

In table 2, are some nutraceuticals of the most used carotenoids.

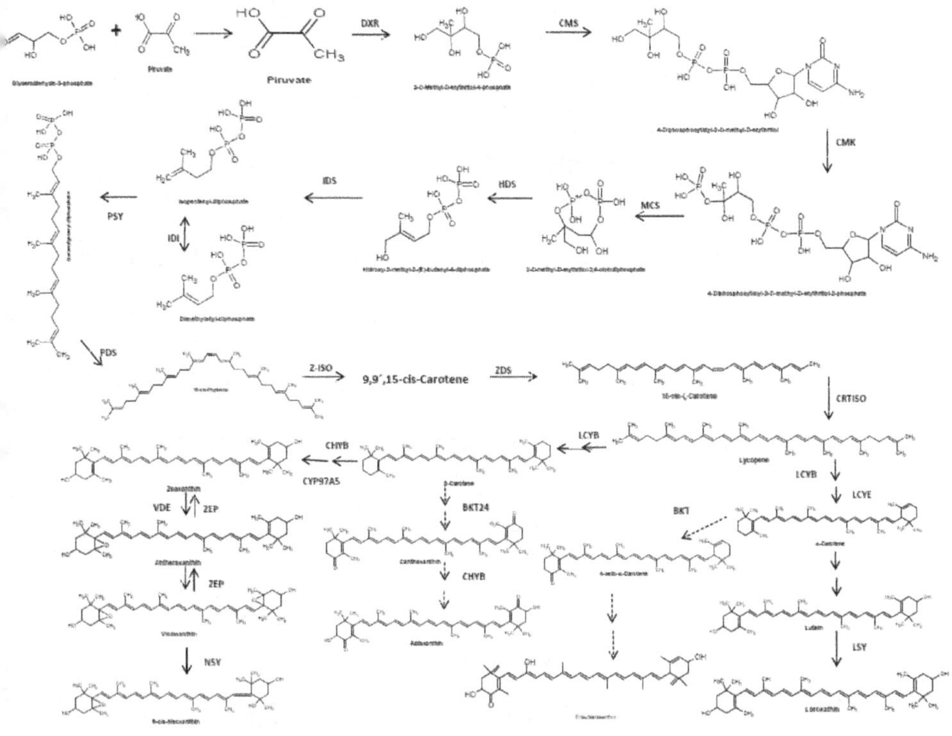

Fig. 2. Putative pathways of carotenoid biosynthesis in *Chlamydomonas*. Hypothetical
zygospore-specific pathways are indicated by dotted arrows. For the enzymes of the
pathways only abbreviations are given. DXS (1-deoxy-D-xylulose-5-phosphate synthetase,
EC 2.2.1.7). DXR (1-deoxy-D-xylulose-5-phosphate reductoisomerase, EC 1.1.1.267). CMS (4-
diphosphocytidil-2-C-methyl-D-erythriol synthase, EC 2.7.7.60). CMK (4-diphosphocytidil-
2-C-methyl-D-erythriol kinase, EC 2.7.1.14.8). MCS (2-C-methyl-D-erythritol-2,4-
cyclophosphate synthase, EC 4.6.1.12). HDS (4-hydroxy-3-methylbut-2-en-1-yl diphosphate
synthase, EC 1.17.7.1). IDS (isopentenyl dimethylallyl diphosphate synthase, EC 1.17.1.2.3).
IDI (isopentenyl-diphosphate delta-isomerase, EC 5.3.3.2). GGPPS (geranylgeranyl-
diphosphate synthase, EC 2.5.1.81). PSY (phytoene synthase, EC 2.5.1.32). PDS (phytoene
desaturase, EC 1.3.5.5). Z-ISO (ζ-Carotene isomerase); ZDS (ζ-carotene desaturase, EC
1.3.5.6). CRTISO (carotenoid isomerase, EC 5.2.1.13). LCYB (lycopene-β-cyclase). LCYE
(lycopene-ε-cyclase), CHYB (carotene-β-hydroxylase, EC 1.14.13.-). CYP97A5 (carotene-β-
hydroxylase, EC 1.14.13.129). CYP97C3 (carotene-ε-hydroxylase, EC 1.14.99.45). ZEP
(zeaxanthin epoxidase, EC 1.14.13.90). VDE (violaxanthin epoxidase, EC 1.10.99.3). NSY
(neoxanthin synthase, EC 5.3.99.9). LSY (loroxanthin synthase), and BKT (carotene-β-
ketolase).

Carotenoid	Nutracetical effect	Reference
Lycopene	Antimutagenic effect	(Polivkova et al., 2010)
	Neuroprotective action	(Sandhir et al., 2010)
	Nephroprotective action	(Sahin et al., 2010)
	Prevent preclampsia	(Banerjee et al., 2009)
	Reduce risk of hip fracture	(Sahni et al., 2009)
	Antioxidant effect	(Erdman et al., 2009)
	Reduce eosinophil influx in asthma	(Wood et al., 2008)
	Cardioprotective effect against doxorubicin–caused damage	(Anjos Ferreira et al., 2007)
	Reduce inflammatory cytokines expression in pancreatitis	(Kim, 2011a; Kim, 2011b)
	Inhibit the growth and progression of colon cancer	(Tang et al., 2011)
	Enhanced antioxidant enzymes and immunity function in gastric cancer	(Luo & Wu, 2011)
	Inhibit NFκB–modulated IL-8 expression in macrophages–cigarette activated	(Simone et al., 2011)
	Reduced oxidative stress in allergic rhinitis	(Li et al., 2011b)
	Attenuated endothelial dysfunction in diabetes	(Zhu et al., 2011)
	Reduce cognitive decline in Parkinson's disease	(Kaur et al., 2011)
	Reduces LDC cholesterol and systolic blood pressure	(Ried & Fakler, 2011)
Asthaxanthin	Reduces endothelial dysfunction in diabetic rats	(Zhao et al., 2011)z
	Produces anxiolytic–like effects in mice	(Nishioka et al., 2011)
	Reduces oxidative stress and mitochondrial dysfunction in brain due MPTP/MPTP⁺	(Lee et al., 2011)
	Reduce IL-6 microglia production	(Kim et al., 2010)
	Reduce blood pressure in hypertensive rats	(Monroy-Ruiz et al., 2011)
	Neuroprotective action against focal ischemia	(Lu et al., 2010)
	Attenuate thrombosis	(Khan et al., 2010)
	Reduce retinal injury in elevated intraocular pressure	(Cort et al., 2010)
	Reduce UVA – induced skins photoaging	(Suganuma et al., 2010)
	Hepatoprotective action	(Curek et al., 2010)

Table 2. Nutraceutical application of lycopene and astaxanthin.

Carotenoids are lipid soluble and in general they follow the same absorption pathway as lipids, however other mechanisms of absorption have been proposed. To learn more, read the review of Kotake-Nara and Nagao (Kotake-Nara & Nagao, 2011). Once in the bloodstream, carotenes are fundamentally ligated to low density lipoprotein (LDL) whereas the xanthophylls are more evenly distributed between high density lipoproteins (HDL) and low density lipoproteins (LDL). Nonpolar carotenoids (lycopene, α-carotene, β-carotene) are located in the hydrophobic core and the polar (xanthophylls) would be, at least in part, on the surface of lipoproteins (Furr & Clark, 1997). For the microalgae, carotenoids are synthesized in high concentrations under several different environmental conditions, and humans exploited these as nutraceuticals in food.

5. Nutraceutical effects of chlorophylls, PUFA and other vitamins

There are other components in microalgae that could modulate redox environment to prevent oxidative stress and can affect intracellular communication. These components are chlorophyll, PUFAs, and vitamins such as vitamin A, B, C, and E.

Microalgae, like all chloroplast-containing photosynthetic eukaryotes, synthesize chlorophyll pigments. In Chlorophyceae chorophylls a and b are the most predominant. The chlorophylls have a porphyrin ring structure similar to heme, but with a central nonreactive magnesium ion instead of iron. To review chlorophyll biosynthesis in microalgae, read the chapter of Beale (Beale, 2008). The information about the biological activities of chlorophyll as nutraceuticals is scarce. They do have antipoliferative (Wu et al., 2010) and antioxidant (Serpeloni et al., 2011) activities. The chlorophyllin-cooper complex, a water-soluble commercial version of chlorophyll, possesses antimutagenic (Chernomorsky et al., 1997) and anticancer activities (Chernomorsky et al., 1997). The other components of microalgae; PUFAs, and vitamins A, B, C, and E, could be a nutraceutical because there is much evidence of how they modulate intracellular signals and act as antioxidants.

6. *Chlorella* genus as nutraceutic

Chlorella species are encountered in all water habitats having cosmopolitan occurrences. The species of this genus have a simple form, a unicellular green alga belonging to the Chlorophyceae family. The *Chlorella* sp. is morphologically classified into four types; a) spherical cells (ratio of the two axes equals one), b) ellipsoidal cells (ratio of the longest axis to the shorter axis 1.45 to 1.60), spherical or ellipsoidal cells, and globular to subspherical cells. Their reproduction is asexual. Each mature cell divides usually producing four or eight (and more rarely 16) autospores, which are freed by rupture or dissolution of the parental walls.

Our research group has used *Chlorella vulgaris* as nutraceutical, particularly against mercury-caused oxidative stress and renal damage. For that we used male mice that were assigned into six groups; 1) a control group that received 100 mM phosphate buffer (PB) ig and 0.9% saline ip, 2) PB + $HgCl_2$ 5 mg/kg ip, 3) PB + 1000 mg/kg *Chlorella vulgaris* ig, and three groups receiving $HgCl_2$ + 250, 500, or 1000 mg/kg *Chlorella vulgaris* ig. The administration of the microalgae or PB was made 30 min before saline or $HgCl_2$ for 5 days. Our results demonstrated that $HgCl_2$ caused oxidative stress and cellular damage, whereas *Chlorella vulgaris* administration prevents oxidative stress (figure 3) and cellular damage (figure 4) in the kidney (Blas-Valdivia et al., 2011). We proposed that *Chlorella vulgaris's* carotenes play an important role in preventing $HgCl_2$-caused lipid peroxidation. Carotenes have a wide pharmacological spectrum of effects. The inhibition of lipid peroxidation may be caused by the free radical scavenging property of these compounds (Miranda et al., 2001) Carotenes can scavenge singlet oxygen and they terminate peroxides by their redox potential because of the hydroxyl group in its structure. Thus, the ROS-steady state is maintained in the kidney damage lower than in animals with mercury intoxication. The biochemical behavior of this microalgae against mercury-caused oxidative stress is similar to the purified component of cyanobacteria such as *Pseudoanabaena tenuis* (Cano-Europa et al. 2010) or *Spirulina maxima* (Sharma et al., 2007).

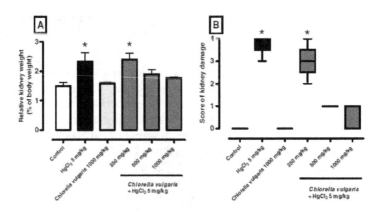

Fig. 3. Quantification of relative kidney weight (A) and the score of kidney damage (B) of mice treated with $HgCl_2$ and *Chlorella vulgaris*. In A each bar represents the mean ± S.E.M. In B each box represents the median ± intercuartilic space.* $P < 0.05$ vs. control. Author right permission. Springer ©.

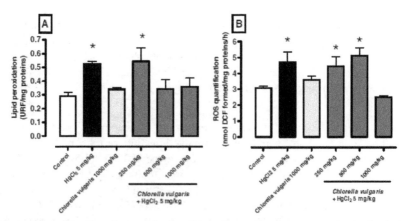

Fig. 4. Quantification of lipid peroxidation (A) and reactive oxygen species in the kidneys of mice treated with $HgCl_2$ and *Chlorella vulgaris*. Bar represents the mean ± S.E.M.* $P < 0.05$ vs. control. Author right permission. Springer ©.

Here are some experiments that demonstrated the nutraceutical use of *Chlorella* (Table 3).

Study	Evidences
The administration of *Chlorella* sp. reduces endotoxemia, intestinal oxidative stress and bacterial traslocation in experimental biliary obstruction (Bedirli et al., 2009)	*Chlorella* administration inhibits bacterial culture and it avoids oxidative stress.
Hot water extract of *Chlorella vulgaris* induced DNA damage and apoptosis (Yusof et al., 2010)	The extract of *Chlorella vulgaris* inhibited DNA synthesis, causing apoptosis and it increases p53, caspase-3, and Bax expression in hepatoma cells (HEpG2)
Attenuating effect of *Chlorella* supplementation on oxidative stress and NFκB. Activation in peritoneal macrophages and liver of C57BL/6 mice fed on atherogenic diet (Lee et al., 2003)	*Chlorella* supplementation decreases the NFκB activation and superoxide anion production and because it increases SOD and catalase activity
Chlorella accelerates dioxin excretion in rats (Morita et al., 1999)	*Chlorella* enhanced dioxin metabolism and excretion by feces
Effect of *Chlorella* and its fractions on blood pressure, cerebral stroke lesions, and life-span in stroke-prone spontaneously hypertensive rats (Sansawa et al., 2006)	A *Chlorella* supplemented diet decreases blood pressure and the incidence rate of cerebral stroke in SHRSP.
Hypocholesterolemic mechanism of *Chlorella*: *Chlorella* and its indigestible fraction enhance hepatic cholesterol 7α-hydroxylase in rats (Shibata et al., 2007)	*Chlorella* powder increases the expression of CYP7A1, a limiting enzyme of the main pathway of the cholesterol catabolism, lowering the concentration of LDL in plasma
Chlorella vulgaris triggers apoptosis in hepatocarcinogenesis-induced rats (Mohd Azamai et al., 2009)	*Chlorella vulgaris* inhibits the anti-apoptotic protein Bcl-2
Effect of *Chlorella vulgaris* on lipid metabolism in Wistar rats fed high fat diet (Lee et al., 2008)	*Chlorella vulgaris* decreases HDL cholesterol concentration by a reduction in the intestinal absorption
Antioxidant effect of the marine algae *Chlorella vulgaris* against naphthalene-induced oxidative stress in the albino rats (Vijayavel et al., 2007)	*Chlorella vulgaris* inhibits production of free radicals, decreasing lipoperoxidation, and increasing the activity of antioxidant enzymes as SOD, catalase, GPX and reduced glutathione, preventing from the toxicity of naftalene
Six-week supplementation with *Chlorella* has favorable impact on antioxidant status in Korean male smokers (Lee et al., 2010)	*Chlorella* supplement exhibits antioxidant activity decreasing ROS and increasing the activity of SOD and catalase
Chlorella pyrenoidosa supplementation reduces the risk of anemia, proteinuria and edema in pregnant women (Nakano et al., 2010)	*Chlorella pyrenoidosa* exhibits an antiinflammatory activity regulated by cytokine. It increased the production of IL-10

Study	Evidences
Effect of *Chlorella* intake on cadmium metabolism in rats (Shim et al., 2009)	*Chlorella* inhibits cadmium absorption and it promotes the excretion through the feces. Also, it stimulates the production of metallothionein in the small intestine.
Isolation of phophorylated polysaccharides from algae: the inmmunostimulatory principle of *Chlorella pyrenoidosa* (Suarez et al., 2010)	The *Chlorella* polysaccharides increases the production of NO in macrophages enhancing the innate immune response, mediated by Toll-like receptors (TLR-4)
Influence of *Chlorella* powder intake during swimming stress in mice (Mizoguchi et al., 2011)	*Chlorella vulgaris* exhibits an antioxidant activity, reducing the lipoperoxidation, avoiding the DNA damage. However it does not show hypoglycemic activity

Table 3. Nutraceutical evidences of *Chlorella*.

7. *Chlamydomonas* genus as nutraceutic

Chlamydomonas spp. are unicellular algae with cell walls and with either two or four flagella. The genus *Chlamydomonas* is of worldwide distribution and is found in a diversity of habitats including temperate, tropical, and polar regions. *Chlamydomonas* species have been isolated from freshwater ponds and lakes, sewage ponds, marine and brackish waters, snow, garden and agricultural soil, forests, deserts, peat bogs, damp walls, sap on a wounded elm tree, an artificial pond on a volcanic island, mattress dust in the Netherlands, roof tiles in India, and a Nicaraguan hog wallow. These algae belong to the family *Chlamydomonadaceae* that consists of approximately 30 genera. DNA sequence analysis clearly demonstrates, however, that this family is composed of multiple phylogenetic lineages that do not correspond to the morphologically defined genera. Although the identities of the species are uncertain, it is noteworthy that the traits in which they differed included body shape, thickness of the cell wall, presence or absence of the apical papilla, lateral vs. basal position of the chloroplast, chloroplast position, and shape of the eyespot, all of which were later used as criteria to separate species. Although cell-body shape and size vary among *Chlamydomonas* species (as defined by morphological criteria), the overall polar structure, with paired apical flagella and basal chloroplast surrounding one or more pyrenoids, is constant. Cells are usually free-swimming in liquid media but on solid substrata may be nonflagellated, and are often seen in gelatinous masses similar to those of the algae *Palmella* or *Gloeocystis* in the order *Tetrasporales*. This condition has been referred to as a palmelloid state. Some species may also form asexual resting spores, or akinetes, in which the original vegetative cell wall becomes much thicker, and carotenoids, starch, and lipids may accumulate (Harris et al., 2008).

Our group has studied the nutraceutical properties of *Chlamydomonas gloeopara*, a microalgae collected from a eutrophic reservoir (La Piedad Lake) in Cuautitlan Izcalli, Mexico. That reservoir is located at 19°39′N (latitude) and 99°14′W (longitude). Our research group has used *Chlamydomonas gloeopara* as a nutraceutical, particularly against mercury-caused oxidative stress and renal damage. For that we used male mice that were assigned into six groups; 1) a control group that received 100 mM phosphate buffer (PB) ig and 0.9% saline ip, 2) PB + HgCl$_2$ 5 mg/kg ip, 3) PB + 1000 mg/kg *Chlamydomonas gloeopara* ig, and three groups receiving HgCl₂ + 250, 500, or 1000 mg/kg *Chlamydomonas gloeopara* ig. The administration of the microalgae or PB was made 30 min before saline or HgCl₂ for 5 days. Our results demonstrated that *Chlamydomonas gloeopara* as well as *Chlorella* prevents renal damage (figure 5, panel A-F) by reducing the oxidative stress of lipid peroxidation (figure 5, panel G).

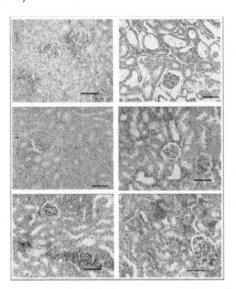
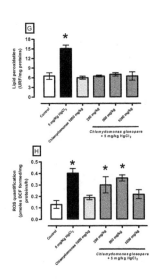

Fig. 5. Effect on *Chlamydomonas gloeopara* administation on HgCl₂-caused renal damage (panel A-F) and oxidative stress (panel G and H). Photomicrographs of renal cortex . Panel A shows control group. Panel B shows group treated with HgCl₂. Panel C shows group treated with *Chlamydomonas gloeopara* 1000 mg/kg . Panels D, E and F show groups treated with *Chlamydomonas gloeopara* 250, 500 and 1000 mg/kg plus HgCl₂. The tissue was stained by hematoxylin-eosin. Treatment with HgCl₂ causes cell atrophy, hyperchromatic nuclei, and edema. Histological alterations were partially ameliorated in groups treated with *Chlamydomonas gloeopara*. *Chlamydomonas gloeopara* administration reduced lipid peroxidation (G) and reactive oxygen species (H) in the kidneys of mice treated with HgCl₂ and *Chlorella vulgaris*. Bar is the mean ± SE * $P < 0.05$ vs. control.

3. *Haematococcus* genus as nutraceutic

Haematococcus are green microalgae; single-celled aquatic organisms. It is known that *Haematococcus* is the primarily source of astaxanthin, a ketocarotenoid that is a natural nutritional component. In the marine environment, astaxanthin is biosynthesized in the food chain, within the microalgae or phytoplankton, at the primary production level. When these algae are exposed to harsh environmental conditions and ultraviolet light, they accumulate the highest level of astaxanthin and in this process, the algae become blood red. Astaxanthin accumulates 2% to 3% of dry weight and constitutes 85% to 88% of the total carotenoids. Chemically it is a ketocarotenoid (3,3´-dihydroxy-β, β-carotene-4,4´dione) and is the principal pigment of salmonoids and shrimp. Astaxanthin has a higher antioxidant activity than lutein, lycopene, α or β-carotene, and α-tocopherol. Astaxanthin has 100 times and 1(times greater antioxidant activity than vitamin E and β-carotene (Guerin, 2003).

Morphological studies have shown that the algae have a life cycle. The, green vegetativ(cells with two flagellae grow autotrophycally in the light and heterotrophically in the dark In culture, *H. pluvialis* has the typical characteristics of a motile stage, with biflagellat(spherical cells. The growth in a bioreactor, with mechanical stirring, favors the occurrence o(more or less mature aplanospores. This stage becomes dominant together with the evolutior of growth.   The aplanospore color turns gradually red, because of the accumulation o(carotenoids in the chloroplast, and especially outside of them in lipid globules (astaxanthin) The red aplanospores are known as haematocysts. This stage may appear under stres(conditions caused by light, high temperature, increased salinity, nutritional limitation, o: change of carbon source. During the growth stage, the cells with a diameter of 30 µm wer(spherical to ellipsoid and enclosed by a cell wall. The cells had 2 flagellae of equal lengtl emerging form an anterior papilla. As they age, the cells ceased to be mobile, yet the cellula: structure remained the same without the flagellae. Under stress conditions, the volume o the cells increased to a diameter of > 40 µm and the cell wall became resistant. Th(maturation of the cyst cells was accompanied by enhanced carotenoid biosynthesis and ¿ gradual change in cell color to red. When the cystic cells were transferred to optimal growtl conditions, daughter cells were released from the cystic cells, and then vegetative cell(regenerated from daughter cells (Cysewski & Todd Lorenz, 2004).

Haematococcus has the potential as a nutraceutical because there is various evidence of this In table 4, we show some articles that employed *Haematococcus* or its astaxanthin.

Study	Evidences
Haematococcus astaxanthin: applications for human health and nutrition (Guerin, 2003)	This is a review about the uses of astaxantin from *Haematococcus* in health
Optimization of microwave-assisted extraction of astaxanthin from *Haematococcus pluvialis* by response surface methodology and antioxidant activities of the extracts (Zhao et al., 2009)	The extracts have a high antioxidant capacity, inhibit peroxidation of linoleic acid, and neutralize free radicals
Cardioprotection and myocardial salvage by a disodium disuccinate astaxanthin derivative (Cardax™) (Gross & Lockwood, 2004)	The astaxanthin is an antioxidant, antiinflammatory, and cardioprotective. reducer of levels of nitric oxide, tumor necrosis factor alpha, and prostaglandin E2
Ulcer preventive and antioxidative properties of astaxanthin from *Haematococcus pluvialis (Kamath et al., 2008)*	The astaxanthin exerts its gastroprotection of gastric ulceration by activation of antioxidant enzyme such as catalase, superoxide dismutase, and glutathione peroxidase. It inhibits the activity pump Na-K ATPase
Safety assessment of astaxanthin-rich microalgae biomass: acute and subchronic toxicity studies in	The administration of astaxanthin has no adverse effects

Study	Evidences
rats(Stewart et al., 2008)	
Astaxanthin, a carotenoid with potential in human health and nutrition (Hussein et al., 2006).	The antihypertensive and neuroprotective potentials of the compound
Protective effects of *Haematococcus* astaxanthin on oxidative stress in healthy smokers (Kim et al., 2011).	The results suggest that ASX supplementation might prevent oxidative damage in smokers by suppressing lipid peroxidation and stimulating the activity of the antioxidant system in smokers
Astaxanthin-rich extract from the green alga *Haematococcus pluvialis* lowers plasma lipid concentrations and enhances antioxidant defense in apolipoprotein E knockout mice (Yang et al., 2011)	It results suggest that supplementation of astaxanthin-rich *Haematococcus* extract improves cholesterol and lipid metabolism as well as antioxidant defense mechanisms, all of which could help mitigate the progression of atherosclerosis.

Table 4. Nutraceutical evidences of *Haematococcus*.

3. *Dunaliella* genus as nutreutic

Dunaliella salina is a unicelular green alga belonging to the Chlorophyceae family. *Dunaliella* cells are ovoid, spherical, pyriform, fusiform, or ellipsoid with sizes varying from 5 to 25 μm in length and from 3 to 13 μm in width. The cells also contain a single chloroplast, which mostly has a central pyrenoid surrounded by starch granules. *Dunaliella* multiplies by lengthwise division, but sexual reproduction does occur rarely by isogametes with a conjugation process. It proliferates in extremely varied salinities from 0.5 to 5.0 M NaCl. The alga cells do not contain a rigid cell wall; instead a thin elastic membrane surrounds them. It is known to accumulate carotenoids under various stress conditions. It possesses a remarkable degree of environmental adaptation by producing an excess of β-carotene and glycerol to maintain its osmotic balance. β-carotene occurs naturally as its isomers, namely, all-*trans*, 9-*cis*, 13-*cis*, and 15-*cis* forms and functions as an accessory light harvesting pigment, thereby protecting the photosynthetic apparatus against photo damage in all green plants including algae. β-carotene, a component of the photosynthetic reaction center is accumulated as lipid globules in the interthylakoid spaces of the chloroplasts of *Dunaliella*. They protect the algae from damage obtained during excessive irradiance by preventing the formation of reactive oxygen species, by quenching the triplet-state chlorophyll, or by reacting with singlet oxygen ($^1O_2$) and also act as a light filter (Ben-Amotz, 2004). *Dunaliella* nutraceutical properties are shown in table

Study	Conclusion
In vivo antioxidant activity of carotenoids from *Dunaliella salina* a green microalga (Chidambara-Murthy et al., 2005)	Carotenoids provide protection against CCl_4-caused hepatic damage by restoring the activity of hepatic enzymes like peroxidase, super oxide dismutase, and catalase, which reduce ROS and lipid peroxidation.

Study	Conclusion
9-*cis* β-carotene-rich powder of the alga *Dunaliella bardawil* increases plasma HDL-cholesterol in fibrate-treated patients (Shaish et al., 2006)	*Dunaliella* treatment increases plasma HDL-cholesterol and lower plasma triglyceride levels
Ethanol extract of *Dunaliella salina* induces cell cycle arrest and apoptosis in A545 human non-small cell lung cancer cells (Sheu et al., 2008)	Ethanol extract of *Dunaliella salina* inhibits cell proliferation and causes apoptosis possibly via p53 and p21 promoting the protein expression of Fas and FasL
Protective effects of *Dunaliella salina* against experimental induced fibrosarcoma on Wistar rats (Raja et al., 2007).	The *chlorophyta* has a protective effect against experimentally caused fibrosarcoma
Bioavailability of the isomer mixture of phytoene and phytofluene-rich alga *Dunaliella bardawil* in rat plasma and tissues (Werman et al., 2002).	9-*cis* phytoene has a stronger antioxidative effect than the all trans isomer
Hypercholesterolemia induced oxidative stress is reduced in rats with diets enriched with supplement from *Dunaliella salina* algae (Bansal & Sapna, 2011).	*Dunaliella salina* components inhibit lipid peroxidation and also increases Type1 5'-iodothyronine deiodinase (5'-DI) expression, which leads to a T$_3$ level increase
Evaluation of carotenoid extract from *Dunaliella salina* against cadmium-induced cytotoxicity and transforming growth factor β1 induced expression of smooth muscle α-actin with rat liver cell lines (Jau-Tien et al., 2011).	Carotenoid extract of *Dunaliella salina* contains abundant *cis* and *trans* β-carotenes. These antioxidants decrease the lipid peroxidation and also inhibit activation of hepatic stellate cells (HSCs).
Protective effects of *Dunaliella salina*- a carotenoids-rich alga, against carbon tetrachloride-induced hepatotoxicity in mice (Hsu et al., 2008).	Carotenoids of *D. salina* inhibit the lipid peroxidation and increases the antioxidant enzyme activity

Table 5. Nutraceutical evidences of *Dunaliella*.

10. Final remarks

The functional food and nutraceutical market is growing. However, to promote health the active compounds must be ingested in high concentration. This is a great problem because sometimes the components such as carotenoids, polyphenols, and chlorophylls are extracted from vegetables or plants. In their production, we are modifying the environment, thus the use of biotechnology of microalgae or other microorganisms like bacteria or fungus could be an alternative because they may be environmentally friendly. The sun can be used as energy source and the medium could be fresh or sea water, with the carbon source as CO_2 and other inorganic or organic sources. In this chapter we show the evidence of some genera, particularly of Chlorophyceae class as *Chlorella, Chlamydomonas, Haematococcus,* and *Dunaliella*. It is evident that their components modulate intracellular communication and they act as antioxidants.

There are many microalgae never used as nutraceuticals that could be used for human or animal health, such as the microalgae used in aquaculture to fed shrimp and fish. Examples of those kinds of microalgi are *Pavlova* and *Tetraselmis* that produce high concentration of PUFAs.

11. Acknowledgement

This study was partially supported by SIP-IPN 20110283 y 20110336. Thanks to Dr. Ellis Glazier for editing this English-language text.

12. References

Actis-Goretta,L., Ottaviani,J.I. & Fraga,C.G. (2006). Inhibition of angiotensin converting enzyme activity by flavanol-rich foods. *Journal Agriculture Food Chemestry* Vol.54, No. 1,(January 2006), pp. 229-234, ISSN 0021-8561

Anjos Ferreira,A.L., Russell,R.M., Rocha,N., Placido Ladeira,M.S., Favero Salvadori,D.M., Oliveira,N., Matsui,M., Carvalho,F.A., Tang,G., Matsubara, L.S. & Matsubara,B.B. (2007). Effect of lycopene on doxorubicin-induced cardiotoxicity: an echocardiographic, histological and morphometrical assessment. *Basic and Clinical Pharmacology and Toxicology* Vol.101, No.1,(July 2007) pp. 16-24, ISSN 1742-7843

Armstrong,G.A. & Hearst,J.E. (1996). Carotenoids 2: Genetics and molecular biology of carotenoid pigment biosynthesis. *FASEB Journal* Vol.10, No.2,(February 1996), pp. 228-237, ISSN 0950-9232

Balk,R. (2011). Roger C. Bone, MD and the evolving paradigms of sepsis. *Contributions to Microbioly* Vol.17, pp.1-11,(January 2011), ISSN 14209519

Banerjee,S., Jeyaseelan,S. & Guleria,R. (2009). Trial of lycopene to prevent pre-eclampsia in healthy primigravidas: results show some adverse effects. *Journal of Obstetrics and Gynaecology Research* Vol.35, No.2, pp.477-482,(June 2009), ISSN (printed).

Bansal,M. & Sapna,J. (2011). Hypercholesterolemia induced oxidative stress is reduced in rats with diets enriched with supplement from Dunaliella salina algae. *Americal Journal of Biomedical Science* Vol.1, pp.196-204, ISSN 14230127

Beale,S.I. (2008). Biosynthesis of chlorophylls and heme. In The *Chlamydomonas* Sourcebook, eds. Harris,E.H., Stern,D.B. & Witman,G.B., USA: Academic Press.

Bedirli,A., Kerem,M., Ofluoglu,E., Salman,B., Katircioglu,H., Bedirli,N., Ylmazer,D., Alper,M., and Pasaoglu,H. Administration of *Chlorella* sp. microalgae reduces endotoxemia, intestinal oxidative stress and bacterial translocation in experimental biliary obstruction. *Clinical nutrition* (Edinburgh, Scotland) Vol.28, No.6, pp.674-678,(December 2009), ISSN 0261-5614

Ben-Amotz,A. (2004). Industrial production of microalgal cell-mass and secondary products. Major industrial species: *Dunaliella*. In *Handbook of microalgal culture: Biotechnology and applied Phycology*, ed. Richmond Amos, pp. 273-280. Australia: Blackwell science.

Bernal,J., Mendiola,J.A., Ibanez,E. & Cifuentes,A. (2011). Advanced analysis of nutraceuticals. *Journal of Pharmaceutical and Biomedical Analysis* Vol.55, No.4, pp.758-774,(June 2011), ISSN 0731-7085

Bertrand,M. (2010). Carotenoid biosynthesis in diatoms. *Photosynthesis Research* Vol.106, No. 1-2, pp.89-102,(November 2010), ISNN 01668595

Biesalski,H.K. (2007). Polyphenols and inflammation: basic interactions. *Current Opinion in Clinical Nutrition and Metabolic Care* Vol.10, No.6, pp.724-728,(November 2007), ISNN 1473-6519.

Blas-Valdivia,V., Ortiz-Butron,R., Pineda-Reynoso,M., Hernandez-Garcia,A. & Cano-Europa,E. (2011). *Chlorella vulgaris* administration prevents $HgCl_2$-caused oxidative stress and cellular damage in the kidney. *Journal of Applied Phycology* Vol.23, No.1, pp.53-58,(February 2011), ISNN 1573-5176

Cano-Europa,E., Ortiz-Butron,R., Blas-Valdivia,V., Pineda-Reynoso,M., Olvera-Ramírez,R. & Franco-Colín,M. (2010). Phycobiliproteins from *Pseudanabaena tenuis* rich in c-phycoerythrin protect against $HgCl_2$-caused oxidative stress and cellular damage in the kidney. Journal of Applied Phycology Vol.22, No. 4, pp.495-501, ISSN 0921-8971

Cazzonelli,C.I. & Pogson,B.J. (2010). Source to sink: regulation of carotenoid biosynthesis in plants. *Trends in Plant Science* Vol.15, No.5, pp.266-274, (May 2010), ISSN 1360-1385

Celiz,G., Daz,M. & Audisio,M.C. (2011). Antibacterial activity of naringin derivatives against pathogenic strains. *Journal of Applied Microbiology* Vol.111, No.3, pp.731-738,(September 2011), ISSN 1364-5072

Chan,C.M., Fang,J.Y., Lin,H.H., Yang,C.Y. & Hung,C.F. (2009). Lycopene inhibits PDGF-BB-induced retinal pigment epithelial cell migration by suppression of PI3K/Akt and MAPK pathways. *Biochemical and Biophysical Research Communications* Vol.388, No.1, pp.172-176,(October 2009), ISSN 0006-291X

Chao,C.L., Weng,C.S., Chang,N.C., Lin,J.S., Kao,S.T. & Ho,F.M. (2010). Naringenin more effectively inhibits inducible nitric oxide synthase and cyclooxygenase-2 expression in macrophages than in microglia. *Nutrition Research* Vol.30, No.12, pp.858-864, (December 2010), ISSN 0271-5317

Chernomorsky,S., Rancourt,R., Virdi,K., Segelman,A. & Poretz,R.D. (1997). Antimutagenicity, cytotoxicity and composition of chlorophyllin copper complex. *Cancer Letters* Vol.120, No.2, pp.141-147, (December 1997), ISSN 0304-3835

Chidambara-Murthy,K., Vanitha,A., Rajesha,J., Mahadeva-Swamy,M., Sowmya,P. & Ravishankar,G. (2005). In vivo antioxidant activity of carotenoids from Dunaliella s salina--a green microalga. *Life Sciences* Vol.76, No.12, pp.1381-1390,(February 2005), ISSN 0024-3205.

Corradini,E., Foglia,P., Giansanti,P., Gubbiotti,R., Samperi,R. & Lagana,A. (2011). Flavonoids: chemical properties and analytical methodologies of identification and quantitation in foods and plants. *Natural Products Research* Vol.25, No.5, pp.469-495, (March 2011), ISSN 1478-6427

Cort,A., Ozturk,N., Akpinar,D., Unal,M., Yucel,G., Ciftcioglu,A., Yargicoglu,P. & Aslan,M. (2010). Suppressive effect of astaxanthin on retinal injury induced by elevated intraocular pressure. *Regulatory Toxicology and Pharmacology* Vol.58, No.1, pp.121-130,(October 2010), ISSN 0273-2300

Curek,G.D., Cort,A., Yucel,G., Demir,N., Ozturk,S., Elpek,G.O., Savas,B. & Aslan,M. (2010). Effect of astaxanthin on hepatocellular injury following ischemia/reperfusion. *Toxicology* Vol.267, No.1-3, pp.147-153, (January 2010), ISSN 0300-483X

Cvorovic,J., Tramer,F., Granzotto,M., Candussio,L., Decorti,G. & Passamonti,S. (2010). Oxidative stress-based cytotoxicity of delphinidin and cyanidin in colon cancer cells. *Archives of Biochemistry and Biophysics* Vol.501, No.1, pp.151-157, (October 2010), ISSN 0003-9861

Cysewski,G.R. & Todd Lorenz,R. (2004). Industrial production of microalgal cell-mass and secondary products. Species of high potential: Haematococcus. In Handbook of Microalgal Culture: Biotechnology and Applied Phycology, ed. Richmond Amos, pp. 281-288

Degenhardt,J., Kollner,T.G. & Gershenzon,J. (2009). Monoterpene and sesquiterpene synthases and the origin of terpene skeletal diversity in plants. Phytochemistry Vol.70, No.15-16, pp.1621-1637, (October 2010), ISNN 0031-9422

Du,W.X., Olsen,C.W., Avena-Bustillos,R.J., Friedman,M. & McHugh,T.H. (2011). Physical and antibacterial properties of edible films formulated with apple skin polyphenols. Journal of Food Science Vol.76, No.2, pp.M149-M155, (March 2011), ISNN 1750-3841

Elmarakby,A.A., Ibrahim,A.S., Faulkner,J., Mozaffari,M.S., Liou,G.I. & Abdelsayed,R. (2011). Tyrosine kinase inhibitor, genistein, reduces renal inflammation and injury in streptozotocin-induced diabetic mice. *Vascular Pharmacology*, (July 2011), In press, ISNN 1537-1891

Erdman,J.W., Balentine,D., Arab,L., Beecher,G., Dwyer,J.T., Folts,J., Harnly,J., Hollman,P., Keen,C.L., Mazza,G., Messina,M., Scalbert,A., Vita,J., illiamson,G. & urrowes,J. (2007). Flavonoids and heart health: Proceedings of the ILSI North America Flavonoids Workshop, May 31-June 1, 2005, Washington, DC. *The Journal of Nutrition* Vol.137, No. 3, pp.718S-737S,(March 2007), ISNN 1541-6100

Erdman,J.W., Ford,N.A. & Lindshield,B.L. (2009). Are the health attributes of lycopene related to its antioxidant function? *Archives of Biochemistry and Biophysics* Vol.483, No.2, pp.229-235, (March 2009), ISNN 003-9861.

Fontana Pereira,D., Cazarolli,L.H., Lavado,C., Mengatto,V., Figueiredo,M.S., Guedes,A., Pizzolatti,M.G. & Silva,F.R. (2011). Effects of flavonoids on α-glucosidase activity: Potential targets for glucose homeostasis. *Nutrition* Epv_ahead of print, (July 2011), ISNN 0899-9007

Fowler,Z.L. & Koffas,M.A. (2009). Biosynthesis and biotechnological production of flavanones: current state and perspectives. *Applied Microbioogyl and Biotechnology* Vol.83, No.5, pp.799-808, (July 2009), ISNN 1432-0614

Funakoshi-Tago,M., Nakamura,K., Tago,K., Mashino,T. & Kasahara,T. (2011). Anti-inflammatory activity of structurally related flavonoids, Apigenin, Luteolin and Fisetin. *International Immunopharmacology* Vol.11, No.9, pp.1150-1159, (September 2011), ISNN 1567-5769

Furr,H.C. & Clark,R.M. (1997). Intestinal absorption and tissue distribution of carotenoids. *Nutrional Biochemistry* Vol.8, pp.364-377, (1997), ISNN 0955-2863

Ganjare,A.B., Nirmal,S.A. & Patil,A.N. (2011). Use of apigenin from Cordia dichotoma in the treatment of colitis. *Fitoterapia* In press (June 2011), ISNN 0367-326X

Gasiorowski,K., Lamer-Zarawska,E., Leszek,J., Parvathaneni,K., Yendluri,B.B., Blach-Olszewska,Z. & Aliev,G. (2011). Flavones from root of Scutellaria baicalensis Georgi: drugs of the future in neurodegenration? *CNS and Neurological Disorders-Drug Targets* Vol. 10, No.2, pp.184-191, (March 2011), ISNN 1871-5273

Grassi,D., Desideri,G., Croce,G., Tiberti,S., Aggio,A. & Ferri,C. (2009). Flavonoids, vascula function and cardiovascular protection. *Current Pharmaceutical Design* Vol. 15, No 10, pp. 1072-1084 (January, 2009) ISSN 1381-6128

Gross,G.J. & Lockwood,S.F. (2004). Cardioprotection and myocardial salvage by a disodiun disuccinate astaxanthin derivative (Cardax™). *Life Sciences* Vol. 75 No. 2 , pp 215 224 (May, 2004) ISSN 0024-3205

Guerin,M. (2003). Haematococcus astaxanthin: applications for human health and nutrition *Trends in Biotechnology* Vol. 21, No. 5, pp 210-216, (May, 2003) ISSN 0167-7799

Hardy,G. (2000). Nutraceuticals and functional foods: introduction and meaning. *Nutritior* Vol. 16, No 7-8,pp 688-689 (July, 2000) ISSN 0899-9007

Hargrove,J.L., Greenspan,P., Hartle,D.K. & Dowd,C. (2011). Inhibition of aromatase and α amylase by flavonoids and proanthocyanidins from *Sorghum bicolor* bran extracts *Journal Medical and Food* Vol. 14,No.7-8 , pp 799-807 (July,2011) ISSN 1557-7600

Harris,E.H., Stern,D.B. & Witman,G.B. (2008). The genus *Chlamydomonas*. In Th *Chlamydomonas* Sourcebook, eds. Harris,E.H., Stern,D.B. & Witman,G.B., USA Academic Press.

Heyman,R.A., Mangelsdorf,D.J., Dyck,J.A., Stein,R.B., Eichele,G., Evans,R.M. & Thaller,C (1992). 9-cis retinoic acid is a high affinity ligand for the retinoid X receptor. *Cel* Vol. 68, No. 2, pp 397-406 (January, 1992) ISSN 0096-8674

Hirose,E., Matsushima,M., Takagi,K., Ota,Y., Ishigami,K., Hirayama,T., Hayashi,Y. Nakamura,T., Hashimoto,N., Imaizumi,K., Baba,K., Hasegawa,Y. & Kawabe,T (2009). Involvement of heme oxygenase-1 in kaempferol-induced anti-allergic actions in RBL-2H3 cells. *Inflammation* Vol. 32, No. 2, pp 99-108 (April, 2009) ISSN 1466-1861

Hsu,C.L. & Yen,G.C. (2006). Induction of cell apoptosis in 3T3-L1 pre-adipocytes by flavonoids is associated with their antioxidant activity. Molecular *Nutrition & Foo Research* Vol. 50, No. 11, pp 1072-1079 (November, 2006) ISSN 1613-4133

Hsu,Y., Tsai,C., Chang,W., Ho,Y., Chen,W. & Lu,F. (2008). Protective effects of *Dunaliell salina*--a carotenoids-rich alga, against carbon tetrachloride-induced hepatotoxicity in mice. *Food and Chemical Toxicology* Vol. 46, No. 10,pp 3311-3317 (October, 2008 ISSN 0278-6915

Hussein,G., Sankawa,U., Goto,H., Matsumoto,K. & Watanabe,H. (2006). Astaxanthin, a carotenoid with potential in human health and nutrition. *Journal Natural Products* Vol. 69, No. 3, pp 443-449 (March, 2003) ISSN 0970-129X

Jackson,H., Braun,C.L. & Ernst,H. (2008). The chemistry of novel xanthophyll carotenoids *American Journal of Cardiology* Vol. 101, no 10A, pp 50D-57D (May, 2008) ISSN 0002 9149

Jang,Y.J., Kim,J., Shim,J., Kim,J., Byun,S., Oak,M.H., Lee,K.W. & Lee,H.J. (2011). Kaempfero Attenuates 4-hydroxynonenal-induced apoptosis in PC12 Cells by directly inhibiting NADPH oxidase. *Journal of Pharmacology And Experimental Therapeutic* vol. 337, No. 3, pp 747-754 (June, 2011) ISSN 0022-3565

Jau-Tien,L., Ying-Chung,L., Chao-chin,H., You-Cheng,S., Fung-Jou,L. & Deng-Jye,Y. (2011) Evaluation of carotenoid extract from *Dunaliella salina* against cadmium-inducec cytotoxicity and transforming qrowth factor b1 induced expression of smootl muscle a-actin with rat liver cell lines. *Journal of Food Science* Vol. 18, pp 301-30€ (2011) ISSN 1750-3841

Kamath,B.S., Srikanta,B.M., Dharmesh,S.M., Sarada,R. & Ravishankar,G.A. (2008). Ulcer preventive and antioxidative properties of astaxanthin from *Haematococcus pluvialis*. *Europe Journal of Pharmacology* Vol. 590, No. 1-3, pp 387-395 (August, 2008) ISSN 0014-2999

Kaur,H., Chauhan,S. & Sandhir,R. (2011). Protective effect of lycopene on oxidative stress and cognitive decline in rotenone induced model of Parkinson's disease. *Neurochemical Reserch* Vol. 36, No.8, pp 1435-1443 (August ,2011) ISSN 1573-6903

Khan,S.K., Malinski,T., Mason,R.P., Kubant,R., Jacob,R.F., Fujioka,K., Denstaedt,S.J., King,T.J., Jackson,H.L., Hieber,A.D., Lockwood,S.F., Goodin,T.H., Pashkow,F.J. & Bodary,P.F. (2010). Novel astaxanthin prodrug (CDX-085) attenuates thrombosis in a mouse model. *Thrombosis Reserch* Vol. 126, No. 4, pp 299-305 (October, 2010) ISSN 0049-3848

Kim,H. (2011). Inhibitory mechanism of lycopene on cytokine expression in experimental pancreatitis. *Annals of the New York Academy of Sciences* Vol. 1229, No 1, pp 99-102 (July, 2011) ISSN 0077-8923

Kim,J.H., Chang,M.D., Choi,H.D., You,Y.K., Kim,J.T., Oh,J.M., and Shin,W.G. Protective effects of *Haematococcus* astaxanthin on oxidative stress in healthy smokers. *Journal of Medicinal Food* In press (2011) ISSN 1557-7600

Kim,J.S., Kwon,C.S. & Son,K.H. (2000). Inhibition of alpha-glucosidase and amylase by luteolin, a flavonoid. *Bioscience, Biotechnology, and Biochemistry* Vol. 64,No. 11, pp 2458-2461 (November, 2000) ISNN 0916-8451

Kim,Y.H., Koh,H.K. & Kim,D.S. (2010). Down-regulation of IL-6 production by astaxanthin via ERK-, MSK-, and NF-κB-mediated signals in activated microglia. *International Immunopharmacology* Vol. 10, No. 12, pp 1560-1572 (December, 2010) ISNN 1567-5769

Kinghorn,A.D., Su,B.N., Jang,D.S., Chang,L.C., Lee,D., Gu,J.Q., Carcache-Blanco,E.J., Pawlus,A.D., Lee,S.K., Park,E.J., Cuendet,M., Gills,J.J., Bhat,K., Park,H.S., Mata-Greenwood,E., Song,L.L., Jang,M. & Pezzuto,J.M. (2004). Natural inhibitors of carcinogenesis. *Planta Medica* Vol. 70, No. 8, pp 691-705 (August, 2004) ISNN 0032-0943

Kontogiorgis,C., Mantzanidou,M. & Hadjipavlou,L. (2008). Chalcones and their potential role in inflammation. *Mini-Reviews in Medicinal Chemistry* Vol. 8, No. 12, pp 1224-1242 (December, 2008) ISSN 1389-5575

Kotake-Nara,E. & Nagao,A. (2011). Absorption and metabolism of xanthophylls. *Marine Drugs* Vol. 9, No.6, pp 1024-1037 (January, 2011) ISSN 1660-3397

Kumar,A., Katiyar,S.B., Agarwal,A. & Chauhan,P.M. (2003). Perspective in antimalarial chemotherapy. *Current Medicinal Chemistry* Vol. 10, No. 13 , pp 1137-1150 (July, 2003) ISSN 1070-3632

Kwon,O., Eck,P., Chen,S., Corpe,C.P., Lee,J.H., Kruhlak,M. & Levine,M. (2007). Inhibition of the intestinal glucose transporter GLUT2 by flavonoids. *FASEB Journal* Vol. 21, No. 2, pp 366-377(February, 2007) ISNN 0717-7712

Lagoa,R., Lopez-Sanchez,C., Samhan-Arias,A.K., Ganan,C.M., Garcia-Martinez,V. & Gutierrez-Merino,C. (2009). Kaempferol protects against rat striatal degeneration induced by 3-nitropropionic acid. *Journal of Neurochemistry* Vol. 111, No. 2, pp 473-487(October, 2009) ISNN 1566-5240

Lee,D.H., Kim,C.S. & Lee,Y.J. (2011). Astaxanthin protects against MPTP/MPP+-induced mitochondrial dysfunction and ROS production in vivo and in vitro. *Food and Chemical Toxicology* Vol. 49, No. 1, pp 271-280 (January, 2011) ISSN 0278-6915

Lee,H.S., Choi,C.Y., Cho,C. & Song,Y. (2003). Attenuating effect of chlorella supplementation on oxidative stress and NFkappaB activation in peritoneal macrophages and liver of C57BL/6 mice fed on an atherogenic diet. *Bioscience Biotechnology, and Biochemistry* Vol. 67, No. , pp 2083-2090 (October, 2003) ISSN 1347-6947

Lee,H.S., Park,H.J. & Kim,M.K. (2008). Effect of Chlorella vulgaris on lipid metabolism in Wistar rats fed high fat diet. *Nutrition Research and Practice* Vol. 2, No. 4, pp 204-210 (December, 2008) ISSN 0271-5317

Lee,J.S. (2006). Effects of soy protein and genistein on blood glucose, antioxidant enzyme activities, and lipid profile in streptozotocin-induced diabetic rats. *Life Sciences* Vol 79, No. 16, pp 1578-1584 (September, 2006) ISSN 0024-3205

Lee,S.H., Kang,H.J., Lee,H.J., Kang,M.H. & Park,Y.K. (2010). Six-week supplementation with *Chlorella* has favorable impact on antioxidant status in Korean male smokers *Nutrition* Vol. 26, No. 2, pp 175-183 (February, 2010) ISSN 0029-6643

Li,W., Frame,L.T., Hoo,K.A., Li,Y., D'Cunha,N. & Cobos,E. (2011a). Genistein inhibited proliferation and induced apoptosis in acute lymphoblastic leukemia, lymphoma and multiple myeloma cells in vitro. *Leukemia and Lymphoma* In press (July, 2011) ISSN 1029-2403

Li,X.L., Zhou,A.G., Zhang,L. & Chen,W.J. (2011b). Antioxidant status and immune activity of glycyrrhizin in allergic rhinitis mice. *International Journal of Molecular Sciences* Vol. 12, No. 2, pp 905-916 (January, 2011) ISSN 1422-0067

Liu,I.M., Tzeng,T.F., Liou,S.S. & Chang,C.J. (2012). Beneficial effect of traditional chinese medicinal formula danggui-shaoyao-san on advanced glycation end-product-mediated renal injury in streptozotocin-diabetic rats. *Evidence-based Complementary and Alternative Medicine* In press(January, 2012) ISSN 1741- 427X

Lohr,M., Im,C.S. & Grossman,A.R. (2005). Genome-based examination of chlorophyll and carotenoid biosynthesis in *Chlamydomonas reinhardtii*. Plant Physiology Vol. 138, No. 1, pp 490-515 (May, 2005) ISSN 1040-4651

Lohr,M. (2008). Carotenoids. In The *Chlamydomonas* Sourcebook, eds. Harris,E.H., Stern,D.B. & Witman,G.B., USA: Academic Press.

Lu,Y.P., Liu,S.Y., Sun,H., Wu,X.M., Li,J.J. & Zhu,L. (2010). Neuroprotective effect of astaxanthin on H2O2-induced neurotoxicity in vitro and on focal cerebral ischemia in vivo. *Brain Research* Vol. 1360, pp 40-48 (November, 2010) ISSN 0006-8993

Luo,C. & Wu,X.G. (2011). Lycopene enhances antioxidant enzyme activities and immunity function in N-Methyl-N'-nitro-N-nitrosoguanidine-induced gastric cancer rats. *International Journal of Molecular Sciences* Vol. 12, No. 5, pp 3340-3351 (January, 2011) ISSN 1422-0067

Mahat,M.Y., Kulkarni,N.M., Vishwakarma,S.L., Khan,F.R., Thippeswamy,B.S., Hebballi,V., Adhyapak,A.A., Benade,V.S., Ashfaque,S.M., Tubachi,S. & Patil,B.M. (2010). Modulation of the cyclooxygenase pathway via inhibition of nitric oxide production contributes to the anti-inflammatory activity of kaempferol. *European Journal of Pharmacology* Vol. 642, No.1-3, pp 169-176 (September, 2010) ISSN 0014-2999

Manach,C., Scalbert,A., Morand,C., Remesy,C. & Jimenez,L. (2004). Polyphenols: food sources and bioavailability. *American Journal of Clinical Nutrition* Vol. 79, No. 5, pp 727-747 (May, 2004) ISSN 0002-9165

Mandel,S., Packer,L., Youdim,M.B. & Weinreb,O. (2005). Proceedings from the "Third International Conference on Mechanism of Action of Nutraceuticals". *The Journal of | Nutritional Biochemistry* Vol. 16,No.9 , pp 513-520 (September, 2005) ISSN 0955-2863

Mena,N.P., Bulteau,A.L., Salazar,J., Hirsch,E.C. & Nunez,M.T. (2011). Effect of mitochondrial complex I inhibition on Fe-S cluster protein activity. *Biochemical and Biophysical Research Communications* Vol. 409, No. 2, pp 241-246 (June, 2011) ISSN 0006-291X

Miranda,C.L., Stevens,J.F., Ivanov,V., McCall,M., Frei,B., Deinzer,M.L. & Buhler,D.R. (2000). Antioxidant and prooxidant actions of prenylated and nonprenylated chalcones and flavanones in vitro. *Journal of Agricultural and Food Chemistry* Vol. 48, No.9 , pp 3876-3884 (September, 2000) ISSN 0021-8561

Miranda,M.S., Sato,S. & Mancini-Filho,J. (2001). Antioxidant activity ono.f the microalga *Chlorella vulgaris* cultered on special conditions. *Bollettino Chimico Farmaceutico* Vol. 140, No.3 , pp 165-168 (May,2001) ISSN 0006-6648

Mirshekar,M., Roghani,M., Khalili,M., Baluchnejadmojara...,T. & Arab Moazzen,S. (2010). Chronic oral pelargonidin alleviates streptozotocin-induced diabetic neuropathic hyperalgesia in rat: involvement of oxidative stress. *The Iranian Biomedical Journal* Vol. 14, No. 1-2, pp 33-39 (January, 2010) ISSN 1028-852X

Mizoguchi,T., Arakawa,Y., Kobayashi,M. & Fujishima,M. (2011). Influence of Chlorella powder intake during swimming stress in mice. *Biochemical and Biophysical Research Communications* Vol. 404, No. 1, pp 121-126 (January, 2011) ISSN 0006-291X

Mohan,N., Banik,N.L. & Ray,S.K. (2011). Combination of N-(4-hydroxyphenyl) retinamide and apigenin suppressed starvation-induced autophagy and promoted apoptosis in malignant neuroblastoma cells. *Neuroscience Letters* Vol. 502, No.1, pp 24-29 (September, 2011) ISSN 0304-3940

Mohd Azamai,E.S., Sulaiman,S., Mohd Habib,S.H., Looi,M.L., Das,S., Abdul Hamid,N.A., Wan Ngah,W.Z. & Mohd Yusof,Y.A. (2009). Chlorella vulgaris triggers apoptosis in hepatocarcinogenesis-induced rats. *Journal of Zheijang University SCIENCE B* Vol. 10, No. 1, pp 14-21 (January, 2009) ISSN 1862-1783

Mojzis,J., Varinska,L., Mojzisova,G., Kostova,I. & Mirossay,L. (2008). Antiangiogenic effects of flavonoids and chalcones. *Pharmacal Research* Vol. 57, No. 4, pp 259-265 (April, 2008) ISSN 1976-3786

Monroy-Ruiz,J., Sevilla,M., Carron,R. & Montero,M.J. (2011). Astaxanthin-enriched-diet reduces blood pressure and improves cardiovascular parameters in spontaneously hypertensive rats. *Pharmacal Research* Vol. 63, No.1, pp 44-50 (January, 2011) ISSN 1976-3786

Morita,K., Matsueda,T., Iida,T. & Hasegawa,T. (1999). Chlorella Accelerates Dioxin Excretion in Rats. *The Journal of Nutrition* Vol 129, No. 9, pp 1731-1736 (September, 1999) ISSN 0022-3166

Nakai,M., Fukui,Y., Asami,S., Toyoda-Ono,Y., Iwashita,T., Shibata,H., Mitsunaga,T., Hashimoto,F. & Kiso,Y. (2005). Inhibitory effects of oolong tea polyphenols on pancreatic lipase in vitro. *The Journal of Agricultural and Food Chemistry* Vol.53, No. 11, pp 4593-4598 (June, 2006) ISSN 0021-8561

Nakano,S., Takekoshi,H. & Nakano,M. (2010). Chlorella pyrenoidosa supplementation reduces the risk of anemia, proteinuria and edema in pregnant women. *Plant Foods for Human Nutrition* Vol. 65, No.1, pp 25-30 (March, 2010) ISSN 0921-9668

Neelakandan,C., Chang,T., Alexander,T., Define,L., Evancho-Chapman,M. & Kyu,T. (2011). In vitro evaluation of antioxidant and anti-inflammatory properties of genistein-modified hemodialysis membranes. *Biomacromolecules* Vol. 12, No. 7, pp 2447-2455 (July, 2011) ISSN 1525-7797

Nishioka,Y., Oyagi,A., Tsuruma,K., Shimazawa,M., Ishibashi,T. & Hara,H. (2011). The antianxiety-like effect of astaxanthin extracted from *Paracoccus carotinifaciens*. *Biofactors* Vol. 37, No. 1, pp 25-30 (January, 2011) ISSN 1872- 8081

Orsolic,N., Gajski,G., Garaj-Vrhovac,V., Dikic,D., Prskalo,Z. & Sirovina,D. (2011). DNA-protective effects of quercetin or naringenin in alloxan-induced diabetic mice. *European Journal of Pharmacology* Vol. 656, No. 1-3, pp 110-118 (April, 2011) ISSN 0014-2999

Oz,H.S. & Ebersole,J.L. (2010). Green tea polyphenols mediated apoptosis in intestinal epithelial cells by a FADD-dependent pathway. *Journal of Cancer Therapy* Vol. 1, No. 3, pp 105-113 (September, 2010) ISSN 2151-1934

Palozza,P., Colangelo,M., Simone,R., Catalano,A., Boninsegna,A., Lanza,P., Monego,G. & Ranelletti,F.O. (2010). Lycopene induces cell growth inhibition by altering mevalonate pathway and Ras signaling in cancer cell lines. *Carcinogenesis* Vol. 31, No. 7-8, pp 1813-1821 (October, 1999) ISSN 0143-3334

Park,J.B. & Levine,M. (2000). Intracellular Accumulation of Ascorbic Acid Is Inhibited by Flavonoids via Blocking of Dehydroascorbic Acid and Ascorbic Acid Uptakes in HL-60, U937 and Jurkat Cells. *The Journal of Nutrition* Vol. 130, No. 5, pp 1297-1302 (May, 2005) ISSN 0022-3166

Peng,G., Du,Y., Wei,Y., Tang,J., Peng,A.Y. & Rao,L. (2011). A new synthesis of fully phosphorylated flavones as potent pancreatic cholesterol esterase inhibitors. *Organic and Biomolecular Chemistry* Vol. 9, No.7, pp 2530-2534 (April, 2011) ISSN 1477-0520

Polier,G., Ding,J., Konkimalla,B.V., Eick,D., Ribeiro,N., Kohler,R., Giaisi,M., Efferth,T., Desaubry,L., Krammer,P.H. & Li-Weber,M. (2011). Wogonin and related natural flavones are inhibitors of CDK9 that induce apoptosis in cancer cells by transcriptional suppression of Mcl-1. Cell Death and Disease Vol. 2,pp e182 (January, 2011) ISSN 2041-4889

Polivkova,Z., merak,P., Demova,H. & Houska,M. (2010). Antimutagenic effects of lycopene and tomato puree. *Journal of Medicinal Food* Vol. 13, No. 6, pp 1443-1450 (December, 2010) ISSN 1096-620X

Prabu,S.M., Shagirtha,K. & Renugadevi,J. (2011). Naringenin in combination with vitamins C and e potentially protects oxidative stress-mediated hepatic injury in cadmium-intoxicated rats. *Journal of nutritional science and vitaminology* (Tokyo) Vol. 57, No. 2, pp 177-185 (January, 2011) ISSN 0301-4800

Qin,L., Jin,L., Lu,L., Lu,X., Zhang,C., Zhang,F. & Liang,W. (2011). Naringenin reduces lung metastasis in a breast cancer resection model. *Protein & Cell* Vol.2, No.6 (June 2011), pp.507-516, ISSN 1674-8018

Raja,R., Hemaiswarya,S., Balasubramanyam,D. & Rengasamy,R. (2007). Protective effect of *Dunaliella salina* (*Volvocales, Chlorophyta*) against experimentally induced fibrosarcoma on wistar rats. *Microbiological Research* Vol.162, No.2 (January 2007), pp.177-184, ISSN 0944-5013

Ried,K. & Fakler,P. (2011). Protective effect of lycopene on serum cholesterol and blood pressure: Meta-analyses of intervention trials. *Maturitas* Vol.68, No.4 (April 2011), pp.299-310, ISSN 0378-5122

Rimbach,G., Melchin,M., Moehring,J. & Wagner,A.E. (2009). Polyphenols from cocoa and vascular health-a critical review. *International ournal off Molecular Science* Vol.10, No.10 (October 2009), pp.4290-4309, ISSN 1422-0057

Roghani,M., Niknam,A., Jalali-Nadoushan,M.R., Kiasalari,Z., Khalili,M. & Baluchnejadmojara...,T. (2010). Oral pelargonidin exerts dose-dependent neuroprotection in 6-hydroxydopamine rat model of hemi-parkinsonism. *Brain Research Bulletin* Vol.82, No.5-6 (July 2010), pp.279-283, ISSN 0361-9230

Ruhl,R. (2005). Induction of PXR-mediated metabolism by beta-carotene. *Biochimica et Biophysica Acta* Vol.1740, No.2 (May 2010), pp.162-169, ISSN 0006-3002

Sabarinathan,D., Mahalakshmi,P. & Vanisree,A.J. (2010). Naringenin promote apoptosis in cerebrally implanted C6 glioma cells. *Molecular and Cellular Biochemestry* Vol.345, No.1-2 (December 2010), pp.215-222, ISSN

Sahin,K., Tuzcu,M., Sahin,N., Ali,S. & Kucuk,O. (2010). Nrf2/HO-1 signaling pathway may be the prime target for chemoprevention of cisplatin-induced nephrotoxicity by lycopene. *Food and Chemical Toxicology* Vol.48, No.10 (October 2010), pp.2670-2674, ISSN 0278-6915

Sahni,S., Hannan,M.T., Blumberg,J., Cupples,L.A., Kiel,D.P. & Tucker,K.L. (2009). Protective effect of total carotenoid and lycopene intake on the risk of hip fracture: a 17-year follow-up from the Framingham Osteoporosis Study. *Neurochemical International* Vol.24, No.6 (Jun 2009), pp.1086-1094, ISSN 0197-0186

Sandhir,R., Mehrotra,A. & Kamboj,S.S. (2010). Lycopene prevents 3-nitropropionic acid-induced mitochondrial oxidative stress and dysfunctions in nervous system. *Neurochemestry International* Vol.57, No.5 (November 2010), pp.579-587, ISSN 0197-0186

Sansawa,H., Takahashi,M., Tsuchikura,S. & Endo,H. (2006). Effect of chlorella and its fractions on blood pressure, cerebral stroke lesions, and life-span in stroke-prone spontaneously hypertensive rats. *Journal of Nutritional Sciences and Vitaminology* (Tokyo) Vol.52, No.6 (December 2006), pp.457-466. ISSN 0301-4800

Sarkar,F.H., Li,Y., Wang,Z. & Kong,D. (2010). The role of nutraceuticals in the regulation of Wnt and Hedgehog signaling in cancer. *Cancer and Metastasis Review Vol.29,* No.3 (September 2010), pp.383-394, ISSN 0167-7659

Scalbert,A., Johnson,L.T. & Saltmarsh,M. (2005). Polyphenols: antioxidants and beyond. *American Journal of Clinical Nutrition* Vol.81, No.1 (January 2005), pp.215S-2217, ISSN 0002-9165

Serpeloni,J.M., Grotto,D., Aissa,A.F., Mercadante,A.Z., Bianchi,M.D. & Antunes,L.M. (2011). An evaluation, using the comet assay and the micronucleus test, of the antigenotoxic effects of chlorophyll b in mice. *Mutation Research*, ISSN 0027-5107

Shahidi,F. (2009). Nutraceutical and functional food: whole versus processed food. Trend *Food Science and Technology* Vol.20, pp. 376-387, ISSN 1365-2621

Shaish,A., Harari,A., Hananshvili,L., Cohen,H., Bitzur,R., Luvish,T., Ulman,E., Golan,M. Ben-Amotz,A., Gavish,D., Rotstein,Z. & Harats,D. (2006). 9-cis beta-carotene-rich powder of the alga Dunaliella bardawil increases plasma HDL-cholesterol ir fibrate-treated patients. *Atherosclerosis* Vol.189, No.1 (November 2010), pp.215-221 ISSN

Sharma,M.K., Sharma,A., Kumar,A. & Kumar,M. (2007). Evaluation of protective efficacy o *Spirulina fusiformis* against mercury induced nephrotoxicity in Swiss albino mice *Food and Chemical Toxicology* Vol.45, No.6 (June 2007), pp.879-887, ISSN 0278-6915

Sheu,M., Huang,G., Wu,C., Chen,J., Hang,H., Chang,S. & Chung,J. (2008). Ethanol Extract o *Dunaliella salina* induces cell cycle arrest and apoptosis in A549 human non- smal cell lung cancer cells. *In Vivo* Vol.22, No.3 (May 2008), pp.369-378, ISSN 0258- 051X

Shibata,S., Hayakawa,K., Egashira,Y. & Sanada,H. (2007). Hypocholesterolemic mechanism of *Chlorella: Chlorella* and its indigestible fraction enhance hepatic cholesterol catabolism through up-regulation of cholesterol 7alpha-hydroxylase in rats *Bioscience Biotechnology and Biochemistry* Vol.71, No.4(April 2007), pp.916-925 ISSN0916-8451

Shim,J.A., Son,Y.A., Park,J.M. & Kim,M.K. (2009). Effect of Chlorella intake on Cadmium metabolism in rats. *Nutrition Research and Practice* Vol.3, No.1 (March 2009), pp.15 22, ISSN 1976-1457

Simone,R.E., Russo,M., Catalano,A., Monego,G., Froehlich,K., Boehm,V. & Palozza,P. (2011) Lycopene inhibits NF-kB-mediated IL-8 expression and changes redox and PPAR signalling in cigarette smoke-stimulated macrophages. *PLoS One* Vol.6, No.. (January 2011), pp.e19652, ISSN 1932-6203

Singab,A.N., Ayoub,N.A., Ali,E.N. & Mostafa,N.M. (2010). Antioxidant anc hepatoprotective activities of Egyptian moraceous plants against carbor tetrachloride-induced oxidative stress and liver damage in rats. *Pharmaceutica Biology* Vol.48, No.11 (November 2010), pp.1255-1264, ISSN 1388-0209

Song,J., Kwon,O., Chen,S., Daruwala,R., Eck,P., Park,J.B. & Levine,M. (2002). Flavonoic inhibition of sodium-dependent vitamin C transporter 1 (SVCT1) and glucose transporter isoform 2 (GLUT2), intestinal transporters for vitamin C and glucose *Journal of Biological Chemistry* Vol.277, No.18 (may 2002), pp.15252-15260, ISSN 0021-9258

Sternberg,Z., Chadha,K., Lieberman,A., Hojnacki,D., Drake,A., Zamboni,P., Rocco,P. Grazioli,E., Weinstock-Guttman,B. & Munschauer,F. (2008). Quercetin anc interferon-beta modulate immune response(s) in peripheral blood mononuclea cells isolated from multiple sclerosis patients. *Journal of Neuroimmunology* Vol.205 No.1-2 (December 2008), pp.142-147 ISSN 0165-5728

Stewart,J.S., Lignell,A., Pettersson,A., Elfving,E. & Soni,M.G. (2008). Safety assessment o astaxanthin-rich microalgae biomass: Acute and subchronic toxicity studies in rats *Food and Chemical Toxicology* Vol.46, No.9 (September 2008), pp.3030-3036, ISSN 0278-6915

Suarez,E.R., Kralovec,J.A. & Grindley,T.B. (2010). Isolation of phosphorylatec polysaccharides from algae: the immunostimulatory principle of Chlorella pyrenoidosa. *Carbohydrate Research* Vol.345 No.9 (June 2010), pp.1190-1204ISSN 0008-6215

Suganuma,K., Nakajima,H., Ohtsuki,M. & Imokawa,G. (2010). Astaxanthin attenuates the UVA-induced up-regulation of matrix-metalloproteinase-1 and skin fibroblast 2010), pp.136-142.

Tang,F.Y., Pai,M.H. & Wang,X.D. (2011). Consumption of lycopene inhibits the growth and progression of colon cancer in a mouse xenograft model. *Journal of Agricultural and Food Chemestry* Vol.59, No.16 (Agust 2011), pp.9011-9021, ISSN 0021-8561

Tokimitsu,I. (2004). Effects of tea catechins on lipid metabolism and body fat accumulation. *Biofactors* Vol.22, No.1-4 (January 2004), pp.141-143, ISSN 0951-6433

Tsuda,T. (2008). Regulation of adipocyte function by anthocyanins; possibility of preventing the metabolic syndrome. *Journal of Agricultural and Food Chemestry* Vol.56, No.3 (February 2008), pp.642-646, ISSN 0021-8561

Vijayavel,K., Anbuselvam,C. & Balasubramanian,M.P. (2007). Antioxidant effect of the marine algae Chlorella vulgaris against naphthalene-induced oxidative stress in the albino rats. *Molecular and Cellular Biochemestry* Vol.303, No.1-2 (September 2007), pp.39-44, ISSN 0300-8177

Wang,L.S., Sun,X.D., Cao,Y., Wang,L., Li,F.J. & Wang,Y.F. (2010). Antioxidant and pro-oxidant properties of acylated pelargonidin derivatives extracted from red radish (*Raphanus sativus* var. *niger, Brassicaceae*). *Food and Chemical Toxicology* Vol.48, No.10 (October 2010), pp.2712-2718, ISSN 0278-6915

Werman,M., Mokady,S. & Ben-Amotz,A. (2002). Bioavailability of the isomer mixture of phytoene and phytofluene-rich alga Dunaliella bardawil in rat plasma and tissues. *Journal of Nutrition and Biochemestry* Vol.13, No.10 (October 2010), pp.585-591, ISSN 0955-2863

Williams,R.J., Spencer,J.P. & Rice-Evans,C. (2004). Flavonoids: antioxidants or signalling molecules? *Free Radical Biology and Medical* Vol.36, No.7 (April 2004), pp.838-849, ISSN 8755-9668

Winkel-Shirley,B. (2001). Flavonoid biosynthesis. A colorful model for Genetics, Biochemistry, Cell Biology, and Biotechnology. *Plant Physiology* Vol.126, No.2 (June 2001), pp.485-493,ISSN 0032-0889

Wolfram,S., Raederstorff,D., Preller,M., Wang,Y., Teixeira,S.R., Riegger,C. & Weber,P. (2006). Epigallocatechin Gallate Supplementation Alleviates Diabetes in Rodents. The *Journal of Nutrition* Vol.136, No.10 (October 2010), pp.2512-2518, ISSN 0022-3166

Wood,L.G., Garg,M.L., Powell,H. & Gibson,P.G. (2008). Lycopene-rich treatments modify noneosinophilic airway inflammation in asthma: proof of concept. *Free Radical Reserch* Vol.42, No.1(January 2008), pp.94-102, ISSN 1029-2470

Wu,S.J., Ng,L.T., Wang,G.H., Huang,Y.J., Chen,J.L. & Sun,F.M. (2010). Chlorophyll a, an active anti-proliferative compound of *Ludwigia octovalvis*, activates the CD95 (APO-1/CD95) system and AMPK pathway in 3T3-L1 cells. *Food and Chemical Toxicology* Vol.48, No.2 (February 2010), pp. 716-721, ISSN 0278-6915

Xi,Y.D., Yu,H.L., Ma,W.W., Ding,B.J., Ding,J., Yuan,L.H., Feng,J.F. & Xiao,R. (2011). Genistein inhibits mitochondrial-targeted oxidative damage induced by beta-amyloid peptide 25-35 in PC12 cells. *Journal of Bioenergetics and Biomembranes* Vol.43, No.4 (Agust 2011), pp. 399-407, ISSN 0145-479X

Yang,L., Takai,H., Utsunomiya,T., Li,X., Li,Z., Wang,Z., Wang,S., Sasaki,Y., Yamamoto,H. & Ogata,Y. (2010). Kaempferol stimulates bone sialoprotein gene transcription and new bone formation. *Journal of Cellular Biochemestry* Vol.110, No.6 (Agust 2010) pp.1342-1355, ISSN 1097-4644

Yang,Y., Seo,J.M., Nguyen,A., Pham,T.X., Park,H.J., Park,Y., Kim,B., Bruno,R.S. & Lee,J (2011). Astaxanthin-rich extract from the green alga *Haematococcus pluvialis* lowers plasma lipid concentrations and enhances antioxidant defense in apolipoprotein E knockout mice. *The Journal of Nutrition* Vol.141, No.9 (September 2011), pp. 1611-1617, ISSN 0022-3166

Yusof,Y.A., Saad,S.M., Makpol,S., Shamaan,N.A. & Ngah,W.Z. (2010). Hot water extract of *Chlorella vulgaris* induced DNA damage and apoptosis. *Clinics* (Sao Paulo) Vol.65, No.1? (January 2010), pp.1371-1377, ISSN 1807-5932

Zeisel,S.H. (1999). Regulation of "Nutraceuticals". *Science* Vol.285, No.5435 (September 1999), pp.1853-1855, ISSN 0036-8075

Zhao,L., Chen,G., Zhao,G. & Hu,X. (2009). Optimization of Microwave-Assisted Extraction of astaxanthin from Haematococcus pluvialis by response surface methodology and antioxidant activities of the extracts. *Separation Science and Technology* Vol.44, No.1 (January 2009), pp. 243-262, ISSN 0149-6395

Zhao,Z.W., Cai,W., Lin,Y.L., Lin,Q.F., Jiang,Q., Lin,Z. & Chen,L.L. (2011). Ameliorative effect of astaxanthin on endothelial dysfunction in streptozotocin-induced diabetes in male rats. *Arzneimittel-Forschung* Vol.61, No.4 (January 2011), pp. 61, 239-246, ISSN 0004-4172

Zhu,C., Bai,C., Sanahuja,G., Yuan,D., Farre,G., Naqvi,S., Shi,L., Capell,T. & Christou,P (2010). The regulation of carotenoid pigmentation in flowers. *Archives of Biochemestry and Biophysics* Vol.504, No.1 (December 2010), pp. 132-141, ISSN 0003-9861

Zhu,J., Wang,C.G. & Xu,Y.G. (2011). Lycopene attenuates endothelial dysfunction in streptozotocin-induced diabetic rats by reducing oxidative stress. *Pharmaceutical Biology* (March 2011) ISSN 1388-0209

Permissions

The contributors of this book come from diverse backgrounds, making this book a truly international effort. This book will bring forth new frontiers with its revolutionizing research information and detailed analysis of the nascent developments around the world.

We would like to thank Prof. Dr. Volodymyr I. Lushchak and Dr. Dmytro V. Gospodaryov, for lending their expertise to make the book truly unique. They have played a crucial role in the development of this book. Without their invaluable contribution this book wouldn't have been possible. They have made vital efforts to compile up to date information on the varied aspects of this subject to make this book a valuable addition to the collection of many professionals and students.

This book was conceptualized with the vision of imparting up-to-date information and advanced data in this field. To ensure the same, a matchless editorial board was set up. Every individual on the board went through rigorous rounds of assessment to prove their worth. After which they invested a large part of their time researching and compiling the most relevant data for our readers. Conferences and sessions were held from time to time between the editorial board and the contributing authors to present the data in the most comprehensible form. The editorial team has worked tirelessly to provide valuable and valid information to help people across the globe.

Every chapter published in this book has been scrutinized by our experts. Their significance has been extensively debated. The topics covered herein carry significant findings which will fuel the growth of the discipline. They may even be implemented as practical applications or may be referred to as a beginning point for another development. Chapters in this book were first published by InTech; hereby published with permission under the Creative Commons Attribution License or equivalent.

The editorial board has been involved in producing this book since its inception. They have spent rigorous hours researching and exploring the diverse topics which have resulted in the successful publishing of this book. They have passed on their knowledge of decades through this book. To expedite this challenging task, the publisher supported the team at every step. A small team of assistant editors was also appointed to further simplify the editing procedure and attain best results for the readers.

Our editorial team has been hand-picked from every corner of the world. Their multi-ethnicity adds dynamic inputs to the discussions which result in innovative outcomes. These outcomes are then further discussed with the researchers and contributors who give their valuable feedback and opinion regarding the same. The feedback is then collaborated with the researches and they are edited in a comprehensive manner to aid the understanding of the subject.

Apart from the editorial board, the designing team has also invested a significant amount of their time in understanding the subject and creating the most relevant covers. They scrutinized every image to scout for the most suitable representation of the subject and create an appropriate cover for the book.

The publishing team has been involved in this book since its early stages. They were actively engaged in every process, be it collecting the data, connecting with the contributors or procuring relevant information. The team has been an ardent support to the editorial, designing and production team. Their endless efforts to recruit the best for this project, has resulted in the accomplishment of this book. They are a veteran in the field of academics and their pool of knowledge is as vast as their experience in printing. Their expertise and guidance has proved useful at every step. Their uncompromising quality standards have made this book an exceptional effort. Their encouragement from time to time has been an inspiration for everyone.

The publisher and the editorial board hope that this book will prove to be a valuable piece of knowledge for researchers, students, practitioners and scholars across the globe.

List of Contributors

Zorica Jovanović
Department of Pathological Physiology, Faculty of Medicine, Kragujevac, Serbia

Anwar Norazit
Eskitis Institute for Cell and Molecular Therapies, Griffith University, Nathan, Queensland, Australia
Department of Molecular Medicine, Faculty of Medicine, University of Kuala Lumpur, Malaysia

George Mellick
Eskitis Institute for Cell and Molecular Therapies, Griffith University, Nathan, Queensland, Australia

Adrian C. B. Meedeniya
Eskitis Institute for Cell and Molecular Therapies, Griffith University, Nathan, Queensland, Australia
Griffith Health Institute, Griffith University, Gold Coast, Queensland, Australia

Martina Škurlová
Department of Normal, Pathological, and Clinical Physiology, Third Faculty of Medicine, Charles University in Prague, Czech Republic

Levente Lázár
1st Departmet of Obstetrics and Gynecology, Semmelweis University Budapest, Hungary

Jolanta Zuwala-Jagiello
Department of Pharmaceutical Biochemistry, Wroclaw Medical University, Poland

Eugenia Murawska-Cialowicz
Department of Physiology and Biochemistry, University of Physical Education, Wroclaw, Poland

Monika Pazgan-Simon
Clinic of Infectious Diseases, Liver Diseases and Acquired Immune Deficiency, Wroclaw Medical University, Poland

Tatsuo Shimosawa, Xu Qingyou and Yutaka Yatomi
Department of Clinical Laboratory, Faculty of Medicine, University of Tokyo, Japan

Tomoyo Kaneko, Mu Shengyu, Hong Wang, Sayoko Ogura, Rika Jimbo, Bohumil Matan, Yuzaburo Uetake, Daigoro Hirohama, Fumiko Kawakami-Mori and Toshiro Fujita
Department of Endocrinology and Nephrology, Faculty of Medicinem, Japan

Yusei Miyamoto
Department of Integrated Biosciences, Graduate School of Frontier Sciences, University o
Tokyo, Japan

Cano-Europa, Blas-Valdivia Vanessa, Franco-Colin Margarita and Ortiz-Butron Rocio
Escuela Nacional de Ciencias Biológicas del Instituto Politécnico Nacional, México

Amy H. Yang and Wenhu Huang
Drug Safety Research & Development, Pfizer Inc., La Jolla Laboratories, USA

Kirsi Ketola and Kristiina Iljin
Medical Biotechnology, VTT Technical Research Centre of Finland and University of Turku
Turku, Finland

Anu Vuoristo and Matej Orešič
VTT Technical Research Centre of Finland, Espoo, Finland

Olli Kallioniemi
Institute for Molecular Medicine, Finland (FIMM), University of Helsinki, Finland

Manuel de Miguel and Mario D. Cordero
Departamento de Citología e Histología Normal y Patológica, Facultad de Medicina, Uni
versidad de Sevilla, Sevilla, Spain

P. Pérez-Matute
HIV and Associated Metabolic Alterations Unit, Infectious Diseases Area, Center for Bio
medical Research of La Rioja (CIBIR), Spain

A.B. Crujeiras
Laboratory of Molecular and Cellular Endocrinology, Instituto de Investigación Sanitaria
Complejo Hospitalario de Santiago and Santiago de Compostela University, Spain

M. Fernández-Galilea and P. Prieto-Hontoria
Department of Nutrition, Food Science, Physiology and Toxicology, University of Navarra
México

Rodriguez-Sanchez Ruth, Torres-Manzo Paola and Hernandez-Garcia Adelaida
Escuela Nacional de Ciencias Biológicas, Instituto Politécnico Nacional Departamento de
Fisiología, México

Printed in the USA
CPSIA information can be obtained
at www.ICGtesting.com
JSHW011503221024
72173JS00005B/1191